Motors, Generators, and Transformers

FORMERLY MULTI-AMP INSTITUTE
AVO INTERNATIONAL

 **Delmar Publishers**
I(T)P An International Thomson Publishing Company

Albany • Bonn • Boston • Cincinnati • Detroit • London • Madrid
Melbourne • Mexico City • New York • Pacific Grove • Paris • San Francisco
Singapore • Tokyo • Toronto • Washington

Notice to the reader

Publisher does not warrant or guarantee any of the products described herein or perform any independent analysis in connection with any of the product information contained herein. Publisher does not assume, and expressly disclaims, any obligation to obtain and include information other than that provided to it by the manufacturer.

The reader is expressly warned to consider and adopt all safety precautions that might be indicated by the activities herein and to avoid all potential hazards. By following the instructions contained herein, the reader willingly assumes all risks in connections with such instructions.

The publisher makes no representation or warranties of any kind, including but not limited to, the warranties of fitness for particular purpose or merchantability, nor are any such representations implied with respect to the material set forth herein, and the publisher takes no responsibility with respect to such material. The publisher shall not be liable for any special, consequential, or exemplary damages resulting, in whole or part, from the readers' use of, or reliance upon, this material.

Delmar Staff
Senior Editor: Mark Huth
Project Manager: Patricia Konczeski
Production Coordinator: Dianne Jensis
Art/Design Coordinator: Cheri Plasse

Copyright © 1995
By Delmar Publishers
a division of International Thomson Publishing Inc.

The ITP logo is a trademark under license

Printed in the United States of America

For more information, contact:

Delmar Publishers
3 Columbia Circle, Box 15015
Albany, New York 12212-5015

International Thomson Publishing Europe
Berkshire House 168 - 173
High Holborn
London WC1V 7AA
England

Thomas Nelson Australia
102 Dodds Street
South Melbourne, 3205
Victoria, Australia

Nelson Canada
1120 Birchmount Road
Scarborough, Ontario
Canada M1K 5G4

The IEEE takes no responsibility for and will assume no liability for damages resulting from the reader's misinterpretation of said information resulting from the placement and context in this publication. Information is reproduced with the permission of the IEEE.

International Thomson Editores
Campos Eliseos 385, Piso 7
Col Polanco
11560 Mexico D F Mexico

International Thomson Publishing GmbH
Königswinterer Strasse 418
53227 Bonn
Germany

International Thomson Publishing Asia
221 Henderson Road
#05 - 10 Henderson Building
Singapore 0315

International Thomson Publishing - Japan
Hirakawacho Kyowa Building, 3F
2-2-1 Hirakawacho
Chiyoda-ku, Tokyo 102
Japan

All rights reserved. No part of this work covered by the copyright hereon may be reproduced or used in any form or by any means—graphic, electronic, or mechanical, including photocopying, recording, taping, or information storage and retrieval systems—without the written permission of the publisher.

1 2 3 4 5 6 7 8 9 10 XXX 01 00 99 98 97 96 95

Library of Congress Cataloging-in-Publication Data
Motors, Generators & Transformers / Multi-Amp Institute.
    p. cm.
Includes index.
ISBN 0-8273-4920-3
—ISBN 0-8273-4921-1 (instructor's guide)

CIP
94-20840

Table of Contents

Part I - Fundamentals — 1
    Chapter 1   Basic Theory and Concepts — 3

Part II - AC Generators — 29
    Chapter 2   AC Generator Theory and Construction — 31
    Chapter 3   AC Generator Protection — 57
    Chapter 4   AC Generator Testing — 79

Part III - AC Motors — 109
    Chapter 5   AC Motor Theory — 111
    Chapter 6   AC Motor Construction and Application — 125
    Chapter 7   AC Motor Control and Starting — 157
    Chapter 8   Protection — 179
    Chapter 9   AC Motor Maintenance and Troubleshooting — 199

Part IV - DC Machines — 227
    Chapter 10   DC Generator Theory and Construction — 229
    Chapter 11   DC Motor Theory and Construction — 255
    Chapter 12   DC Machine Maintenance — 285

Part V - Transformers — 313
    Chapter 13   Transformer Theory — 315
    Chapter 14   Transformer Construction — 335
    Chapter 15   Transformer Protection — 367

Foreword

AVO International Training Institute, formerly Multi-Amp Institute, is America's leading provider of electrical testing, maintenance and safety training. The Institute has provided hands-on training programs for over 75,000 engineers and technicians in the areas of circuit breakers, relays and transformer testing and maintenance, cable splicing, testing and fault location and all parts of the OSHA Federal Register which apply to electrical safety. The training center headquarters is located in Dallas, Texas with regional learning centers in Orlando, Philadelphia and Portland. The Institute offers regularly scheduled open-enrollment courses at these locations or can tailor any of our courses for presentation at the client's facility.

In addition to our on-site and open enrollment courses, the Institute offers several additional training forums for our clients to choose from. These include maintenance training material packages, electrical safety videos and vocational/technical textbooks.

AVO International is the leading provider of test equipment and measuring instruments for electrical power applications. At operations in the United States, Canada, France and the United Kingdom, **AVO** International designs, manufactures, markets and supports more than 1000 products under the brand names the electrical power industry has come to depend on: Multi-Amp®, Biddle™, and Megger®.

AVO International provides calibration, certification, warranty repair services at our operations in the United States, Canada, France and United Kingdom. Metrology services are traceable to national standards.

To request your free catalog or additional information on products, please call (214) 333-3201.

If you desire additional information or any training services, please call (214) 330-3522. We will be glad to mail you detailed information on any of our training programs or products.

Preface

The generation, transmission, distribution, and use of electric power are dependent upon devices that employ magnetic fields. Generators are used to convert mechanical energy to electrical energy by rotating conductors through a magnetic field. Transformers are used to change voltage from one level to another by creating a varying magnetic field that cuts through a winding, thus creating an output voltage. Motors are used to convert electrical energy to mechanical energy by the creation of a rotating magnetic field that cuts through windings, creating secondary mechanical motion.

This book is designed to introduce or refresh the student in the operation, construction, and maintenance of generators, motors, and transformers. It presents a practical approach to the understanding of these key electrical components.

The book starts with a minimal introduction of fundamentals. While some math is used, it is kept to a minimum in favor of practical, intuitive explanations.

The actual presentation of motors and generators takes a different approach in this text. Historically, authors have presented the principles of direct current equipment first, possibly because DC motors and generators were developed first. This text takes a more logical approach and presents AC systems first. This approach is based on twenty years of classroom experience which has shown that students learn the material more readily when AC is presented first.

For each of the equipment types, the same basic approach has been taken. Operating principles are discussed first so that students can grasp the fundamentals. Next, the text covers the practical construction features of "real world" generators, motors, and transformers. In these sections, the student learns how real equipment is constructed; materials, structure, and operation are all covered. In many sections, the basic NEMA standards are identified and, in some cases, actually reproduced so that the student sees how things are actually done.

The final information covered by the text includes protection and maintenance of the equipment. While not intended to be a detailed coverage of such topics, enough information is provided to introduce the student and to prepare him or her for more advanced studies of field experiences.

Acknowledgments

AVO International Training Institute wishes to thank all those companies and individuals for their invaluable assistance in supplying technical reviews, information, and art for this book.

Glenn W. Armstrong, Jr.
Doble Engineering Company

Dean Bartlett
AVO International Training Institute

Dennis J. Berry
National Fire Protection Association

Robert Blakely
Mississippi Gulf Coast Community College

Raybon Bone
Halifax Community College

Gary L. Brisbin
Onan Corporation

John Cadick
Cadick Professional Services

Charles C. Campbell
North Wayne County Vocational School

G. Reynolds Clark
Westinghouse Electric Corporation

Cynthia Corbett
Square D Company

Jim Delaney
Cutler-Hammer

Terry Duchaine
Ideal Industries, Inc.

Bruce Eichenlaub
Savannah Technical Institute

Frank Esper
Kalamazoo Community College

Alden S. Fletcher
The Lincoln Electric Company

Alan Mark Franks
AVO International Training Institute

David Gehlauf
Tri-County Vocational School

Robert N. Gerin
The Gerin Corporation

Richard R. Heddleston
S. D. Myers, Inc.

Kathe Hooper
ASTM

James Hutto
MacArthur State Technical College

O'Neill Jaynes
Texas Electric Motors, Inc.

Ralph E. Krisher, Jr.
General Electric Company

Robert C. Lampe, Jr.
General Electric Company

Jane Lee
GE Electrical Distribution and Control

Robert L. Meltzer
ASTM

Joseph B. McCunney
GE Protection and Control

Val Mosticon
Dean Institute of Technology

Margaret S. Newton
The National Electrical Manufacturers Association

Larry Phillips
Minneapolis Technical College

Stan Phipps
S K Enterprises

Rita Simonetta
ABB Power T&D Company, Inc.

James E. Teague
General Electric Company

Jerome T. Walker
The Institute of Electrical and Electronics Engineers, Inc.

James R. White
Director, AVO International Training Institute

Norris Woodruff
Multilin Inc.

Sandy Young
Educational Resource Manager

A special thanks for his efforts to John Cadick, PE, Cadick Professional Services for his vast knowledge and expertise and the entire staff of AVO International Training Institute for their support during the development of this book.

Part I

Fundamentals

Basic Theory and Concepts 1

OBJECTIVES

After studying this chapter, the student should be able to:

- *Describe the nature and effects of magnetism.*
- *Describe the nature and effects of electricity.*
- *Describe how electricity and magnetism are related.*
- *Describe Ampere's Law and how it relates to electric motors.*
- *Describe Faraday's Law of electromagnetic induction as it relates to electric motors and generators.*

INTRODUCTION

This chapter introduces several key concepts that are essential to understanding motors, generators, and transformers. The two principal concepts are Faraday's law and Ampere's law.

Ampere's law defines the force that a magnetic field has on a current-carrying conductor. Faraday's law defines the voltage that is induced into a conductor when it passes through a magnetic field. These two laws, along with Lenz's law (also covered in this chapter), make up the bulk of the theory upon which motors, generators, and transformers operate.

MAGNETISM

Fundamental Properties

Since ancient times, man has known of the existence of certain naturally occurring substances, called *lodestones*, which possess an attraction for iron and other similar metals. The force exerted by these lodestones is called *magnetism*. Many objects other than lodestones may exhibit magnetic characteristics. Such objects are called *magnets*.

A number of materials, such as cobalt and nickel, have magnetic properties similar to, but weaker than, iron and steel. Such materials are called *ferromagnetic*. Other substances have a very weak magnetic effect and are called *paramagnetic*. A few substances are actually slightly repelled by a magnet and are called *diamagnetic*.

The ancients discovered that the earth is also a magnet. Furthermore, they found that magnets have two poles (Figure 1–1.) When suspended

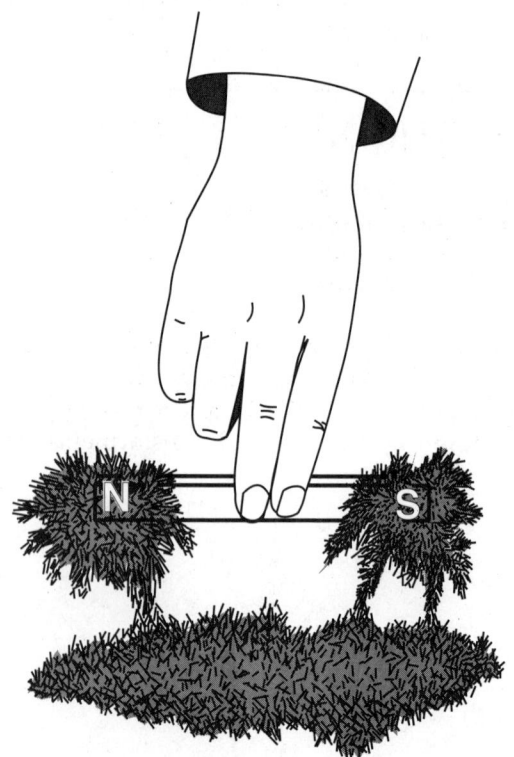

FIGURE 1–1
Magnetic poles

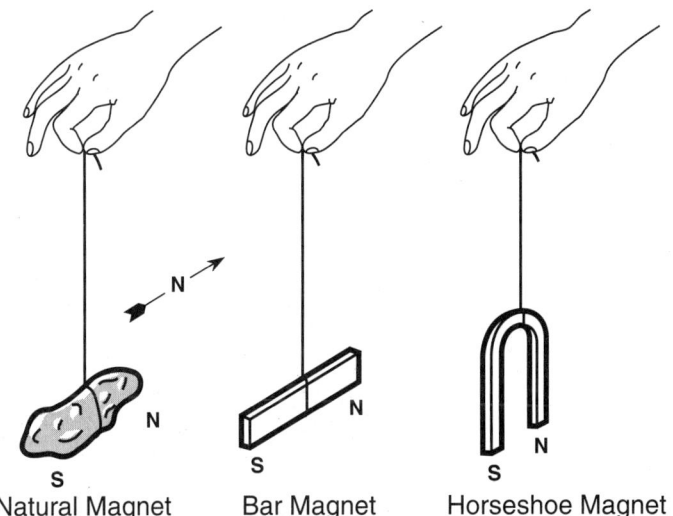

FIGURE 1–2
Magnetic alignment with the earth's field

and allowed to turn freely, a magnet will always align itself so that one end points to the north pole of the earth and the other to the south (Figure 1–2). A north pointing pole repels another north pointing pole and attracts another south pointing pole. Thus the rule that *opposite poles attract and like poles repel* (Figure 1–3). The magnetic force is actually concentrated near the poles. The north pointing pole is called the *north pole* and the south pointing pole is called the *south pole*.

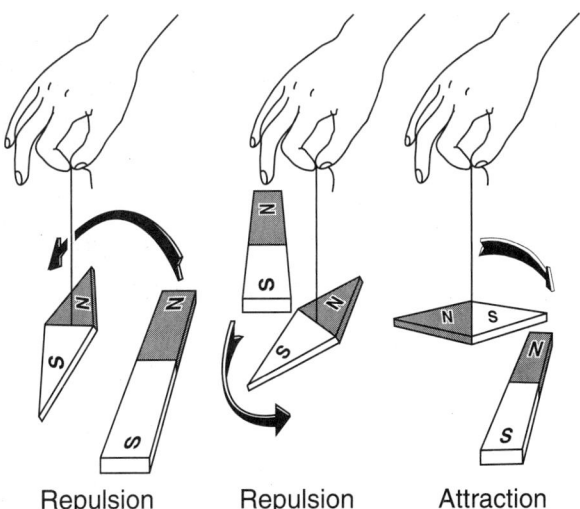

FIGURE 1–3
Laws of attraction and repulsion

Magnetic force can be induced into ferromagnetic materials by stroking them with another magnet, or by placing them in the direct current induced field of a coil (Figure 1–4). After the magnetism has been induced in this manner, a certain portion of it will be retained by the material. The amount retained is called *residual magnetism.* The ability of a material to retain magnetism is called its *retentivity.*

FIGURE 1–4
Methods of inducing magnetic force into ferromagnetic materials

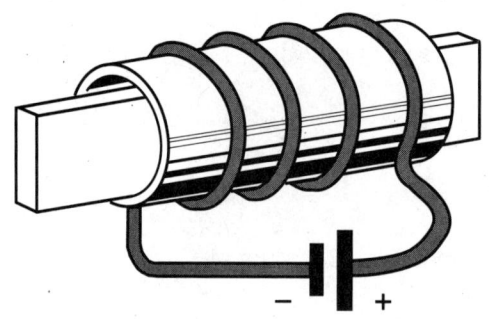

Coil Method

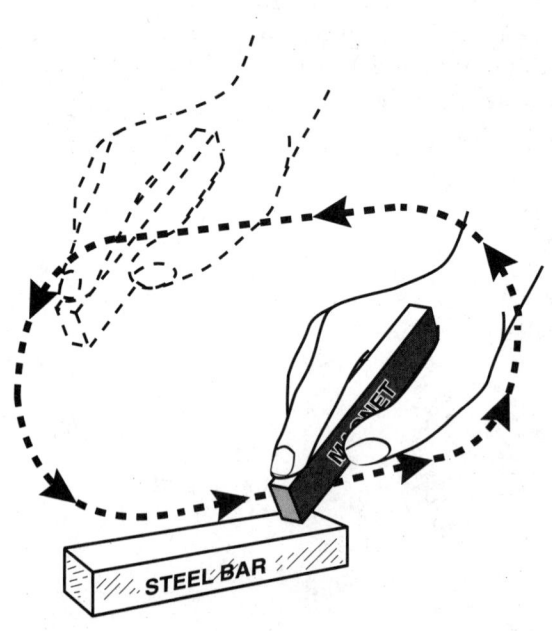

Stroking Method

Force Between Magnetic Poles

The French physicist, Charles A. Coulomb, discovered the basic law of force between two magnetic poles. He discovered that the amount of force depends on two things: (1) the distance between them, and (2) their charge. Coulomb's Law states that two charged bodies will attract or repel each other with a force that is directly proportional to the product of their charges, and inversely proportional to the square of the distance between them.

$$F = \frac{m_1 + m_2}{\mu \times r^2} \qquad (1)$$

Where:
F = Force between two magnetic poles
m_1 = Magnetic strength of magnet number 1
m_2 = Magnetic strength of magnet number 2
m = Permeability of the material
r = distance between the two poles
μ = the permeability of the material in which the poles are located. Permeability may be thought of as the relative ease of magnetization of a material.

In other words, the greater the charge and distance between them, the smaller the force to attract or repel.

Notice that the force is inversely proportional to the square of the distance. This means that if the distance is doubled, the amount of force is only one-fourth as great. From this, you can see why the distance of air gaps is so important in motor, generator, and transformer construction.

Magnetic Fields and Flux

The area around a magnet in which the magnetic force acts is called a *magnetic field*. This field can be thought of as a force. The magnetic field and *lines of force* that surround a magnet are illustrated in Figure 1–5. By observing the small compasses, the direction of the magnetic lines of force may be observed. These magnetic lines of force do not actually exist. They are imaginary lines used to illustrate the pattern of the magnetic field. Keep in mind that the compass is also a magnet. Notice that the north end of the compass needle *always* points toward the south pole, which indicates the direction of the field of force. In other words, the lines of force proceed from the north pole of the magnet, enter the south pole, and again return to the north pole, thus creating a closed loop.

FIGURE 1–5
Magnetic lines of force (flux)

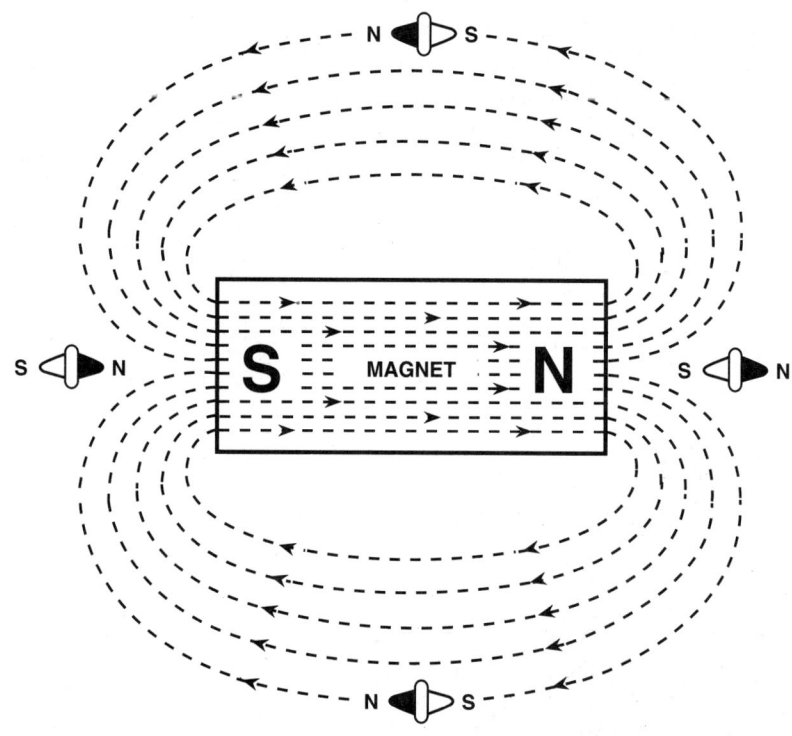

The magnetic lines of force surrounding the magnet are called *magnetic flux*. Flux in a magnetic circuit can be related to gallons of water flowing through a water pipe. The number of lines of force per unit area is called *flux density*, and is measured in lines per square inch or lines per square centimeter.

By studying Figure 1–5, the following observations are made:

1. The magnetic lines of force are continuous and form a closed loop.
2. The magnetic lines of force will never cross each other.
3. Magnetic lines of force that are parallel and traveling in the same direction repel each other. On the other hand, if the magnetic lines of force are traveling in opposite directions, they seem to join together and form into single lines traveling in a direction determined by the magnetic poles.
4. Magnetic lines of force seem to shorten themselves. Therefore, the force between unlike poles causes the poles to be pulled together.
5. Magnetic lines of force will pass through all materials, magnetic and nonmagnetic.

Basic Theory and Concepts

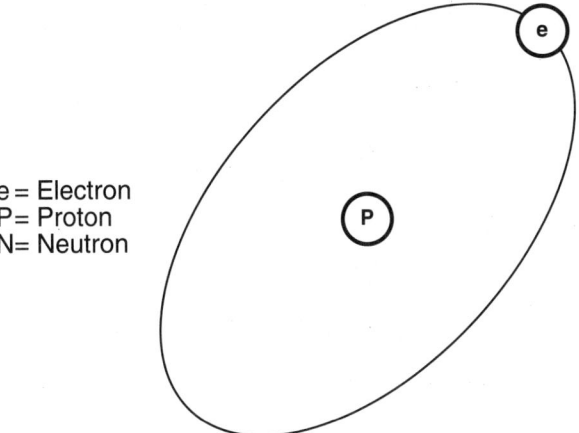

FIGURE 1–6
Hydrogen atom

e = Electron
P = Proton
N = Neutron

ELECTRICITY

Electron Theory

All matter is composed of atoms. Atoms in turn are composed of three major components: *protons, neutrons,* and *electrons.* Figures 1–6, 1–7, and 1–8 are stylized drawings of the hydrogen, helium, and carbon atoms respectively. Notice that the neutrons and protons reside together at the center of the atom. This structure is referred to as the *nucleus* of the atom. The electrons orbit around the nucleus in a sort of "electron cloud."

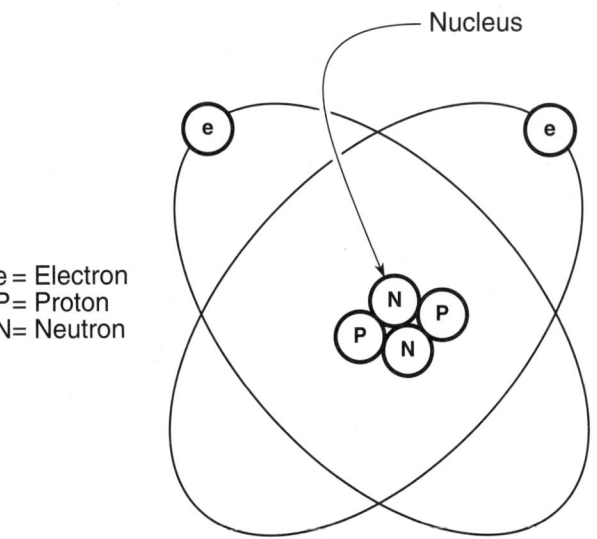

FIGURE 1–7
Helium atom

e = Electron
P = Proton
N = Neutron

FIGURE 1–8
Carbon atom

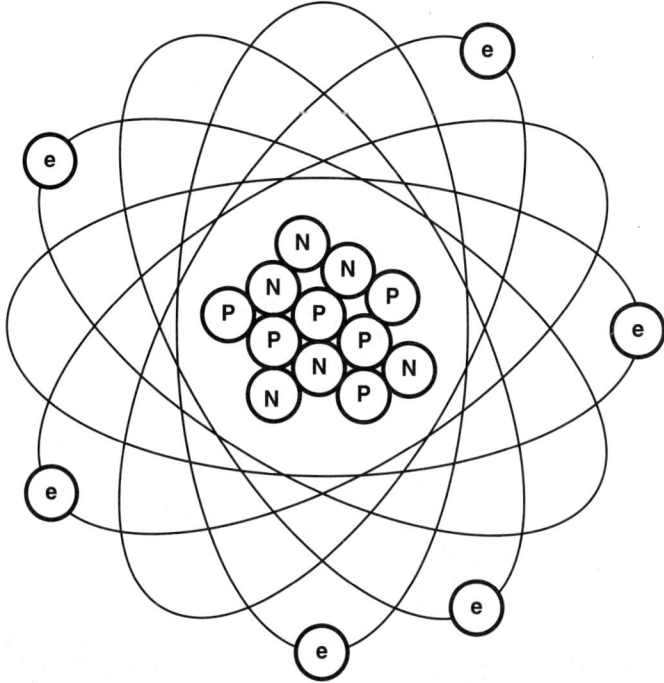

The electron is the principal element in electricity. It exhibits an electric force field that is analogous to the magnetic field discussed earlier. The proton exhibits a similar electric force but with an opposite polarity. The electron has a negative electric force and the proton has a positive electric force. These two forces, also called *charges,* behave in much the same way that magnetic poles do (Figure 1–9). Like charges repel and unlike charges attract. Thus, two electrons will repel, while an electron and a proton have an electrostatic attraction for each other.

Voltage, Current, and Resistance

Voltage Electrons can be broken free from their atoms and moved through a conductor. The force that is used to move electrons is called *voltage.* There are a number of sources of voltage, batteries, Figure 1–10, and generators, Figure 1–11. The amount of pressure to move electrons is measured in the unit of *volts* (V).

Voltage may be either *direct voltage* or *alternating voltage.* Direct voltage is a constant, unidirectional pressure that pushes the electrons in one direction only. Alternating voltage pushes first in one direction and then in the other. Thus, the electrical pressure alternates. The battery shown in Figure 1–10 is a source of direct voltage. The generator shown in Figure 1–11 is a source of alternating voltage. Some generators are sources of direct voltage.

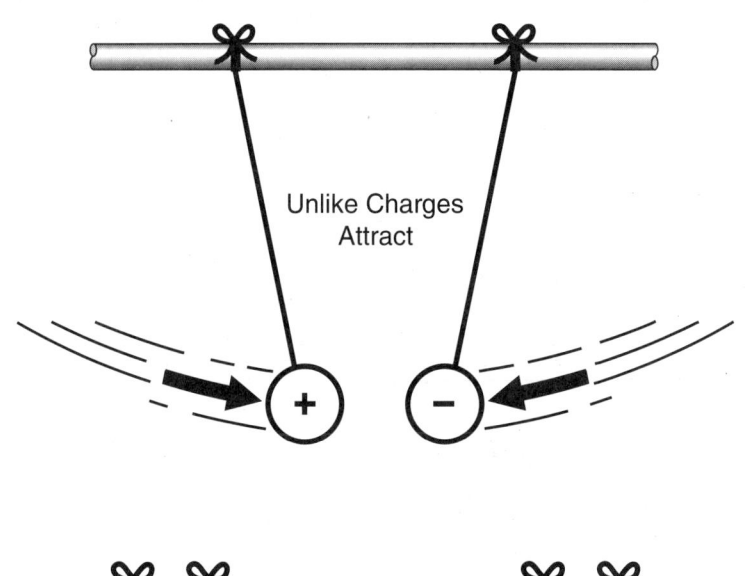

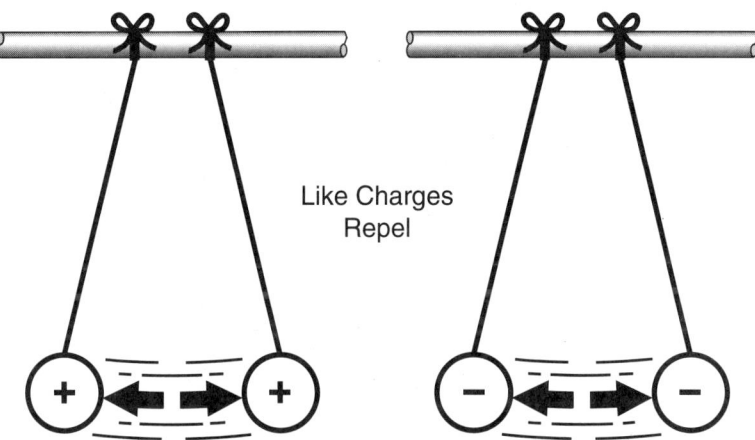

FIGURE 1–9
Reaction between positive and negative charges

Current The flow of electrons through a conductor is called *current* and is measured in *amperes* (A). When the electrons move, they "carry" energy from the source. If a "load" which is capable of using the electrical energy is connected to the wires, useful work can be accomplished. Figure 1–12, p. 14, shows the general concept of electron current flow, and Figure 1–13, p. 14, shows the connection of a source of voltage, some wires, and a load.

Frequency If the source of voltage is a direct source, such as in Figure 1–13, the electron flow will be *direct current* or DC. Conversely, if the voltage is an alternating source, the current flow will be an *alternating current* or AC.

FIGURE 1–10
Batteries

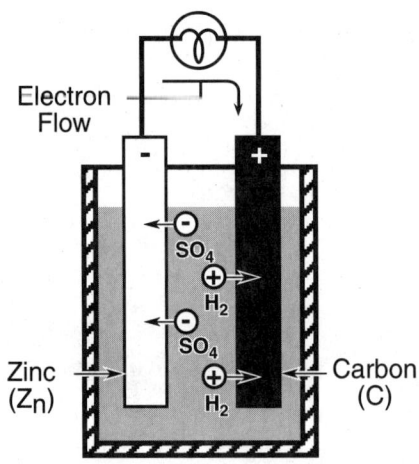

Simple Voltage Cell

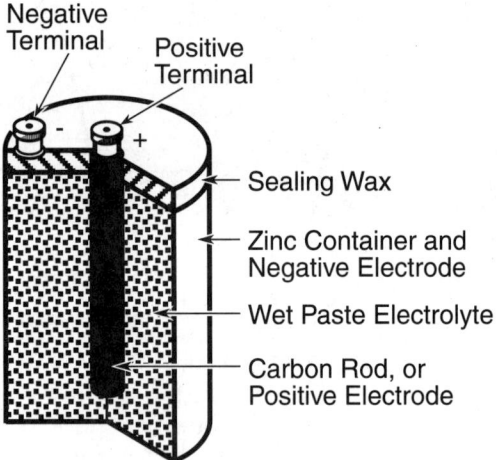

Dry Cell; Cross-Sectional View

Basic Theory and Concepts

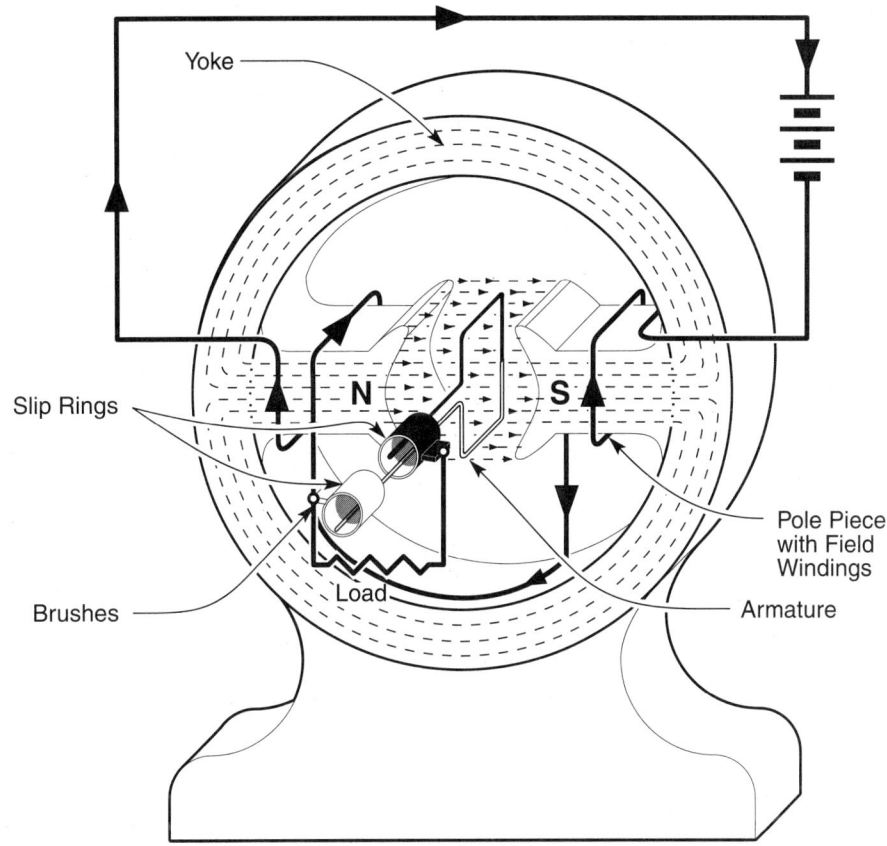

FIGURE 1–11
Simple electrical generator

Figure 1–14 is a time diagram of an alternating current flow. Notice that this current passes through zero, rises to a maximum current flow, decreases back to zero, continues decreasing to a negative maximum, and completes by returning to zero. One complete alternation is called a *cycle*. The number of cycles per second determines the system *frequency*. The unit of frequency is *Hertz* (*Hz*), which is one cycle per second. AC power systems use 60 or 50 Hz, depending on where they are located. The United States uses primarily 60 Hz.

Resistance and Impedance   As electrons move through the circuit, they encounter opposition. This opposition is called *resistance* or *impedance*. Resistance is the opposition that is created by the friction of the movement of

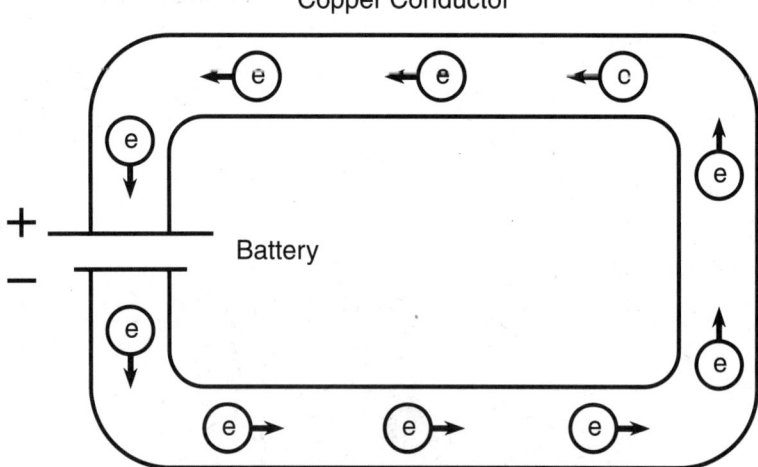

FIGURE 1–12
Current flow

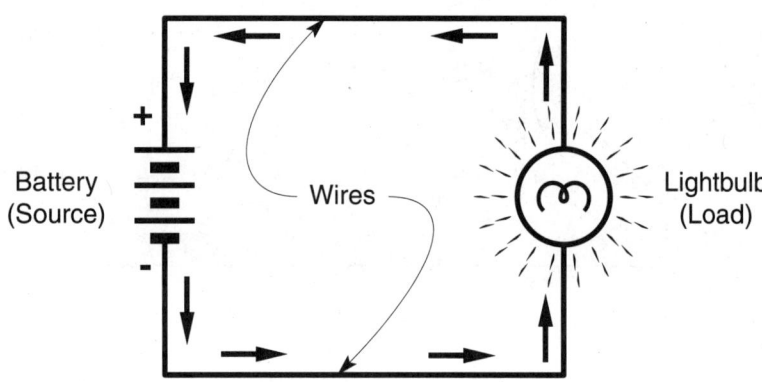

FIGURE 1–13
Direct current
(Courtesy of Cadick
Professional Services)

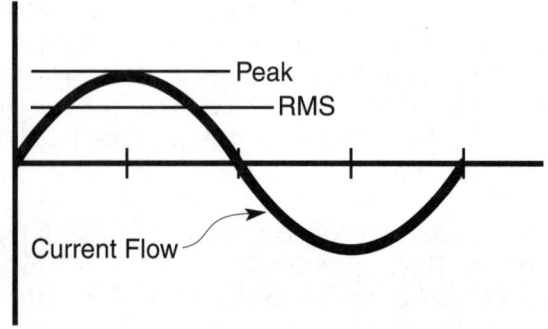

FIGURE 1–14
Alternating current
(Courtesy of Cadick
Professional Services)

the electrons or by the performance of real work. Impedance *includes* resistance plus the opposition created by the formation of magnetic and electrostatic fields. Direct current is opposed only by resistance, while alternating current is opposed by impedance.

Ohm's Law George Simon Ohm discovered a fundamental relationship between voltage, current, and resistance. This relationship is called *Ohm's Law* and is shown below:

$$E = I \times R \tag{2}$$

Where:
E is the circuit voltage expressed in volts
I is the current in the circuit expressed in amperes
R is the resistance of the circuit expressed in ohms (Ω)

If the system uses alternating voltage and current, Ohm's Law is expressed as:

$$E = I \times Z \tag{3}$$

Where:
Z is the impedance of the circuit expressed in ohms

Note that for the alternating current circuit, the voltage and current magnitudes are given in the *RMS* or root-mean-square values. RMS values are used to give an alternating voltage or current their equivalent DC values. Thus, an alternating current which peaks at 169.68 V has an effective or RMS value of 120 V. This means that the 169.68 V peak alternating current has the same effect as a 120 V direct current. In normal, non-distorted power systems, the peak of the AC wave is equal to 1.414 times the RMS value.

Single-Phase and Three-Phase Systems

Figure 1–14 is a *single-phase* system. This means there is only one voltage and one current developed. For a number of reasons it is more economical to generate voltages in groups of more than one phase. Most modern power systems use *three-phase* generation and transmission of electric power. In such a system the three voltages are equal in magnitude and displaced by 120 electrical degrees. Since one full cycle takes 360 electrical degrees, the three phases of a three-phase system are displaced by 1/3 of a cycle from each other. Figure 1–15 shows the waveforms for a three-phase system.

FIGURE 1–15
Three-phase system

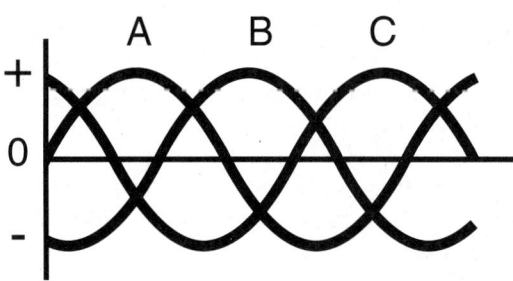

When equal loads are connected to each of the voltages of a three-phase power system, the currents that flow will be equal and 120° apart from each other.

The voltages measured between the three wires of a three-phase system, called the line-to-line voltage, will be equal in magnitude but different in phase angle (120°). It can be shown that the line-to-line voltage in a balanced three-phase system is equal to the line-to-ground voltage multiplied by the square root of three. This is shown mathematically as:

$$E_{LL} = E_{LG} \times \sqrt{3} \qquad (4)$$

Example:

In a 480-V system, the voltage between wires is 480 V. The voltage from wire to ground is

$$\frac{480}{\sqrt{3}} = 277 \text{ V} \qquad (5)$$

Power

Real Power The purpose of electrical power systems is to transfer energy from one place to another so that useful work can be accomplished. This may be in the form of light, heat, or motion. In a DC system, the so-called *real power* in *watts* (W) can be calculated by multiplying the current by the voltage or:

$$P \text{ (watts)} = E \times I \qquad (6)$$

Consider a 120-V DC heater which draws 3 A. Formula (6) tells us that the power drawn is

$$120 \times 3 = 360 \text{ W} \tag{7}$$

In AC systems, the impedance may cause the voltage and current to be out-of-phase with each other. That is, they may not hit their peaks at the same time, but may be displaced by some fraction of a cycle. When this occurs, the real power is not equal to the voltage times the current. Rather it is equal to:

$$P \text{ (watts)} = E \times I \times \cos(\theta) \tag{8}$$

Where:
cos is the trigonometric cosine function
θ is the angle between the current and the voltage

Note that the *cos* (θ) is called the *power factor* of the system. Power factor will be discussed again later.

Reactive Power When an alternating voltage is applied to an inductor or a capacitor, the resulting current flow lags (inductor) or leads (capacitor) by 90°. (Remember that 90° equals 1/4 of one full cycle since the cosine of 90° is zero. You can see from equation (8) that the real power in such a situation is zero, since:

$$\cos 90° = 0 \tag{9}$$

Actually, the inductor or capacitor takes energy from the system during the buildup of the magnetic (inductor) or electric (capacitor) field, and then returns the energy during the decay of the field. Because of this action, the energy is given by and returned to the system each 1/4 cycle. That is, the energy is absorbed by the field during the current rise, and is returned to the system during the current decay. This happens on both the positive and negative half cycles.

The amount of power that is being swapped in and out of the inductor or capacitor can be calculated by multiplying the voltage by the current. This is the same calculation as that illustrated by equation (4) except the current and voltage are 90° out-of-phase with one another. To differentiate between real power and reactive power, reactive power uses the unit of *VAR*. VAR is an acronym for *volt-amps reactive*. VAR can be calculated from the equation:

$$\text{VAR} = E \times I \times \sin(\theta) \tag{10}$$

Since DC systems have no phase angle between the current and voltage, reactive power does not exist in them.

Apparent Power The sum of the real power and the reactive power is referred to as the *apparent power*. The apparent power can be calculated by using equation (4) and ignoring the phase angle. It can also be calculated by adding the watts to the VARS. Since the watts and VARS are 90° out-of-phase, the addition must be performed vectorially as shown in Figure 1–16.

The apparent power can be calculated from the formula:

$$VA = E \times I \tag{11}$$

As mentioned earlier, power factor is numerically equal to the cos (θ) where θ is the angle between the current and the voltage. θ is also the angle between the real power and the apparent power. In fact, the actual definition of power factor is the ratio of real power to apparent power or:

$$\text{Power Factor} = \frac{\text{watts}}{\text{volt-amperes}} \tag{12}$$

Example of power factor:
What is the power factor of a motor which is operating at 400 W and 500 VA?

$$\text{Power Factor} = \frac{400}{500} = 0.8 \text{ or } 80\% \tag{13}$$

Power factor may be expressed as a percent by multiplying equation (12) by 100.

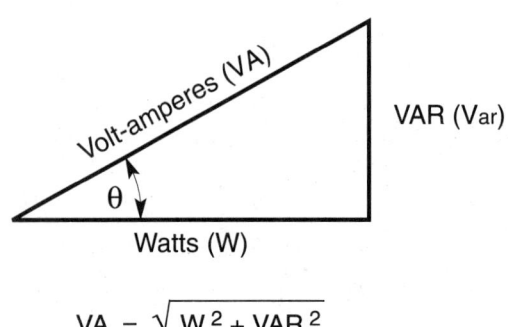

FIGURE 1–16 Relationship between Watts, VARS, and Volt-Amperes

$$VA = \sqrt{W^2 + VAR^2}$$

Basic Theory and Concepts

Power in Three-Phase Systems True power, reactive power, and apparent power may be calculated for three-phase systems with only slight modifications of the formulas already presented. Multiplying the single-phase values by three, and remembering from formula (4) that the line-to-line voltage is equal to the line-to-ground voltage times the square root of three, the following formulas apply:

$$P \text{ (3-phase watts)} = E_{LL} \times I \times \sqrt{3} \times \cos(\theta) \tag{14}$$

and

$$VAR = E_{LL} \times I \times \sqrt{3} \times \sin(\theta) \tag{15}$$

and

$$VA_{\text{three-phase}} = E_{LL} \times I \sqrt{3} \tag{16}$$

Example:

Consider a motor operating in a three-phase power system. The phase-to-phase voltage is 480 volts, the single-phase current is 20 amperes, and the phase angle (θ) is 30°. The three-phase power, var, and volt amps can be calculated as follows:

$$P = 480 \times 20 \times 1.732 \times \cos(30°) = 14{,}400 \text{ W} \tag{17}$$
$$VAR = 480 \times 20 \times 1.732 \times \sin(30°) = 8314 \text{ W}$$
$$\text{Volt amps} = 480 \times 1.732 = 16{,}628$$

$$\text{Power Factor} = \frac{14{,}400}{16{,}628} = 0.866 \text{ or } 86.6\%$$

ELECTROMAGNETISM

Electric Current and Magnetism

In 1820 Hans Christian Oersted discovered that a compass was deflected from its normal north-south orientation when placed close to a current carrying wire, Figure 1–17. When current flows up in the north-south oriented wire, the north pole of the compass is deflected 90° and points east. When current flows down, or south, the compass is deflected to the west.

Oersted also discovered that the effect is strongest close to the wire. When the compass is moved around the wire, Figure 1–18, the compass

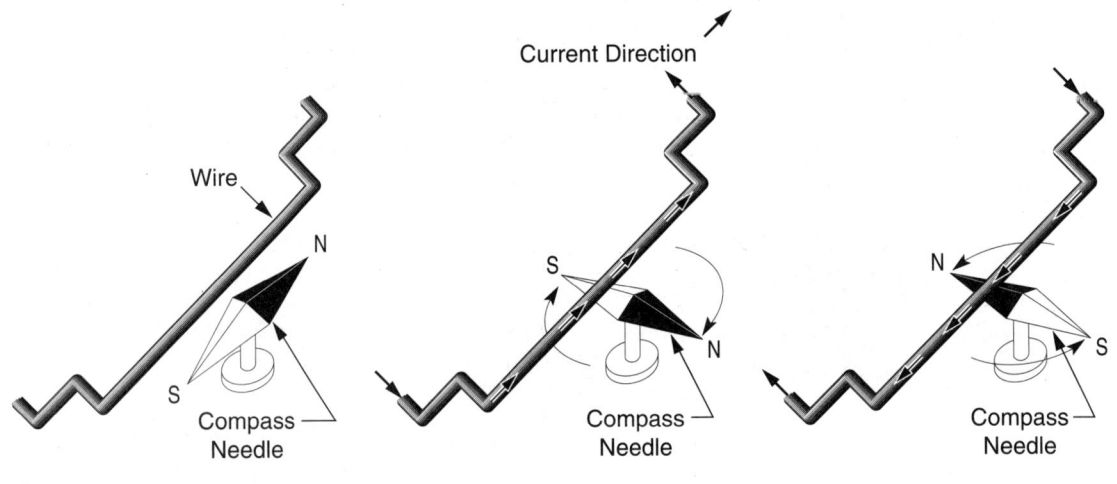

FIGURE 1–17
Oersted's discovery.
Oersted discovered
that a current carrying
wire will deflect a
compass needle. Arrow
shows direction of
compass rotation.

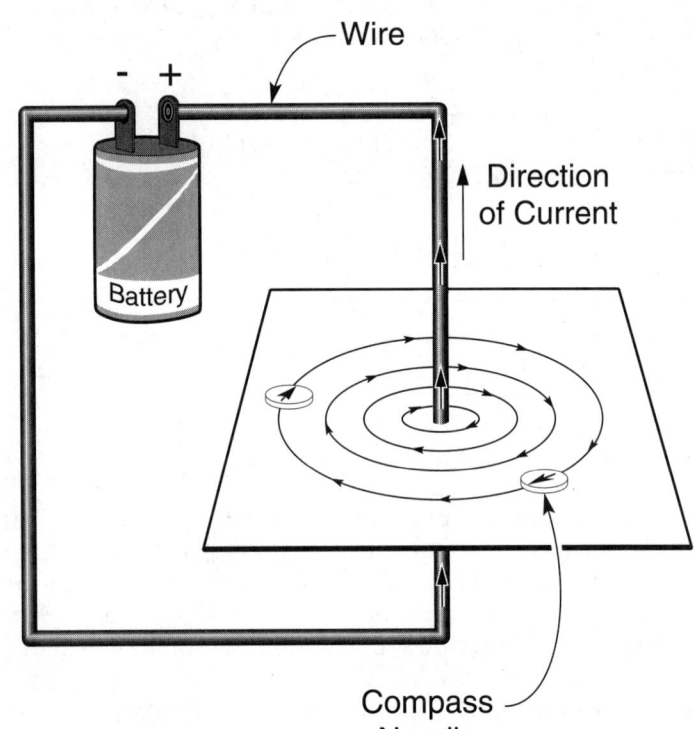

FIGURE 1–18
Magnetic field set up
around a current
carrying wire
(Courtesy of Cadick
Professional Services)

Basic Theory and Concepts

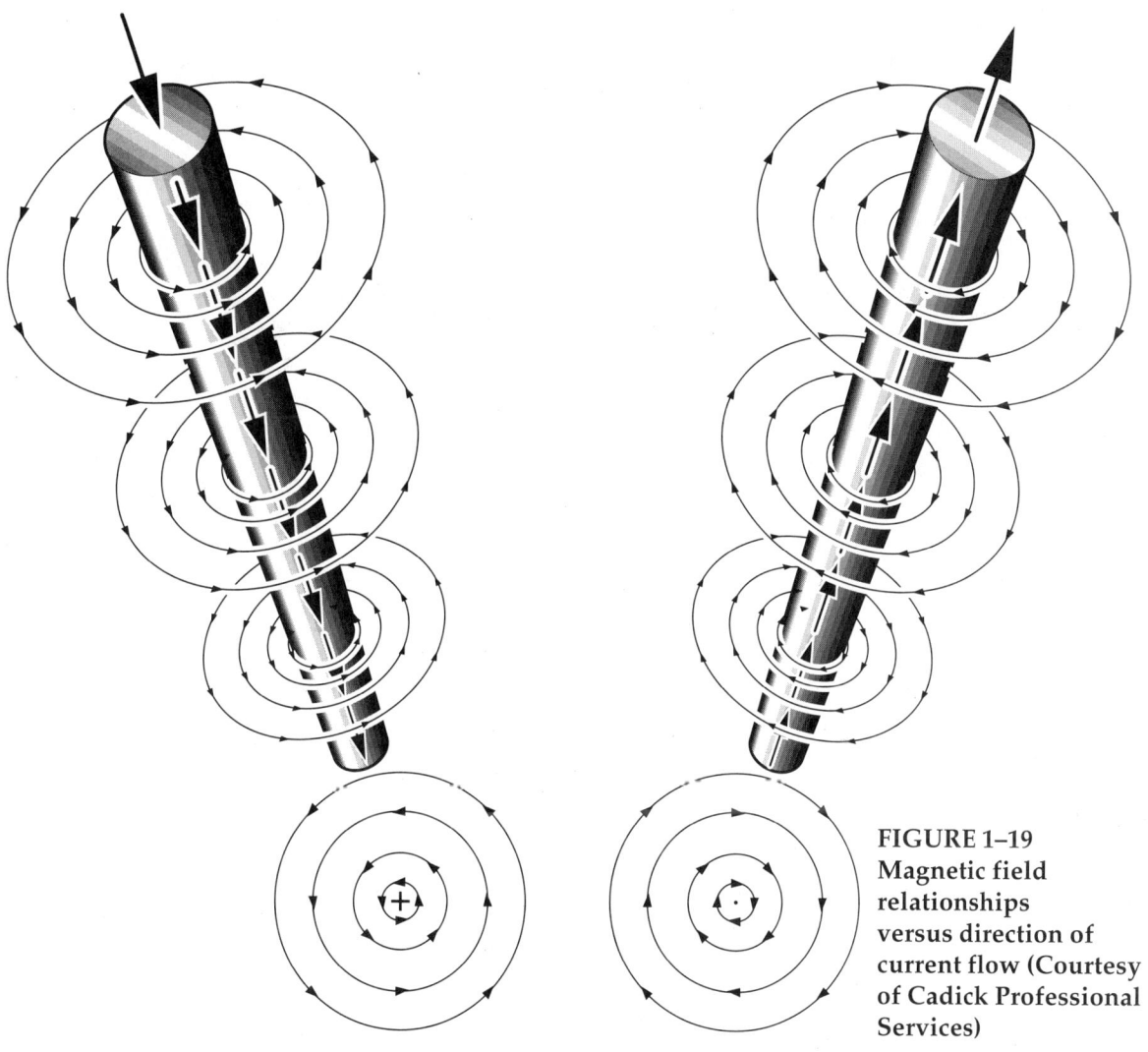

FIGURE 1–19 Magnetic field relationships versus direction of current flow (Courtesy of Cadick Professional Services)

needle traces a circle; the closer to the wire, the smaller the diameter of the circle.

Clearly, the current flow is producing magnetic flux lines that are perpendicular to the current flow. Furthermore, the relative polarity of the magnetic field can be shown as in Figure 1–19. This relationship led to the development of the familiar *left-hand rule,* Figure 1–20, used for determining the polarity of a magnetic field.

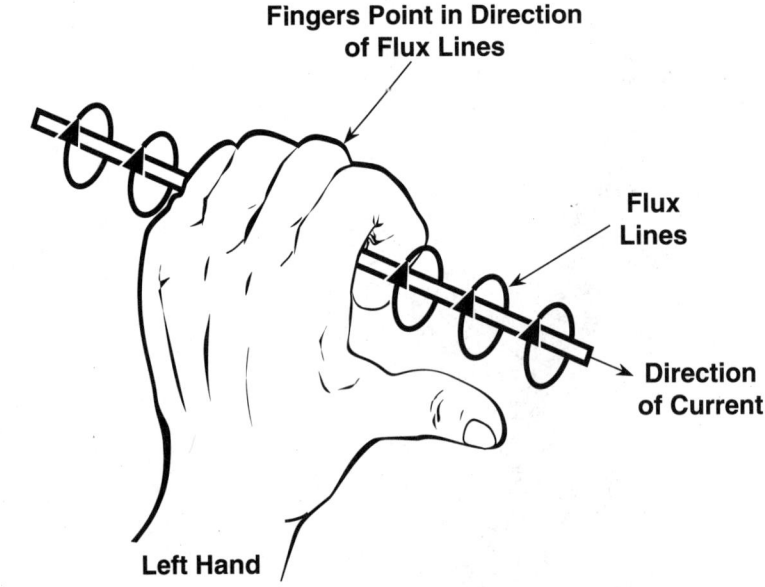

FIGURE 1–20 The left-hand rule for determining electromagnetic field polarity

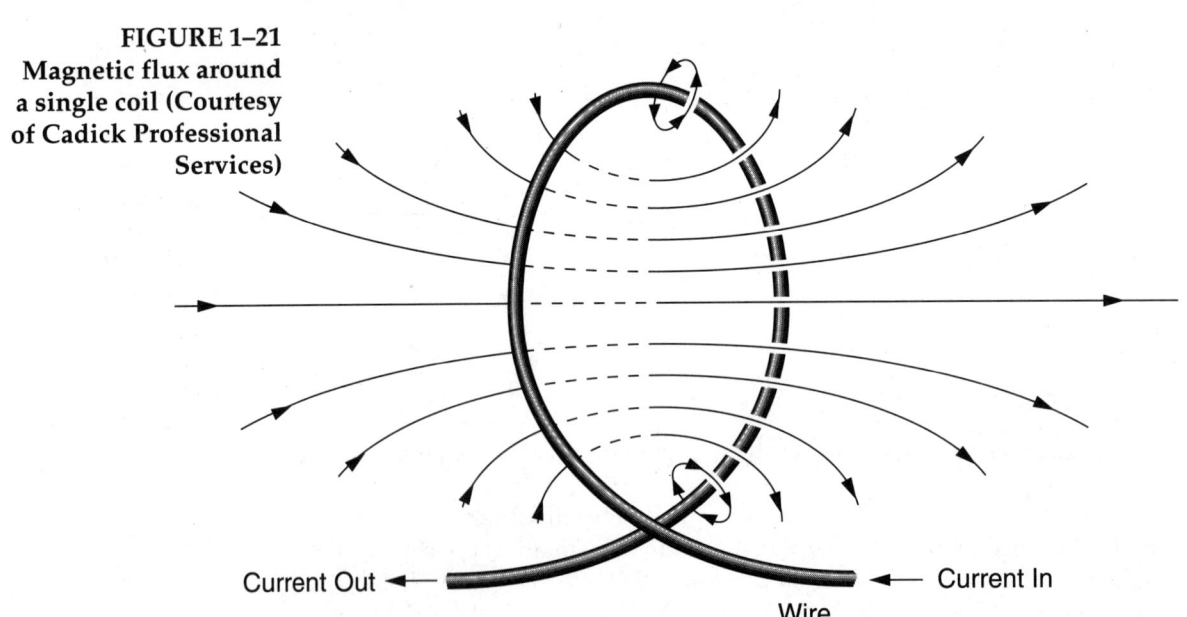

FIGURE 1–21 Magnetic flux around a single coil (Courtesy of Cadick Professional Services)

Basic Theory and Concepts

Coils (Inductors)

An electromagnetic field can be strengthened by coiling or looping the wire, as shown in Figure 1–21. The magnetic fluxes from the two sides of the loop add and strengthen each other; the more turns, the stronger the field, Figure 1–22. In fact, the strength of the magnetic field is proportional to the number of turns multiplied by the current flow. The field intensity of a coil can be determined by using the equation:

$$H = \frac{2 \times \pi \times N \times I}{10 \times r} \tag{18}$$

Where:
H is the field intensity in oersteds
N is the number of turns on the coil

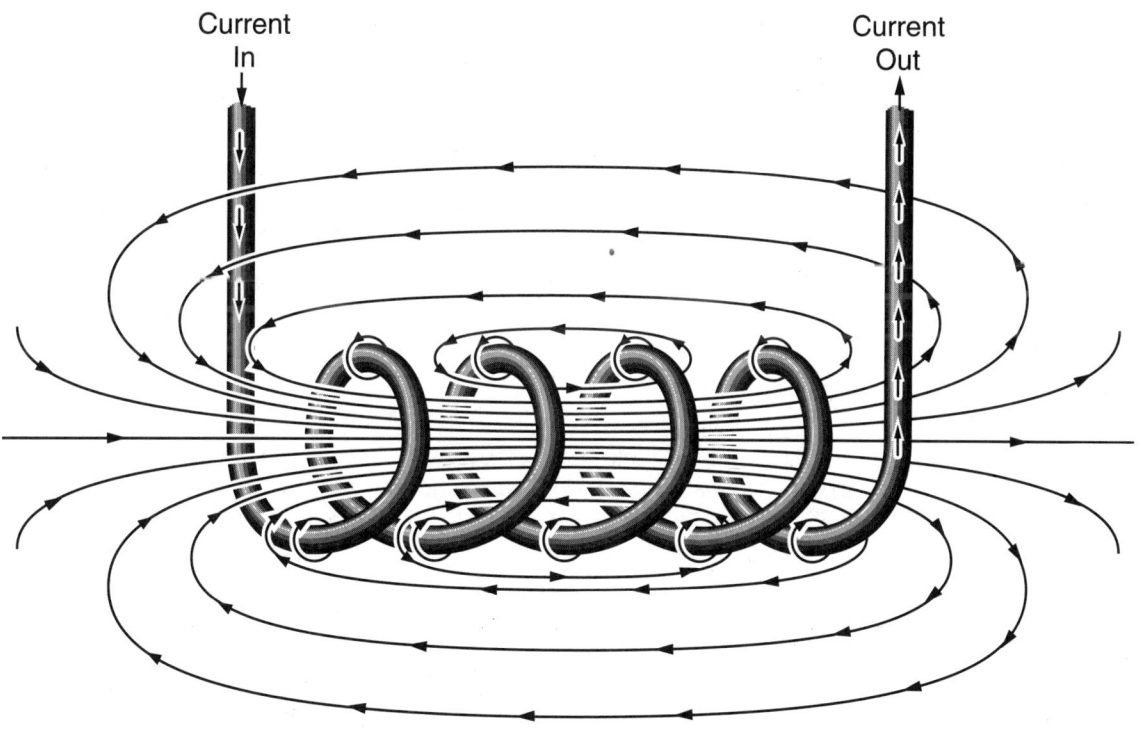

FIGURE 1–22
Magnetic field strengthened by using multiple turns of wire (Courtesy of Cadick Professional Services)

FIGURE 1–23 Left-hand rule for a coil

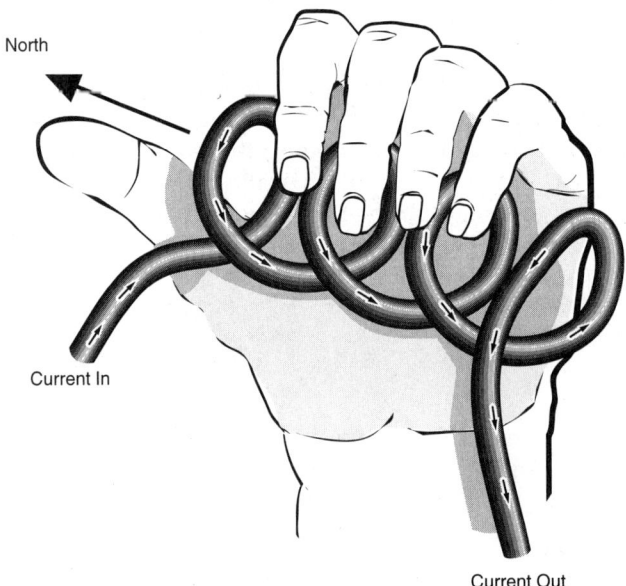

I is the current flow in amperes
r is the radius of the coil turns in centimeters

The left-hand rule can also be used to determine the polarity of a coil's magnetic field, Figure 1–23.

AMPERE'S LAW

Ampere's law, developed by André M. Ampere, states that *any current-carrying conductor located in a magnetic field at right angles to the lines of force will be pushed by a force that is directly proportional to the flux density, the current, and the length of the conductor.*

This law serves as the fundamental, underlying principle upon which all electric motors work. The direction of the force can be determined by using the right-hand motor rule as shown in Figure 1–24. First the thumb, index finger, and middle finger of the right hand are all positioned at right angles to each other. If the index finger is then pointed in the direction of the flux (north to south) and the middle finger pointed in the direction of the electron current flow, the thumb will point in the direction of the force. The right-hand rule can be used to determine the direction that a motor will turn. This will be covered in more detail later.

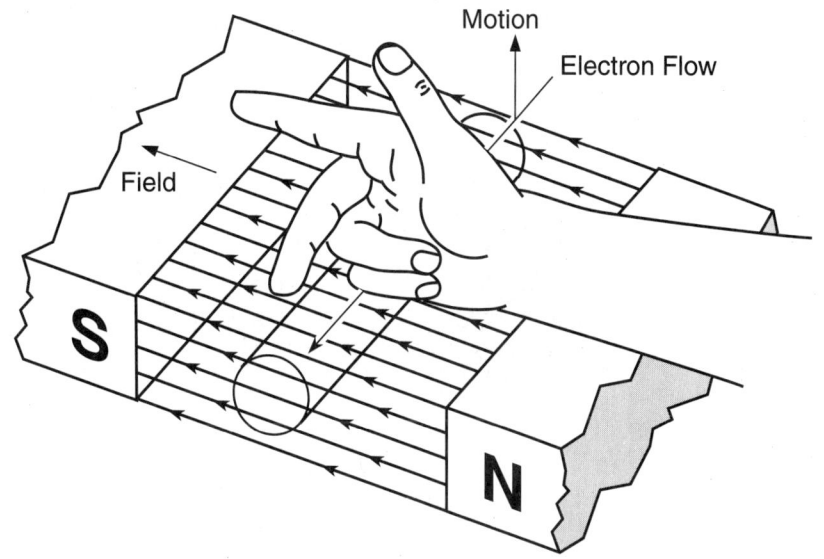

FIGURE 1–24
The right-hand rule for force on a conductor. (Also called the Right-Hand Motor Rule)

ELECTROMAGNETIC INDUCTION

Faraday's Law

Michael Faraday discovered and detailed the basic law of electromagnetic induction. This law, illustrated in Figure 1–25, shows that as a conductor is moved through a magnetic field, a voltage is generated across the conductor. If the conductor is connected to a complete circuit, a current will flow. If the conductor is moved in the opposite direction, a voltage and current of the opposite polarity will be generated. The polarity of the induced voltage can be determined by using the left-hand generator rule shown in Figure 1–26.

Note that it does not matter if the conductor moves and the field is stationary, or the field moves and the conductor is stationary. Because of this fact, a voltage will be induced in the secondary winding of the transformer, Figure 1–27, since the expanding and collapsing magnetic field is cutting through the windings on the secondary and, therefore, creating an output voltage.

Faraday's law defines the magnitude of the voltage generated when magnetic lines of flux are cut by a conductor, and it states that: *the magnitude of the generated voltage is proportional to the number of lines of flux that are being cut each second.*

The output voltage can be increased by increasing the rate of cutting, the speed of the conductor, or the number of lines by increasing the magnetic strength.

FIGURE 1–25
Creating a voltage with electromagnetic induction

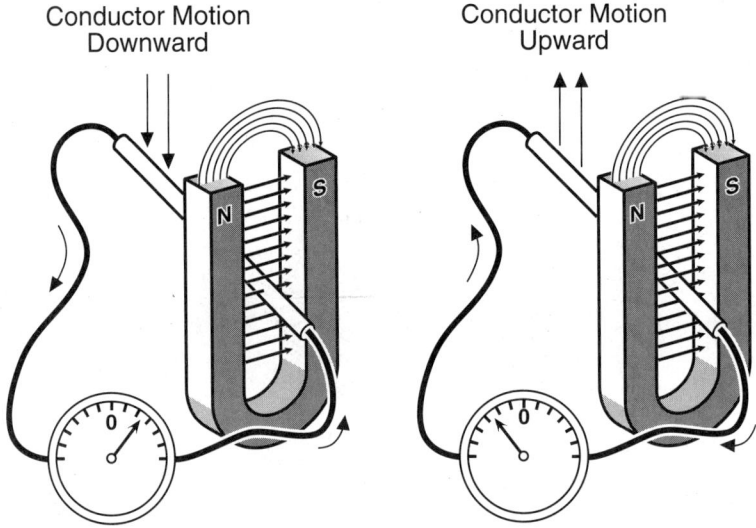

The law of electromagnetic induction is the fundamental principle upon which all generators work.

Lenz's Law

Another important principle was first developed by H.F. Emil Lenz in 1834. Lenz's law states that: *A current set up by a voltage induced due to the motion of*

FIGURE 1–26
The generator left-hand rule

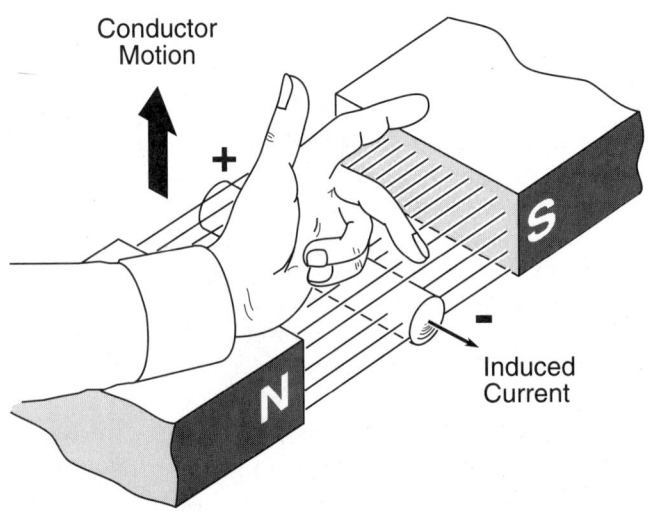

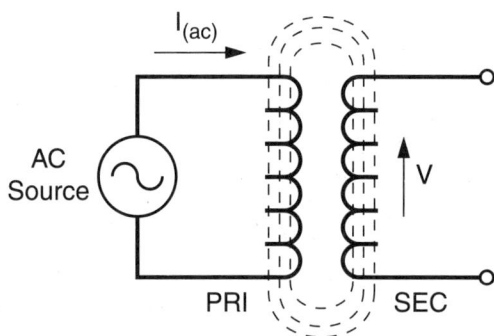

FIGURE 1–27
An expanding and collapsing field cuts through the secondary windings and creates an output voltage (Courtesy of Cadick Professional Services)

a closed circuit conductor will create a magnetic field which **opposes** *the field which caused the current in the first place.*

This means that as the conductor moves through the magnetic field set up by the magnet, the current flow in the conductor creates a field which opposes its motion, Figure 1–25. In other words, a resistant force will be felt as the conductor is pushed through the field. This is the reason why it is harder to turn a generator as the load increases on it. The output current increases, which increases the magnetic field, and therefore increases the push.

SUMMARY

Magnetism is the attraction that certain substances, called magnets, have for ferrous materials such as iron. Magnetism was first observed in naturally occurring rocks called lodestones; nevertheless, magnetism's greatest use is found in its ability to be induced into other substances by physical or electrical means. The magnetic force of a magnet is concentrated in two areas called poles. These poles are of equal strength and attract iron with the same force. Poles are of two types—north poles and south poles. Like poles repel each other and unlike poles attract.

The basic unit of electricity is found in sub-atomic particles called electrons and protons. These particles exhibit electrical properties of equal magnitude and opposite polarity. Electrons, the negative charged electrical particles, can be pushed free from their atoms and caused to move or drift from one place to another. This action is called electric current. Energy can be moved from one location to another with an electric current.

Magnetism and electricity are related to each other. Electric current flow creates magnetism, and magnetic lines of flux cutting through a conductor will create an electric current. This relationship is the foundational principle for motors, generators, and transformers.

Two laws quantify the relationship between electricity and magnetism and a third law explains the fundamental behavior of the two when they interrelate. These laws are Ampere's Law, Faraday's Law, and Lenz's Law.

- Ampere's Law
  Any current-carrying conductor located in a magnetic field at right angles to the lines of force will be pushed by a force that is directly proportional to the flux density, the current, and the length of the conductor.
- Faraday's Law
  The magnitude of the generated voltage is proportional to the number of lines of flux that are being cut each second.
- Lenz's Law
  *A current set up by a voltage induced due to the motion of a closed circuit conductor will create a magnetic field which **opposes** the field which caused the current in the first place.*

REVIEW QUESTIONS

1. Magnetic objects which occur in nature are called _____.

2. Magnetic like poles _____ and unlike poles _____.

3. Magnetic force can be expressed as magnetic _____ of force.

4. The ability of a material to retain magnetism is called _____.

5. A certain coil is carrying 5 A and has a magnetic field intensity of 20 oersteds. The coil has 40 turns. You may use the value of 3.14 for the following three questions:

    What is the radius of the coil in centimeters? _____

    What is the field intensity if the current is doubled? _____

    What is the field intensity if the current is 20 A and the number of turns is reduced to 5? _____

6. A certain generator is producing 120 V output. What will the voltage output be if the magnetic field strength is doubled? _____

7. Measurements of a three-phase power system show an apparent power of 3000 VA and a reactive power of 2000 VAR. What is the true power in the circuit? (Refer to Figure 1–16.) _____

8. A generator is connected to a bicycle mechanism. The electrical output of the generator is connected to several loads which may be switched on and off at will. As more load is connected to the generator, the bicycle becomes more difficult to pedal. This is an illustration of _____ law.

9. Electric motors take advantage of the fact that a force is produced on a current-carrying conductor when it is located in a magnetic field. This principle is called _____ law.

10. A certain power system has an apparent power of 5000 VA, a reactive power of 3000 VAR, and a true power of 4000 W. What is the power factor, in percent, of this system? _____

Part II

AC Generators

AC Generator Theory and Construction 2

OBJECTIVES

After studying this chapter, the student should be able to:

- *Describe the construction of the basic AC generator.*
- *Describe how an AC generator creates an alternating voltage.*
- *Determine the operating frequency of an alternating current generator.*
- *Describe the basic construction of real AC generators.*
- *Explain the basic differences between small AC generators and large AC generators.*
- *Describe the construction of large AC generators, including rotors, excitation systems, and cooling systems.*
- *Explain the basic principles of AC generator control.*
- *Diagram typical single-phase and three-phase connections for AC generators.*

INTRODUCTION

The AC generator is at the very heart of the modern power system. Until the mid-1970s, virtually all power was generated in electric utility generating stations. Most of the few generators that were used in places other than electric utilities were for emergency power or backup purposes. Since that time, more and more industrial, commercial, and even residential generators are being used.

The simplest of all generators is the single-phase, AC generator, Figure 2–1. The most complicated machines are built using this simple structure as a basic building block. This chapter describes the basic operation of the AC generator and explains how it creates a voltage.

PRINCIPLES OF OPERATION

The simplest AC generator, Figure 2–1, is composed of a loop of wire that is free to rotate in a magnetic field. The loop of wire is called the *armature,* and the magnetic field is called the *field.* The armature is turned by a device called the prime mover. The prime mover can be water, steam, or wind turbines, a reciprocating engine, or an electric motor depending on the application. For simplicity, the prime mover is not shown in Figure 2–1.

The armature loop is connected to two *slip rings.* Each slip ring has an electrical conducting *brush* that slips over the surface of the ring as the armature rotates. A galvanometer, or voltmeter, is connected to the brushes.

As the armature rotates in the field, a voltage will be generated. This is the basic premise of Faraday's law. Note that the voltages which are created in the vertical portions of the loop cancel out; only the horizontal portions add to the total output voltage. Figures 2–2 and 2–3 can be used to describe the operation of the AC generator as follows:

- The armature rotates in a counterclockwise direction, as viewed from the front.
- In position A, the horizontal loops are not cutting through any flux lines, since they are moving in parallel with them.
- As the armature moves away from its A position, its windings start to cut through the flux lines. The farther from position A it moves, the more lines it cuts per second; therefore, the more output voltage is created.

FIGURE 2–1
Simplified AC generator

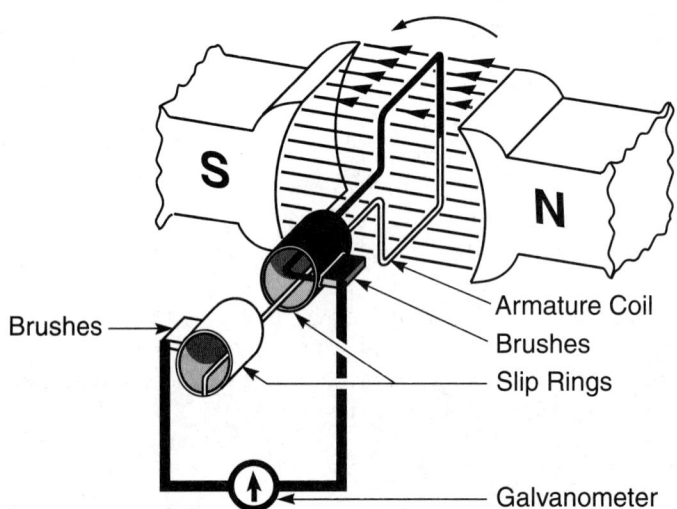

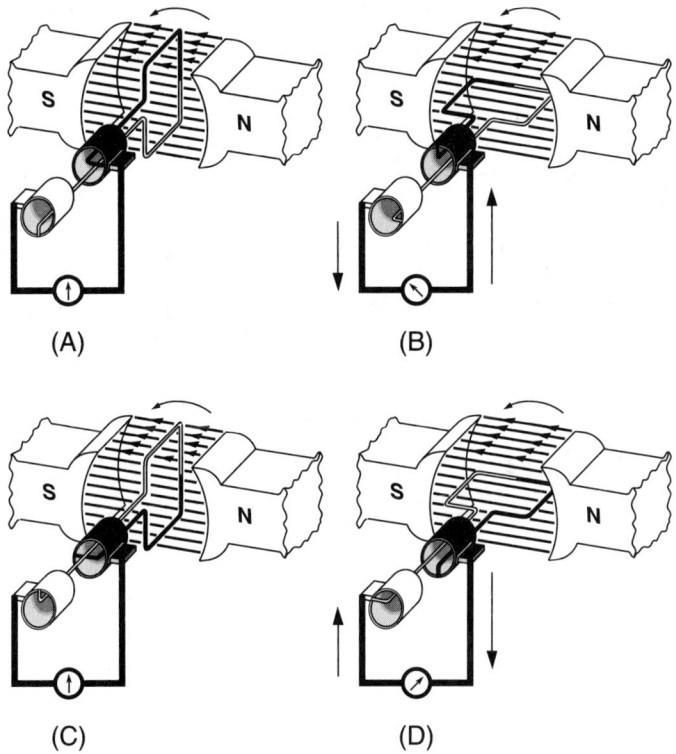

FIGURE 2–2
Generation of an alternating voltage

- When the armature reaches position B, its loops are moving perpendicularly to the field flux; therefore, they are cutting the maximum number of lines per second. According to Faraday's law, the maximum voltage is created.
- As the loop continues past position B, the voltage starts to drop off again since the loops are no longer perpendicular to the flux.
- Eventually, the armature reaches position C, where its motion is again parallel to the flux lines. Here, again, the output voltage is zero.
- As the armature moves from position C to D, the voltage again builds up to a maximum value. This time, however, the voltage is of opposite polarity, since the relative motion is opposite.
- In its final quarter turn, the armature voltage drops from its negative maximum to zero which completes the entire output.

The resulting voltage is plotted in Figure 2–3. This characteristic shape is called a sine wave.

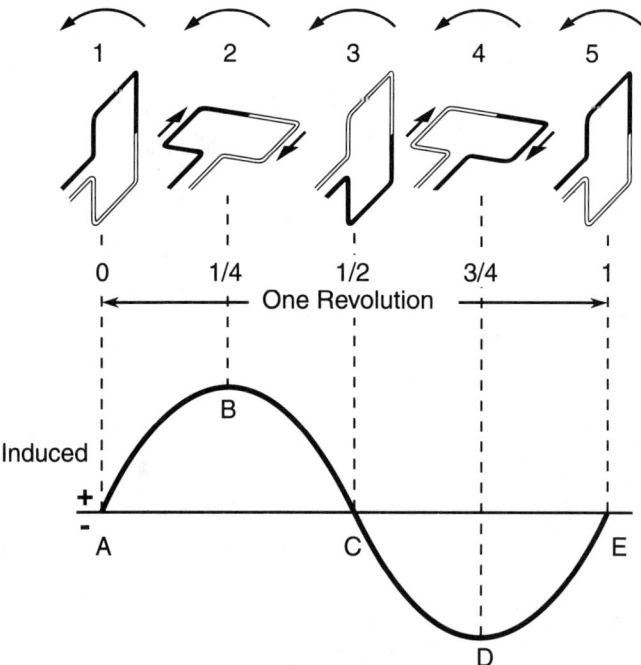

FIGURE 2–3
An alternating voltage sine wave

GENERATOR FIELD AND ARMATURE

With the exception of very, very small generators, magnetic fields are developed using electromagnets. Small generators have a stationary magnetic field and a revolving armature as shown in Figure 2–4. Larger generators eliminate the high current slip ring and brush requirements by employing a stationary armature, Figure 2–5.

DC field current may be supplied to the field by batteries, as shown in Figure 2–4; however, some form of DC generator or rectified AC supply, called an exciter, is normally employed.

The rotating shaft is called the *rotor* and the stationary coils are called the *stator*. These definitions apply regardless of whether the *armature* or the *field* is stationary. Note that the terms *armature* and *field* are electrical terms. In a generator (AC or DC), the armature windings are those windings that are connected to the load. The field windings are those that are used to create the magnetic field. The terms rotor and stator refer to the physical characteristics of the generator; the rotor *always* rotates and the stator is *always* stationary.

Three-phase alternating current can be produced by placing three separate windings on the stationary armature, Figure 2–6. The windings are placed 120° apart, but are otherwise identical. The magnetic field sweeps

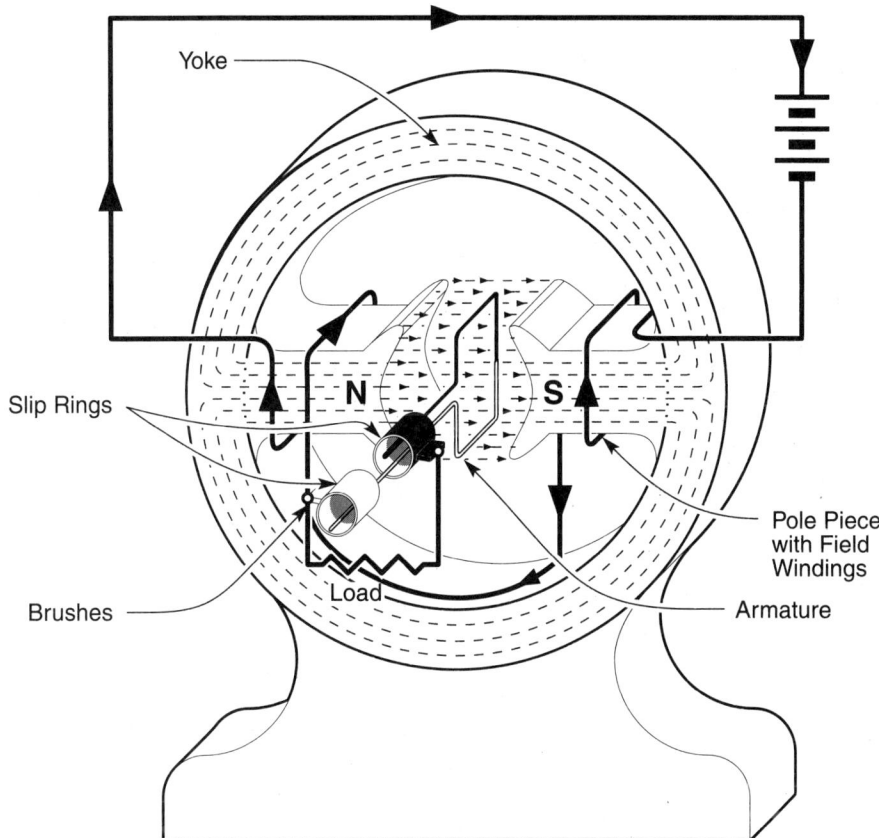

**FIGURE 2–4
AC generator with rotating armature**

across each armature coil as it turns, and with each full revolution it produces the same number of cycles in each winding. The output voltages from each winding will be 120° out-of-phase.

In some generators, the field can be constructed with more than one pair of magnetic poles: that is, two north poles and two south poles, Figure 2–7. The poles are constructed so that north and south poles are installed at alternating positions around the rotor circumference. A four-pole rotor causes a stator coil to be swept by two complete cycles of north and south magnetic fields in one revolution. Thus, the armature produces two full alternations with each revolution. A four-pole rotor can, therefore, rotate at half the speed of a two-pole rotor and produce the same AC frequency. In order to make use of all the magnetic fields of the rotor, the stator windings must contain coils for each pair of poles in each phase. Thus, a four-pole, three-phase AC generator will have two stator coils mounted 90° apart for each of its three-phases.

The frequency of an AC generator can be shown to be

$$f = \frac{S \times P}{120} \tag{1}$$

Where:
f is the output frequency in hertz
S is the generator speed in revolutions per minute (rpm)
P is the total number of north and south poles

Thus, a two-pole, AC generator operating at 3600 rpm will generate a 60-Hz output. Likewise, a four-pole generator will produce a 60-Hz output when it rotates at 1800 rpm since:

$$60 = \frac{3600 \times 2}{120} \text{ and } 60 = \frac{1800 \times 4}{120} \tag{2}$$

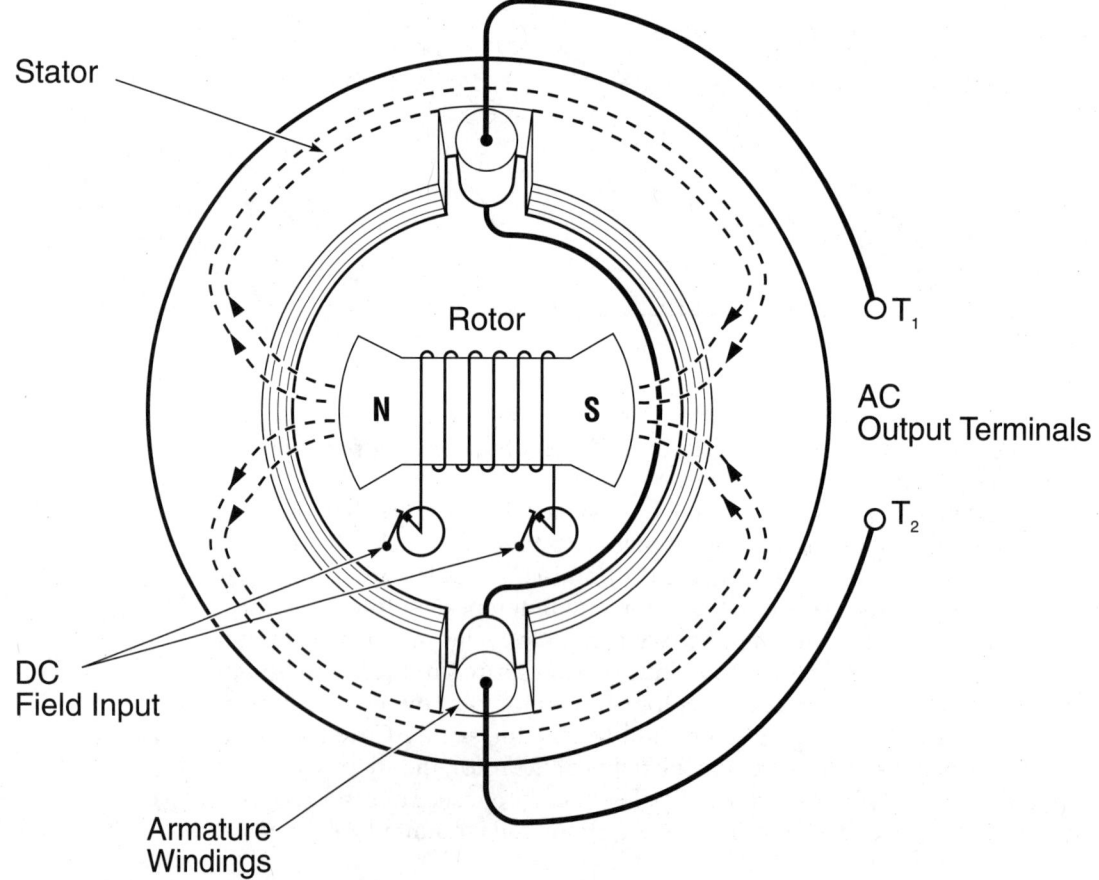

FIGURE 2–5 AC generator with stationary armature

AC Generator Theory and Construction

CONSTRUCTION OF SMALL GENERATORS

General

A small generator is used for local or emergency power generation. Usually less than 15 kilowatts in capacity, they are not normally applied for central station generation by electric utilities. Figures 2–8 and 2–9 are examples of such generators, complete with their prime movers.

Many small generators have revolving armatures and require slip rings and brushes; however, revolving fields have become more common owing to their superior electrical and mechanical design features.

Excitation Systems

General The excitation system supplies the DC current to the field of the generator. All small generator systems are equipped with integrally-mounted excitation systems. The degree of sophistication of the supply is

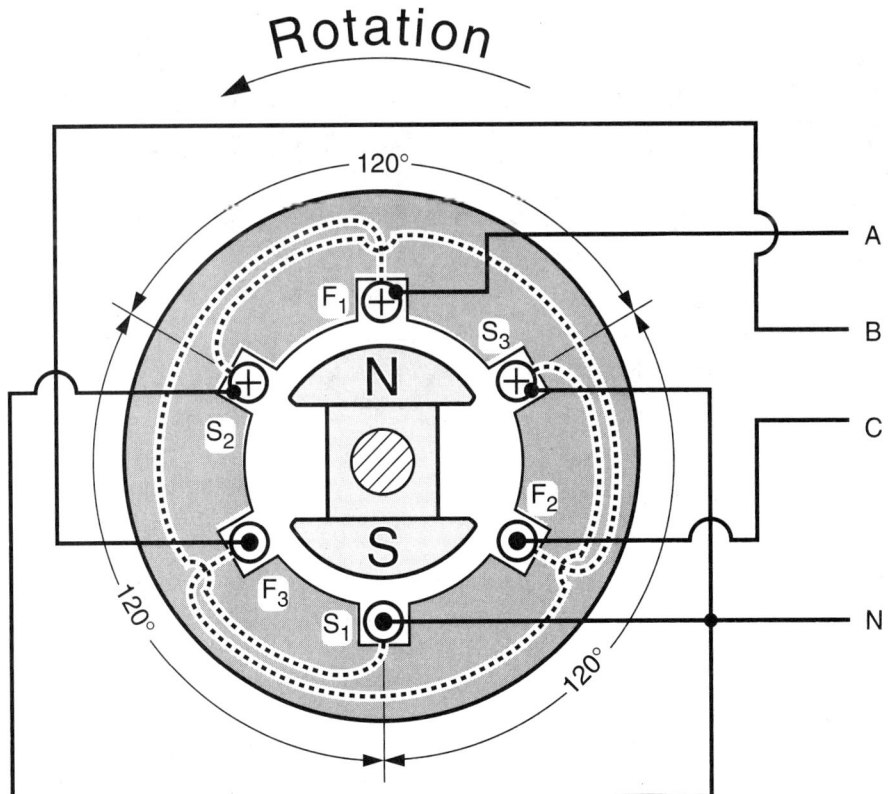

FIGURE 2–6
Three-phase, two-pole AC generator connections (Courtesy of Cadick Professional Services)

FIGURE 2–7
A four-pole rotating armature (Courtesy of Onan Corporation)

determined by the applications requirements. The following paragraphs describe the most common types of excitation systems.

Batteries Some smaller generators use a storage battery system to supply the DC current to the field winding, Figure 2–10. The DC current is supplied to the field through an optional excitation control circuit. The output of the armature is connected to the load and to the charger which supplies charging current to the battery.

Although battery systems provide a very stable voltage supply to the generator field, they are not commonly used because of their relatively high maintenance requirements. Also, modern electronic rectifiers and reg-

AC Generator Theory and Construction

FIGURE 2–8
2.5 kW gasoline engine powered rotating armature, emergency generator with remote start capabilities (Courtesy of Onan Corporation)

ulators are capable of such excellent regulation and ripple elimination that the constant output of the battery is not required.

Battery excitation systems of this type are not commonly used in modern designs.

Permanent Magnet Excitation Permanent magnet generators have two generators on the same shaft. One of the generators, the main generator, is the one that is used to supply load current. The other smaller generator is of the permanent magnet variety as described previously in this chapter. As the prime mover starts to turn, the permanent magnet generator produces an output which is applied to the field winding of the main generator.

Permanent magnet generators (PMG) are constructed in different ways. In the *brushless* type of PMG, the magnet is rotated and the armature is stationary. The output of the armature is fed to a regulator which, in turn, is connected to the field of the main generator.

FIGURE 2–9
1100 kW diesel engine powered rotating field generator (Courtesy of Onan Corporation)

Some small PMGs are constructed with the magnet on the stator and the armature rotating. The main generator is of the revolving field type. The output of the exciter is connected to a rotating bridge diode system which connects to the field of the main generator. These types of systems depend on the native regulation capability of the main generator to control the output voltage and phase angle.

PMGs are among the more commonly used in modern designs.

Self-excited Self-excitation systems are simpler than previously discussed types since they do not require a second generator. The output of the main generator is fed to a regulated, excitation control circuit. The output of the excitation control is connected to the field of the generator, Figure 2–11.

FIGURE 2–10
Battery excitation system (Courtesy of Cadick Professional Services)

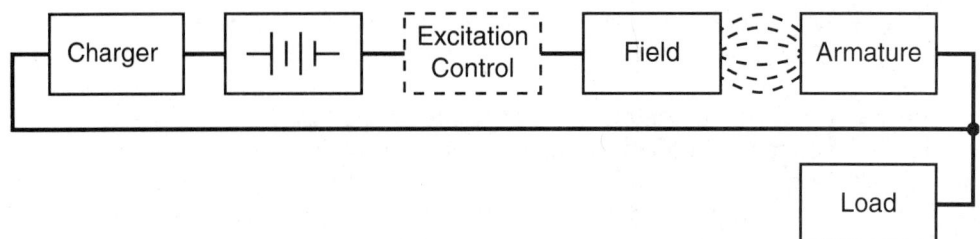

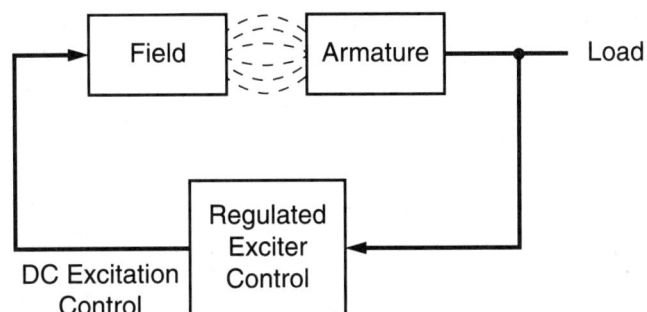

FIGURE 2–11
Self-Excited Generator
(Courtesy of Cadick Professional Services)

If a self-excited system is the only generator connected to the load, it must have some way to develop output when it is first started. Several different techniques have been employed. *Flashing* is a term that is used to describe the momentary application of a voltage to the field. When this occurs, the short burst of current establishes a brief but sufficient magnetic field. This short burst of field creates an armature output that provides input to the exciter control, which maintains the field. Some engine-driven generators have a flashing circuit built into them which uses the starting batteries for the supply.

More recent, self-excited designs depend on the retentivity of the generator's laminations. The residual magnetism in the field laminations is sufficient to create an output voltage. This voltage is applied to the exciter control and the resulting process is identical to that described for flashing. Should a problem ever occur which causes the residual magnetism to drop below the value required for starting, the field can be flashed using a DC supply, such as a battery and a momentary contact switch. A general approach to this procedure is shown in Figure 2–12. The current limiting resistor may or may not be required, depending upon the battery voltage and the field coil current rating. Refer to the manufacturer's literature for specific information.

Induction Generators The generators described so far are called *synchronous* generators, because they have their own source of excitation, and are driven at synchronous speed. *Induction* generators do not have a separate excitation source. Instead, the field windings are short-circuited. This type of generator employs a revolving field. When a voltage is applied to the armature, a *rotating magnetic field* is established in the air gap between the rotor and the stator. The prime mover turns the rotor faster than the rotating magnetic field. This action causes excitation current to be induced into the field coils.

Obviously, an induction generator draws excitation current from the system to which it is connected. This means that an induction generator must *always* be connected to a system that has a synchronous generator to supply the excitation current.

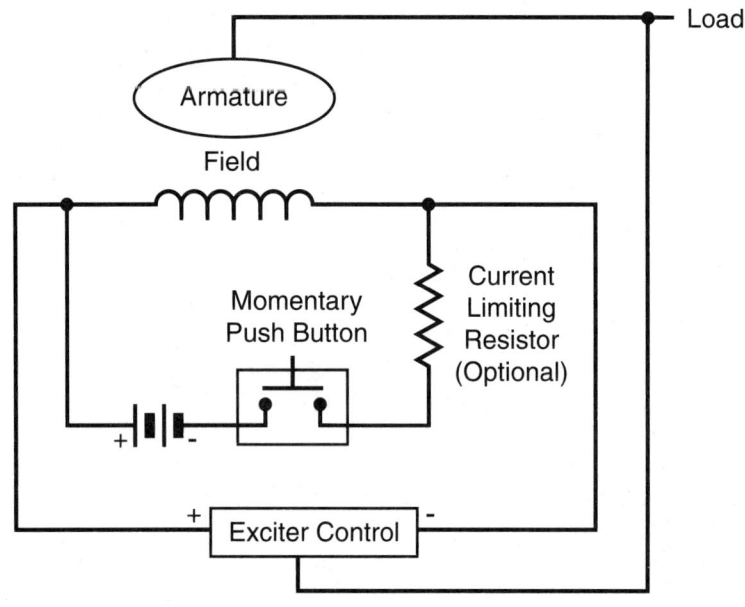

FIGURE 2–12 Circuit for flashing the field (Courtesy of Cadick Professional Services)

CONSTRUCTION OF LARGE GENERATORS

Rotors

Salient-pole   All large generators employ a revolving field. Figure 2–13 is an example of a *salient-pole* rotor. This type of rotor is used when the prime mover is relatively slow, such as a water turbine. The field poles are formed by fastening a number of steel laminations to a spoked frame, also known as a *spider*. The heavy pole pieces produce a flywheel effect on the slow speed rotor. Some salient-pole rotors weigh 300 tons or more. This helps to keep the angular speed constant and reduces variations in the voltage and frequency of the generator output.

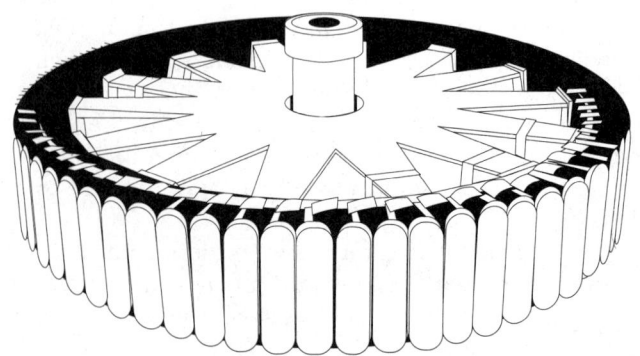

FIGURE 2–13 Salient-pole rotor for large hydroelectric generator

AC Generator Theory and Construction

For large, low-speed salient-pole rotors, the field coils are wound and placed around the poles before they are fastened onto the rotor. The field windings can be insulated with spiral windings of cotton, silk, paper, or fiberglass tape. These materials are impregnated with an insulating compound that fills the spaces between the fibers of the material. Another form of insulation is a coating of a compound that is applied directly to the wire and baked on.

Note that many small generators also use salient-pole rotors, even though they are operated at 1800 or 3600 rpm. Figure 2–7 is a photograph of such a small, salient-pole rotor.

Turbo-Rotors (Round Rotors) Large, high-speed gas or steam turbines are not well suited for the very heavy salient-pole rotors. The weight and inertia of the salient-pole construction would cause the rotor to fly apart at high speeds. In addition, the windage loss of such a rotor becomes excessively high at high speeds.

The turbo-rotor, Figures 2–14 and 2–15, is used in high speed generators for two reasons:

1. The turbo-rotor has less windage loss because of its smooth design.
2. The windings on a turbo-rotor are installed so they can handle the centrifugal forces at high speed.

High speed turbo-rotors are a solid steel forging or a number of steel discs fastened together. The copper field coils are placed in slots in such a way that they distribute the flux evenly around the rotor. Because of the high speed and temperatures at which the round rotor operates, the field windings are often insulated with mica.

Excitation Systems

Rotating Exciters *Rotating exciters* are generators that are used to supply the DC excitation current to the main generator. The rotating exciters may be DC generators, as described later in the text, or AC generators with rectified output. A rotating exciter may be driven by the same shaft as the

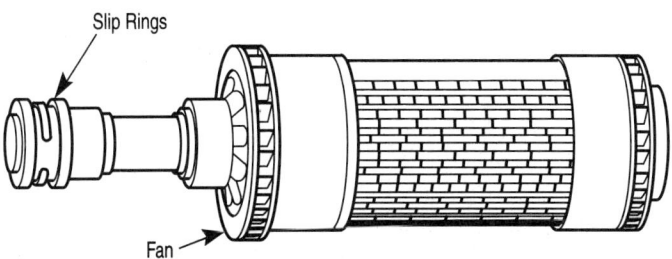

FIGURE 2–14
Turbo Rotors (Round Rotors)

FIGURE 2–15
Close-up view of a turbo-rotor (Courtesy of General Electric Company)

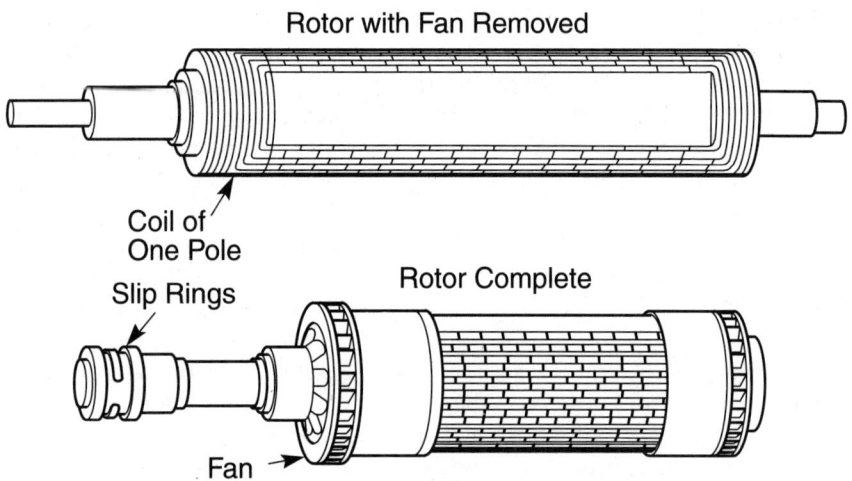

main generator, or by a separate prime mover. In either case, their operation is much like the exciters described previously in the small generators sections.

Static Exciters Because a *static exciter* has no moving parts, it generally has less stringent maintenance requirements, Figure 2–16. Still, the static exciter presents more of a protection problem for overcurrent situations. When a short circuit occurs on the generator output terminals, the voltage decays to the exciter. This causes the generator output to drop very quickly, resulting in a reduced current output.

A characteristic of this type is called a generator decrement curve, Figure 2–17. This curve shows the change in generator current as a function of

FIGURE 2–16
Static exciter system (Courtesy of Cadick Professional Services)

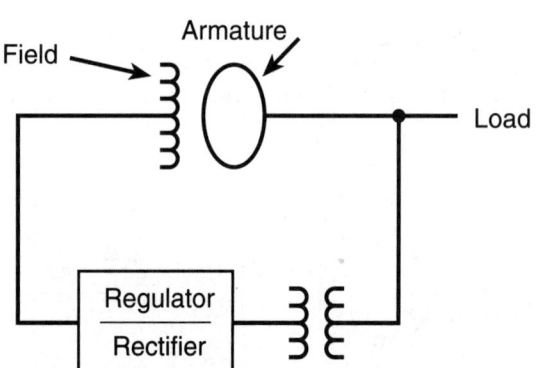

AC Generator Theory and Construction

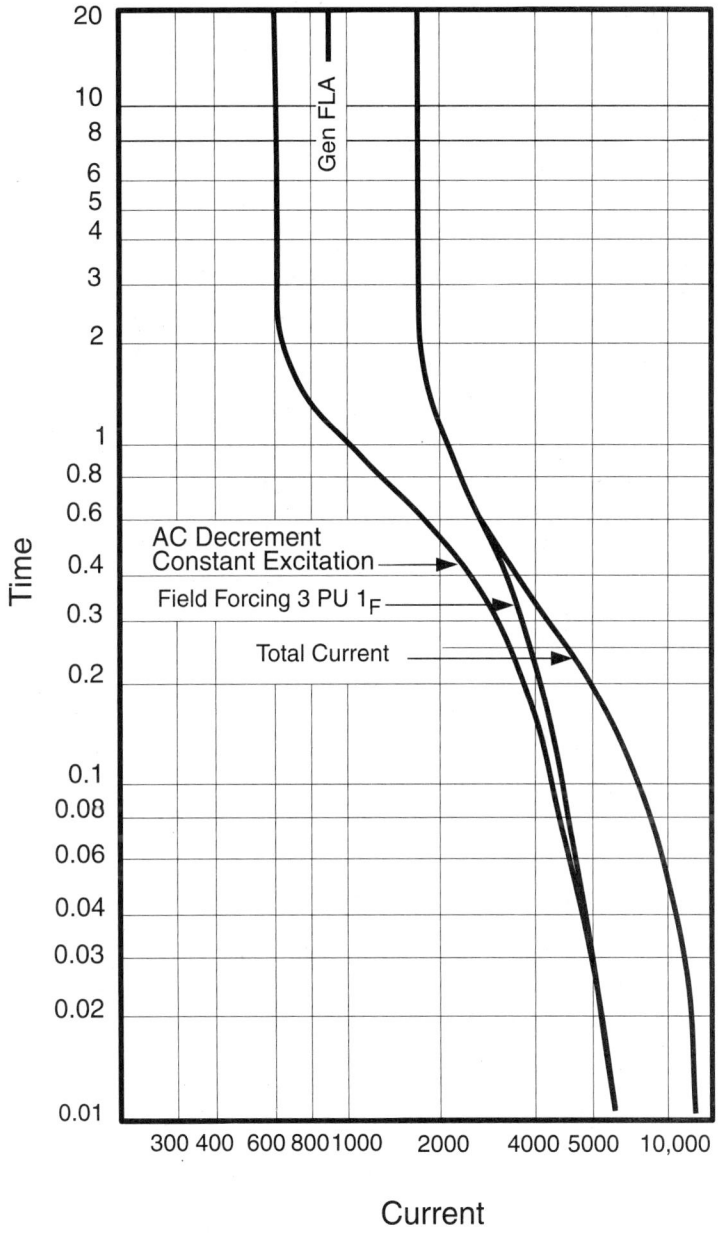

FIGURE 2–17
Generator decrement curve for a 19,500 kVA 12.47 kV generator
(Courtesy of IEEE Std 242-1986, IEEE Recommended Practices for Protection and Coordination of Industrial and Commercial Power Systems)

time after the initiation of a short circuit. The constant excitation curve drops to a value of 600 A in about three seconds. Full load current for the generator is 903 A. Overcurrent protection on such a system can be very difficult.

The actual operating curve may even be worse than shown by Figure 2–17 because the excitation will probably not be constant. Static excitation systems tend to decrement much more quickly than rotating systems, because they have no inertia. Rotating systems are more commonly used for large generators.

Cooling Systems

Air-Cooled Many small to medium-sized generator systems are open, air-cooled systems. This means that the stator and rotor windings are open to the atmosphere. Air is forced over the windings, usually by means of a fan mounted on the rotor, Figure 2–14.

Other systems use a filter to remove small particles from the air. In these generators, the air is drawn through the filter by the fan. The air is cleansed of particles and then passed through the air gap.

Very large high speed systems do not normally employ these types of cooling for three reasons:

1. Air is only a moderately effective coolant. It cools by convection only, and is not extremely efficient.
2. If the system is open to the air, dirt, dust, and liquid sprays can get into the air gap and cause abrasion. If this becomes severe, the windings can be damaged and cause a major failure of the system.
3. At 3600 rpm, or even 1800 rpm, air creates significant drag (windage) on the rotor as it turns. This results in higher losses, greater heating, and less efficiency.

Figure 2–18 shows an example of an air-cooled generator.

Air/Water-Cooled Systems An air/water-cooled system is closed and sealed to the environment. Air is circulated over the rotor and through ventilation tubes that are located in the stator windings. The air is then pumped to a heat exchanger where it is cooled by circulating water. Air/water-cooled systems greatly improve the efficiency of the cooling and eliminate the problem of foreign particles in the winding. However, the air is still a relatively inefficient cooler, and still has a windage problem on very large systems. Figure 2–19 shows an air/water-cooled generator.

Hydrogen-Cooled Systems A hydrogen-cooled system is similar to an air/water-cooled system except hydrogen is substituted for the air. Because hydrogen has such a light molecule, windage is virtually eliminated. Moreover, hydrogen conducts very well; therefore, the cooling effect is both convective and conductive.

AC Generator Theory and Construction

FIGURE 2–18
A General Electric 7A6 air-cooled generator during assembly. (Courtesy of General Electric Company)

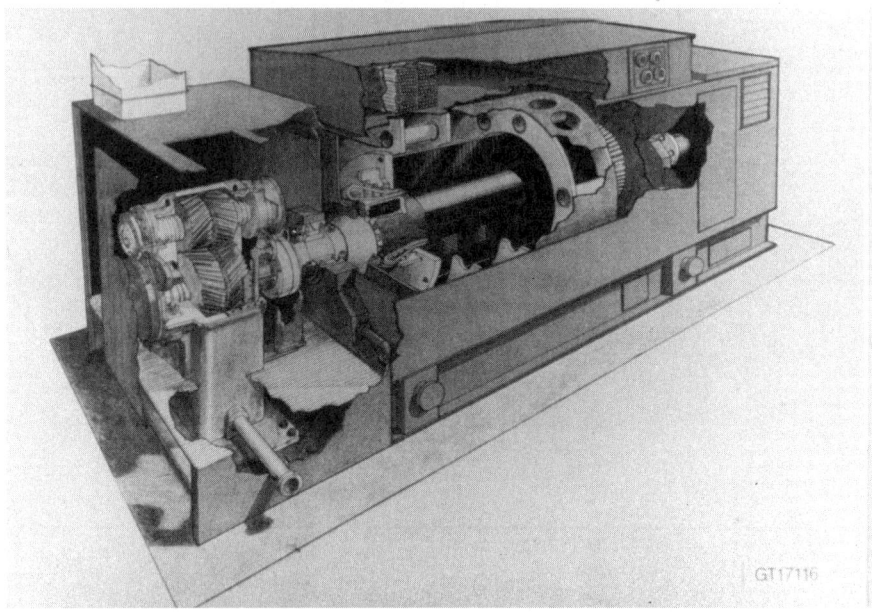

FIGURE 2–19
A cutaway drawing of a General Electric design for a totally-enclosed water-to-air cooled (TEWAC) generator. (Courtesy of General Electric Company)

Hydrogen is explosive in an oxygen atmosphere; nevertheless, if concentrations are kept at least 75% pure and preferably above 95%, hydrogen is extremely stable and unlikely to explode. The hydrogen atmosphere inside the generator is monitored with special equipment. Should a leak develop which reduces the concentration below acceptable values, the alarm system alerts system operators and the system is shut down. Figure 2–20 shows a large, hydrogen-cooled generator.

The seals at the end of the rotor use a layer of oil to trap any hydrogen that leaks through. The oil is circulated through a filtering system where any hydrogen that is present is removed and vented to the atmosphere. The carbon dioxide is used to purge the system of air before the hydrogen is applied.

GENERATOR CONTROLS

General

Three generalizations may be made about the control of an AC generator's output:

1. When additional load is added to a generator, the increased current causes an opposing magnetic field (Lenz's law) which tries to slow the machine. If the machine slows, the frequency will go down. To maintain

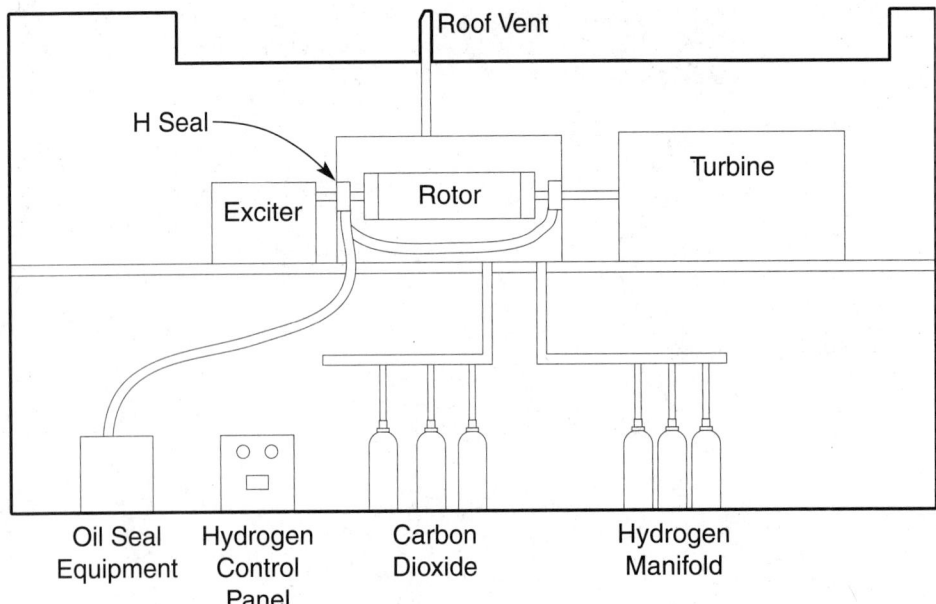

FIGURE 2–20
Typical hydrogen cooling system (water system not shown) (Courtesy of Cadick Professional Services

AC Generator Theory and Construction

speed and frequency, the input power to the prime mover must be increased. If the input power to the prime mover is increased with no additional load added, the generator frequency will increase.
2. Voltage changes are created by a change in flux (Faraday's law). If the field is increased, the output voltage increases. Therefore, voltage is controlled by the excitation level.
3. Reactive power, and therefore phase angle, is also controlled by the exciter.

Single, Isolated Generators

The control and operation of a single generator is relatively simple. The following points cover most requirements, Figure 2–21:

- The engine or turbine controls generator speed and, therefore, frequency.
- As the load varies, the generator governor must open or close the throttle to keep the generator speed constant.
- Generator output voltage is controlled by the generator field current.
- As the load current varies, the generator voltage regulator automatically adjusts the field current to keep the generator voltage constant.
- As the load VAR requirements vary, the regulator automatically adjusts the field current.
- Since only one generator is being used, it supplies the entire load. Because of this, the load determines how many watts and VARS the unit must deliver.

Multiple Generators

Multiple generators make the control picture somewhat more complex, Figure 2–22. Any change in load must be compensated by one or a combination of all generators. Any change in output of watts or vars from one generator must be compensated by other generators in the system. Assume that initially generator G1 is carrying all of the load and generator G2 is coming on line.

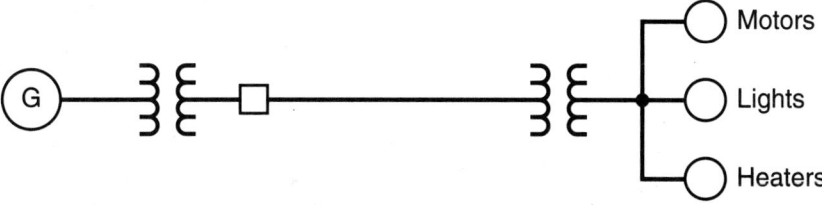

FIGURE 2–21
Single generator supplying load
(Courtesy of Cadick Professional Services)

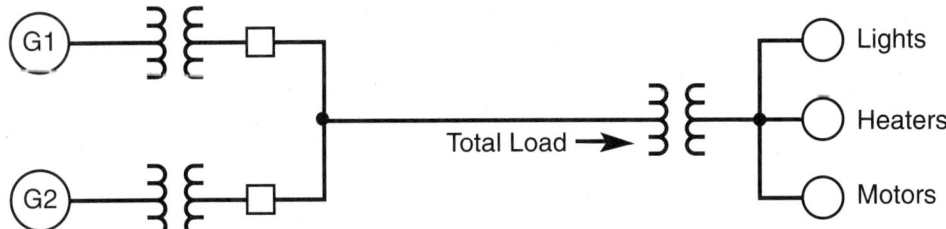

FIGURE 2–22
Two generators in parallel (Courtesy of Cadick Professional Services)

- Initially G1 is operating as a single generator and the previously described control steps apply.
- When G2 is started and synchronized, it will carry 0 W and 0 VARS.
- As the turbine throttle is opened on G2, it begins to supply watts to the load. Unless the throttle on G1 is closed slightly, both units will speed up.
- As G2 picks up load, it will draw excitation current from G1. To compensate for this, the G2 operator must increase the field current. If the excitation is increased to compensate just for the required reactive power, the VAR output of G2 will again be 0.
- If the G2 operator continues to increase the excitation current, G2 will try to supply additional VAR to the system. Since the load does not require the additional VAR, the system voltage will increase. This increase can be compensated by reducing either G1 or G2 excitation so the machines supply the total VAR required by the load.

ELECTRICAL CONNECTION OF GENERATORS

Single-Phase

Single-phase generators are used when only small loads are served. Single-phase generator voltages are usually less than 480 V and typically in the 120 or 240 V category. Figure 2–23 shows the most common single-phase connections. In Figure 2–23a, the armature winding is tapped in the center. If the entire winding supplies 240 V, each leg to the center tap will supply 120 V.

If only a single-phase is required, the connection shown in Figure 2–23b is used.

Three-phase Generators

Figures 2–24 and 2–25 show the most common three-phase connections. In Figure 2–24 the ends of each of the three armature windings are connected together to form a delta configuration. In this connection, the phase-to-

AC Generator Theory and Construction

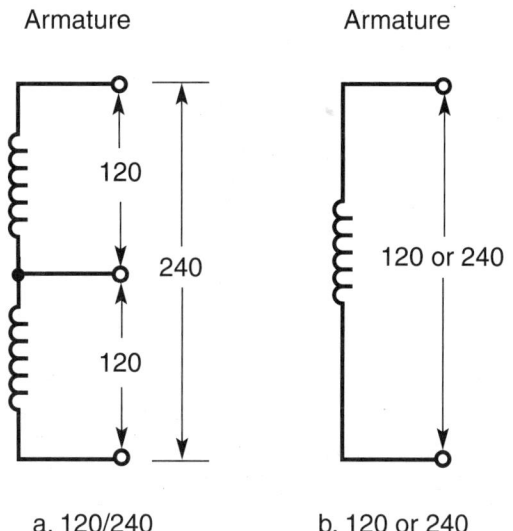

FIGURE 2–23
Single phase 120, 240 or 120/240 V output (Courtesy of Cadick Professional Services)

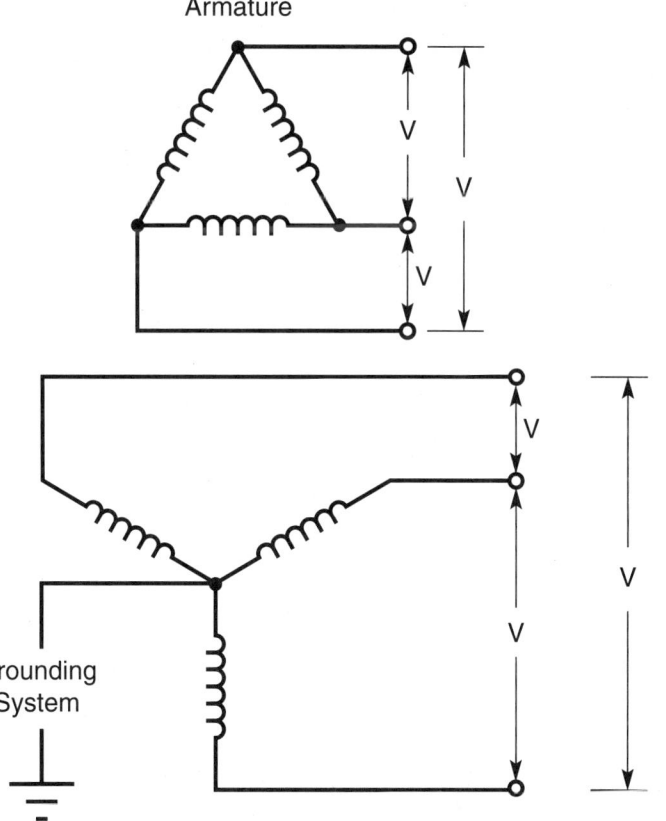

FIGURE 2–24
Delta output connection for voltages from 120 to 22,000 V (Courtesy of Cadick Professional Services)

FIGURE 2–25
Wye output connection for voltages from 120 to 22,000 V (Courtesy of Cadick Professional Services)

phase voltage and the winding voltage are equal to each other. The phase current, however, is equal to the phase current divided by the square root of three.

Figure 2–25 shows the more common wye type connection. In this connection, the winding voltages are equal to the phase voltages divided by the square root of three. The winding currents are equal to the phase currents.

Generator Grounding

Grounding one of the wires of a power system allows static charges to drain, reduces transient overvoltages, and stabilizes the system voltages. Figure 2–26 shows three ways that generators are grounded.

Figure 2–26a is the solidly-grounded system. This method is rarely used except on the smallest of generators. Generators can create very substantial third harmonic voltages (180 Hz in a 60-Hz generator). Large portions of these currents are in-phase in each of the three phases. If the neutral is solidly grounded, third harmonic currents can flow and create overheating problems in connected equipment.

If one phase becomes grounded on a solidly-grounded generator, the fault current in the faulted phase can actually exceed the current if all three phases are tied together. Because the generator is built and braced for the three-phase magnitude, it may be damaged by the single-phase fault.

Because of these reasons, generators are grounded using a circuit, as shown in Figure 2–26b or 2–26c. Figure 2–26b is commonly used on low-

FIGURE 2–26 Common generator grounding methods (Courtesy of Cadick Professional Services)

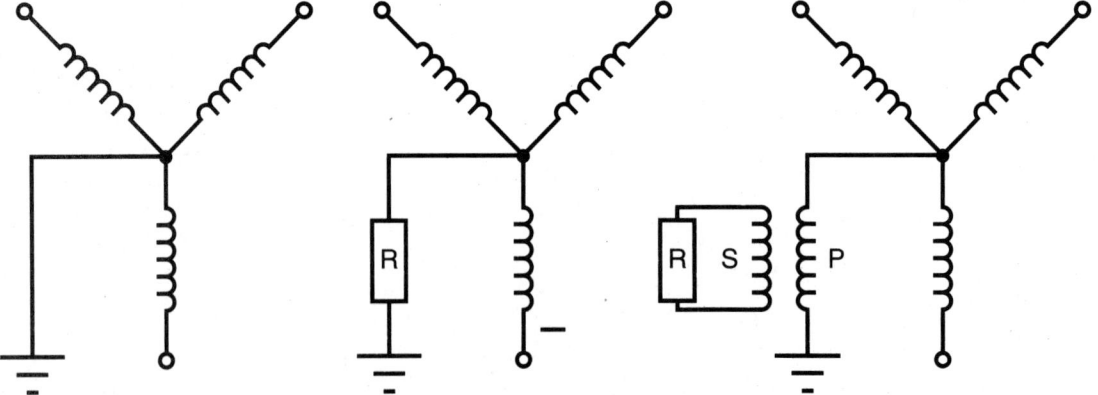

a. Solidly grounded (not preferred)

b. Resistance grounded

c. Resistance grounded with distribution transformer

voltage (less than 480 V) generators. The resistor will limit both third harmonic currents and ground fault currents to acceptable levels.

Figure 2–26c is used for medium-voltage generators up to 22,000 V. The transformer is a distribution class with a primary voltage (P) equal to the generator phase-to-phase voltage and a secondary voltage of 480 V. During a ground fault, the voltage on the primary is equal to the generator phase-to-ground voltage, and the secondary voltage is 277 V.

The resistor limits the current flow in the secondary and, therefore, limits the current in the primary. This method allows lower voltage resistors to be used, and also allows ground fault protection to be employed by connecting an overvoltage relay on the secondary side of the transformer. Such protection will be discussed in the next chapter.

SUMMARY

In their simplest form, AC generators consist of a loop of wire, called the armature, which is rotated in a magnetic field. By Faraday's law, a voltage is created in the armature which can then be connected to an electrical load. The output of such a generator is taken from a set of slip rings which are contacted by electrical brushes. The frequency of output is determined by equation (1).

AC generators may have a revolving armature or a revolving field. The revolving armature is used primarily for smaller generators with minimal load currents. The revolving field is more common and is used on all sizes of generators, especially large ones.

Small generators are usually of single-phase construction and use diesels or gas turbines for prime movers. Larger machines are driven by water, gas, or steam turbines. The excitation systems used by large generators are usually more sophisticated. Small generators often use permanent magnet generators to provide the excitation current.

Large generators are equipped with either salient-pole or turbo-rotors. The salient-pole types are used on slow rotation water turbines, and the turbo-rotor types are used on the high speed steam or gas turbines. Large generators are all of the rotating field types.

Excitation systems of large generators are usually of the rotating type; however, some use a static type of exciter. Cooling systems may be open, filtered air, air/water, or hydrogen-cooled. The smaller generators may use the open or filtered air methods, and the larger ones use hydrogen pumped through cooling tubes in the stator. Hydrogen is also blown through the gap between the stator and rotor. Hydrogen concentrations must be kept above 75% to assure safe operation.

True power output of a generator is controlled by the prime mover. When more load power is required, the prime mover throttle is opened and the generator produces the needed current. If the prime mover is increased when no additional load is added, the generator will speed up and increase its frequency.

Voltage output and power factor of the generator can be controlled by the excitation current. In general, the greater the excitation, the higher the output voltage and the more lagging the power factor.

Generator connections may be either single-phase or three-phase. Three-phase connections may be delta or wye configured. Most generators use a wye configuration with the neutral point grounded through a resistance.

REVIEW QUESTIONS

1. Assume that the loop has rotated 315° from its starting position (A). If the loop is connected to a load, will the electrons be flowing into or out of the black lead?

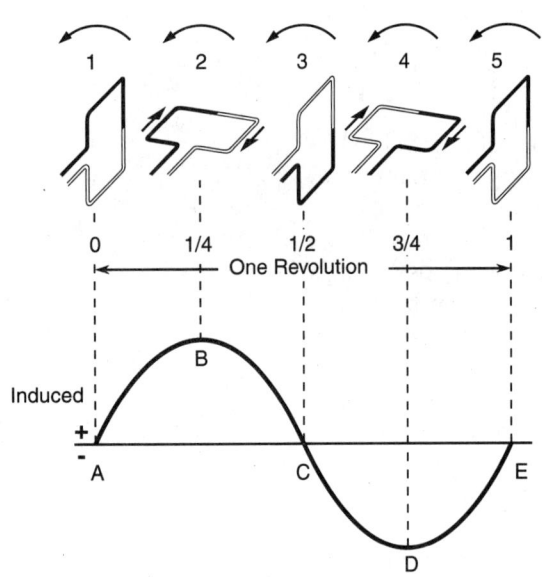

   a. into
   b. out of

2. In Figure 2–4, if twice as many turns were wrapped on the field windings, and the armature is rotated at the same speed, would the armature voltage increase? Why or why not?

3. The rotating portion of a generator is called the _____ and the stationary portion is called the _____ .

4. The _____ of the generator is the winding that is connected to the load. The _____ is used to create the magnetic flux required to create the output voltage.

5. At what speed must a 72-pole generator turn to produce a 60-Hz output? How many poles must a 50-Hz generator have if it is to rotate at 120 rpm?

6. What is the principal advantage of a rotating field generator as compared to a rotating armature generator?

7. Why must a self-excited, small generator be made of magnetic material with relatively high retentivity?

8. Why is a salient-pole generator normally used on a low speed machine?

9. List two reasons for using a turbo-rotor on high speed machines.

10. Why doesn't the hydrogen explode in a hydrogen-cooled generator?

11. How are the power output, voltage, and VAR contribution of a generator determined?

AC Generator Theory and Construction

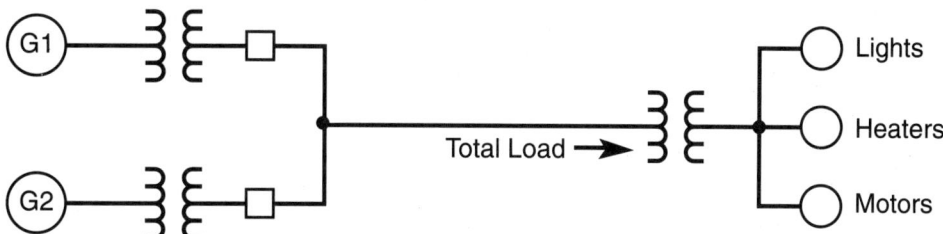

12. Assume that G1 and G2 are each contributing equally to the load. If the operator at G2 suddenly increases the excitation current, and the operator at G1 does nothing, what will happen to the system?

13. Assume that the transformer has a 13,800-V primary and a 480-V secondary. If the secondary resistor is 1 Ω, what resistance is seen by the primary circuit? (The resistance will transfer by the square of the turns ratio with the higher voltage side seeing the higher resistance.) What will the fault current be limited to in the generator? (The transformer primary voltage is 7900 V when a phase-to-ground fault occurs.)

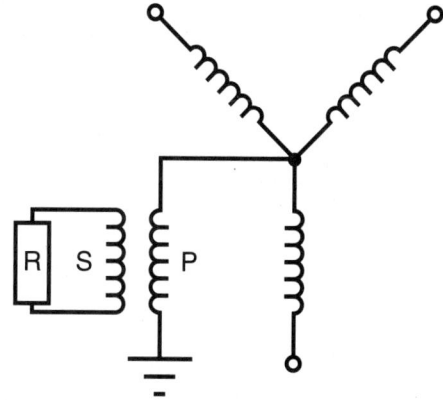

c. Resistance grounded with distribution transformer

AC Generator Protection   3

OBJECTIVES

After studying this chapter, the student should be able to:

- *Describe the types of problems that can damage AC generators.*
- *Describe the types of systems that can be used to protect generators.*
- *Draw simple, one-line diagrams of protective schemes for AC generators of various sizes.*

INTRODUCTION

Even the best designed power systems experience short circuits, open circuits, unbalanced loads, and other such problems. Because of the large investment in a generator, sophisticated protective schemes are warranted. This chapter describes the types of problems that can be experienced with generators, and explains the types of protective schemes that are employed to protect the generator.

Two types of faults can occur—internal faults and external faults. Internal faults are problems that occur inside the generator. Short circuits between two phases or one phase and ground, shorts between stator laminations, and shorted turns and other such problems, fall into the category of internal faults. External faults include unbalanced loads, overloads, and external short circuits.

The nature of the protective schemes used is determined by the type of faults that may occur and the relative size and cost of the generator being protected. Not every protective scheme is used on every generator. Small generators may be protected only by fuses, while larger more sophisticated systems have a full complement of protective systems.

To simplify references to protective relay schemes, the Institute of Electrical and Electronics Engineers (IEEE) and the American National Standards Institute (ANSI) have assigned standard number designations to different relay systems. Time overcurrent relays, for example, are designated as type 51, circuit breakers are type 52, and differential relays are type 87.

PROTECTION SCHEMES

Overload Protection

Generator Overload   All equipment has a maximum loading capacity. If this capacity is exceeded, the equipment will overheat. If the overheating/overload is not corrected quickly enough, the equipment may be permanently damaged.

Generator overload protection may be provided by overcurrent devices, such as fuses or overload relays, or by measuring the temperature of the windings directly using resistance temperature detectors. The following paragraphs describe each of those systems.

Overcurrent Protection   Overcurrent protection may be provided by fuses on very small generators; nevertheless, most larger types use protective relays, Figure 3–1. In this system, the current transformers step the generator current down to a lower level, and apply it to the phase overcurrent relays—51A, 51B, and 51C. The relay pickup currents are set to a nominal value above full load. Using a setting 25% above full load allows the generator to carry its full current without nuisance trips.

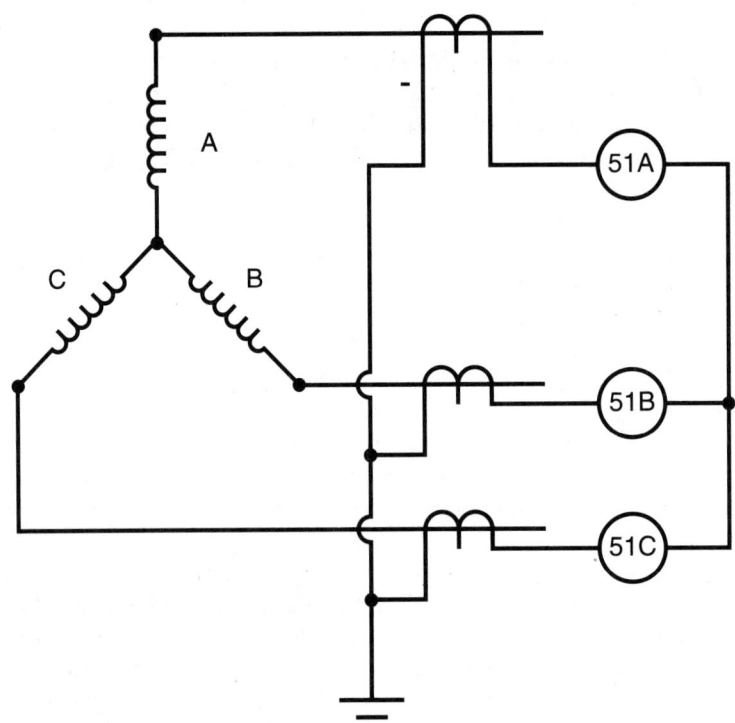

FIGURE 3–1
Generator overcurrent relays (Courtesy of Cadick Professional Services)

AC Generator Protection

The overcurrent relays are connected to trip a circuit breaker on the generator output side. When the current rises above the pickup setting, the relays will operate and disconnect the generator from the line. The relays are time delay units. This allows the generator to carry momentary surges, such as transformer inrush currents or motor starting currents.

Such a scheme will provide overload protection, but may not provide adequate backup protection for external, close-in short circuits. When a close-in short circuit occurs, the generator excitation system may not be able to sustain the generator output voltage. In this situation, as the excitation decreases, so does the output voltage and, therefore, the output current. During a short circuit, the output current of some generators may actually drop to values below full load current.

The voltage-restrained and voltage-controlled overcurrent relays are designed to compensate for such a condition. Such relays have both a current coil and a potential coil, and are connected to a current transformer and a voltage transformer, Figure 3–2. The voltage-controlled overcurrent relay, Figure 3–3, has an undervoltage coil and an overcurrent coil. The overcurrent unit will not operate until the undervoltage coil has dropped out. This type of relay will operate for the short circuit condition, but does not protect for the overload.

The voltage-restrained overcurrent relay, Figure 3–4, has a voltage coil that opposes the current coil. As the voltage decreases, the relay becomes more sensitive to overcurrent. The voltage-restrained overcurrent relay provides protection for both overload and short circuit conditions.

Both voltage-controlled and voltage-restrained overcurrent relays are assigned the ANSI/IEEE number 51V.

Overtemperature Protection The resistance temperature device (RTD) is a temperature sensitive resistor with a very linear temperature to resistance coefficient. As the temperature increases, the resistance increases in a manner similar to that shown in Figure 3–5. The RTD is connected in a relay with a bridge circuit, Figure 3–6. When the temperature rises above an allowable resistance (temperature), the bridge is unbalanced, and the relay operates.

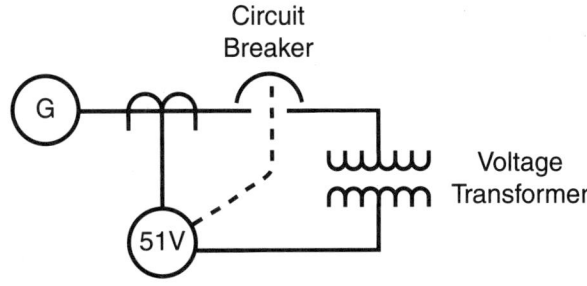

FIGURE 3–2
Generator backup relays (Courtesy of Cadick Professional Services)

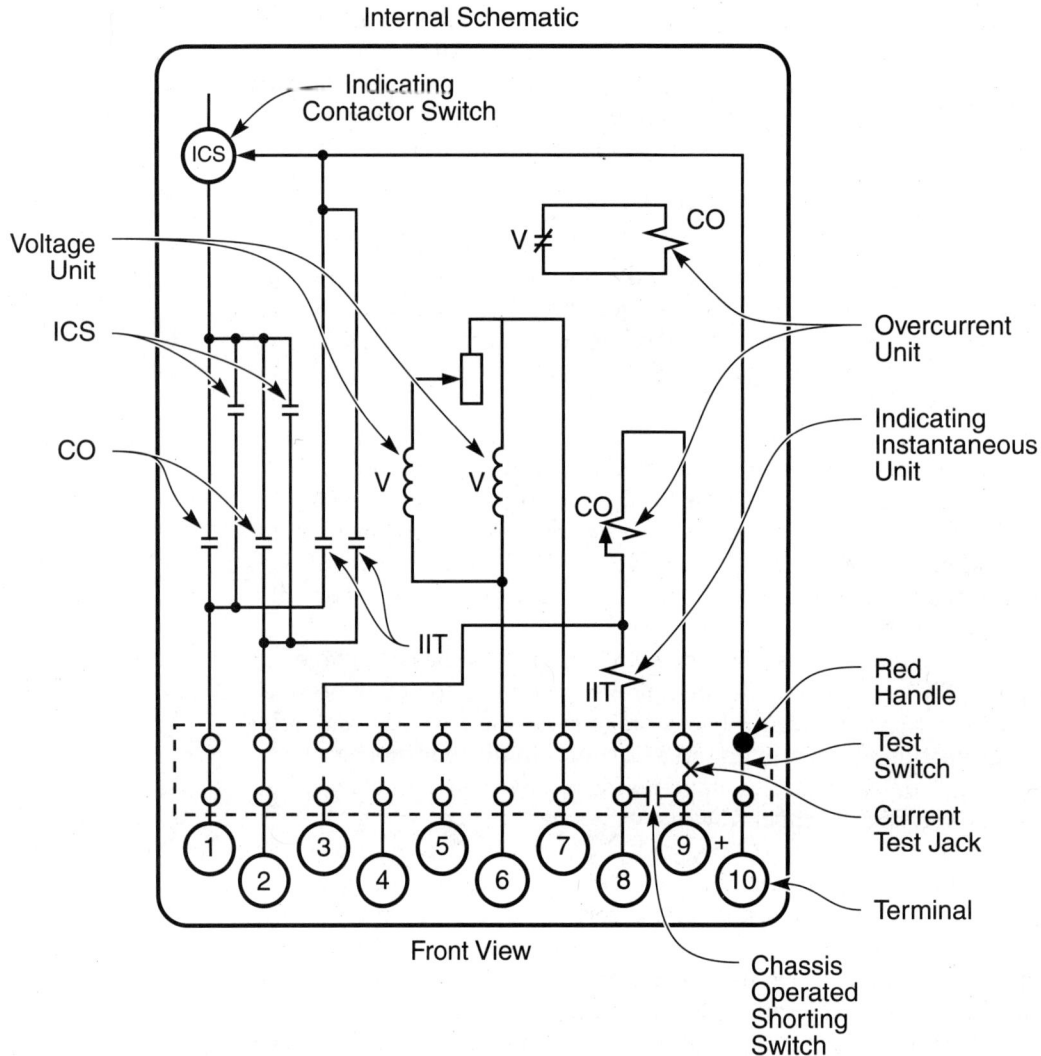

FIGURE 3-3
Voltage-controlled overcurrent relay (Courtesy of ABB Power T & D Company, Coral Spring, FL)

AC Generator Protection

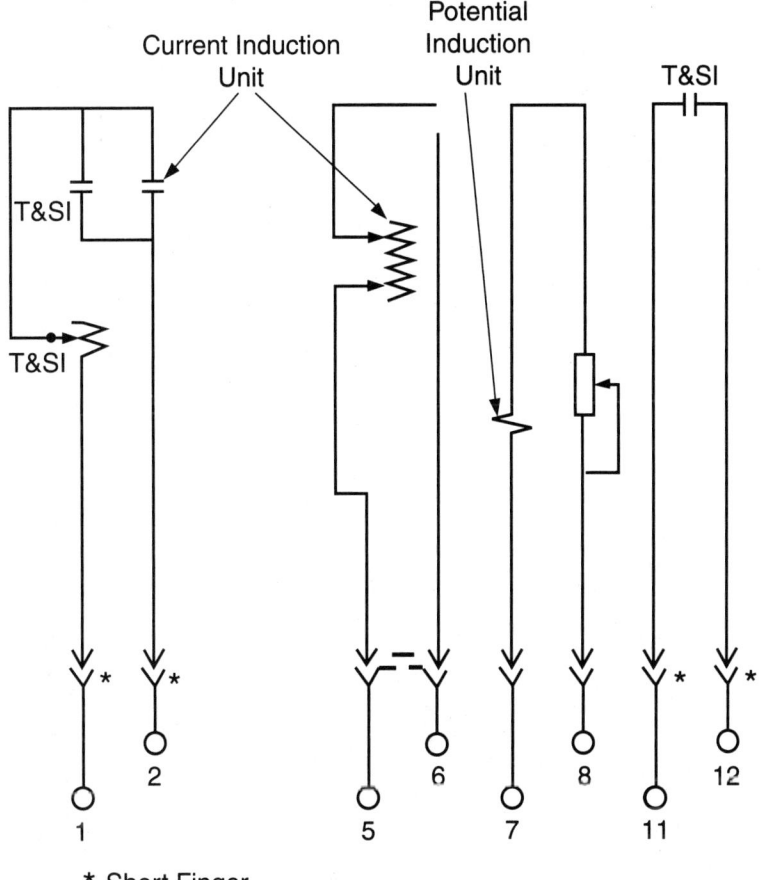

FIGURE 3–4
Voltage-restrained overcurrent relay (Courtesy of GE Protection and Control, Malvern, PA)

RTDs are installed deep in the stator windings of the generator where their temperature is essentially the same as the stator windings and iron. The temperature relay is calibrated to start its operation when the stator temperature rises above some critical value.

Overtemperature relays are assigned ANSI/IEEE type 49.

Differential Protection

Differential protection (ANSI/IEEE type 87) is among the fastest of all protective relay types. Designed to operate only for internal short circuits, differential relays compare the amount of current entering a zone to the amount leaving the same zone, Figure 3–7.

With this connection, an internal short circuit, Figure 3–8, creates operating current in the relays, while an external short circuit creates no

FIGURE 3–5
RTD temperature/resistance curve (Courtesy of General Electric)

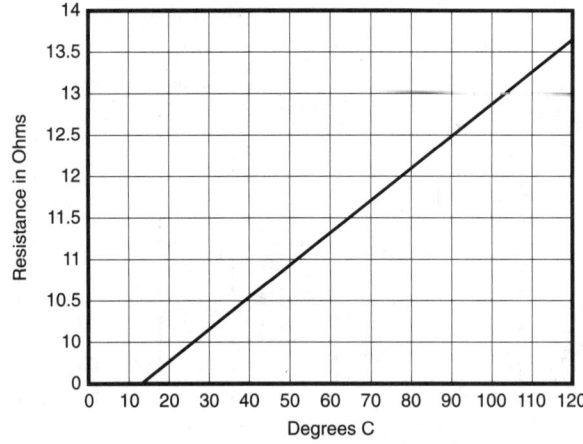

FIGURE 3–6a
The General Electric IRT temperature relay (Courtesy of GE Protection and Control Malvern, PA)

AC Generator Protection

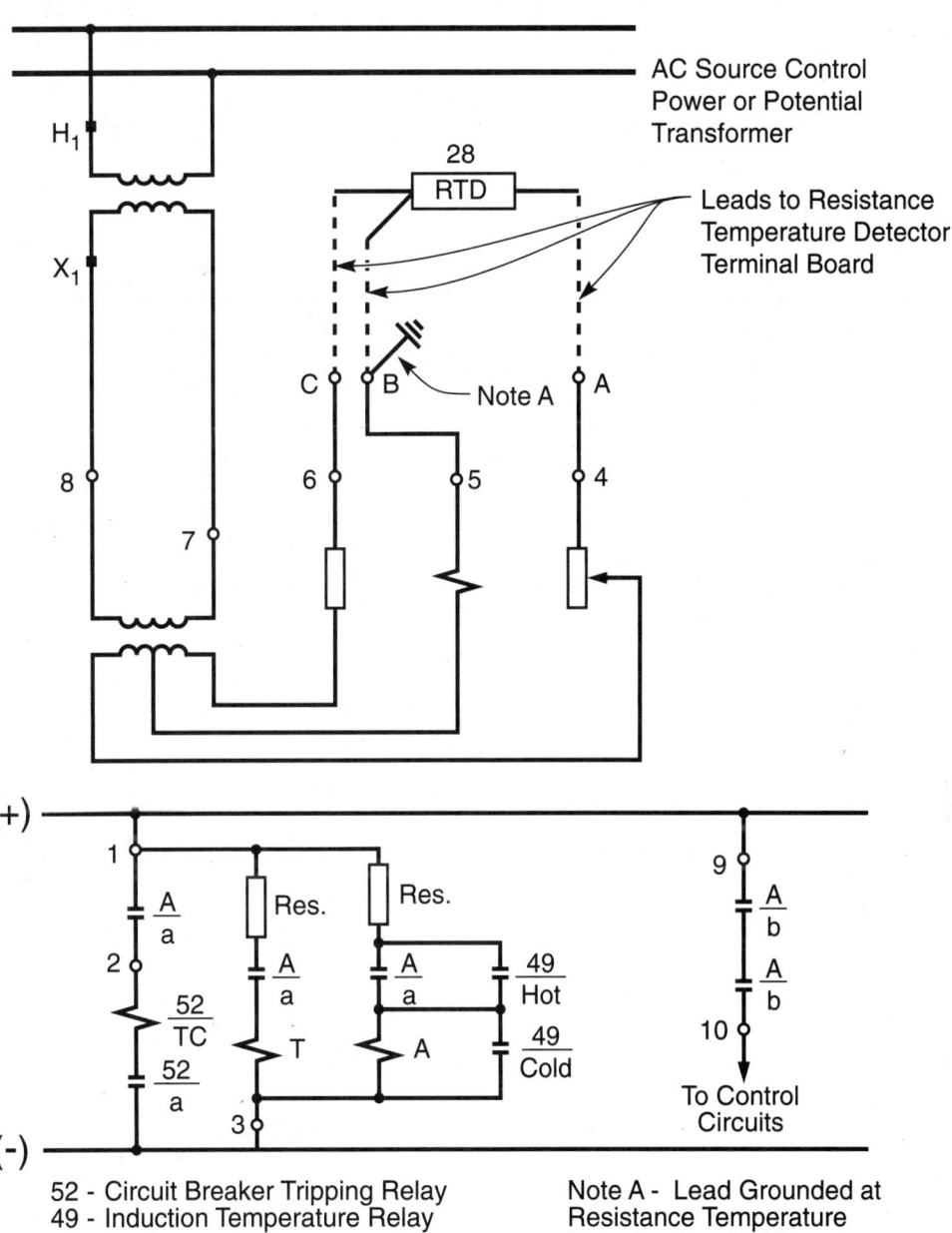

FIGURE 3–6b
IRT connection diagram (Courtesy of GE Protection and Control, Malvern, PA)

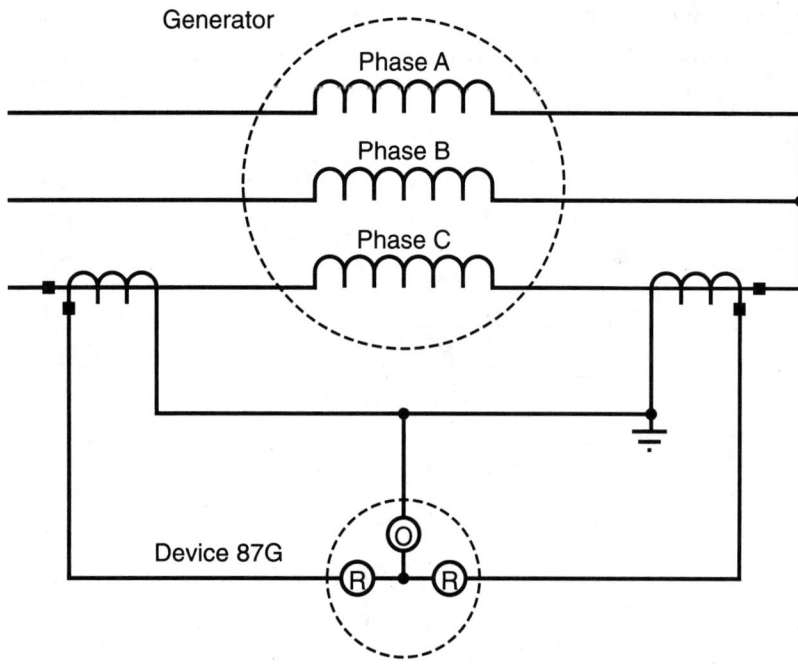

FIGURE 3–7 Generator differential relay system

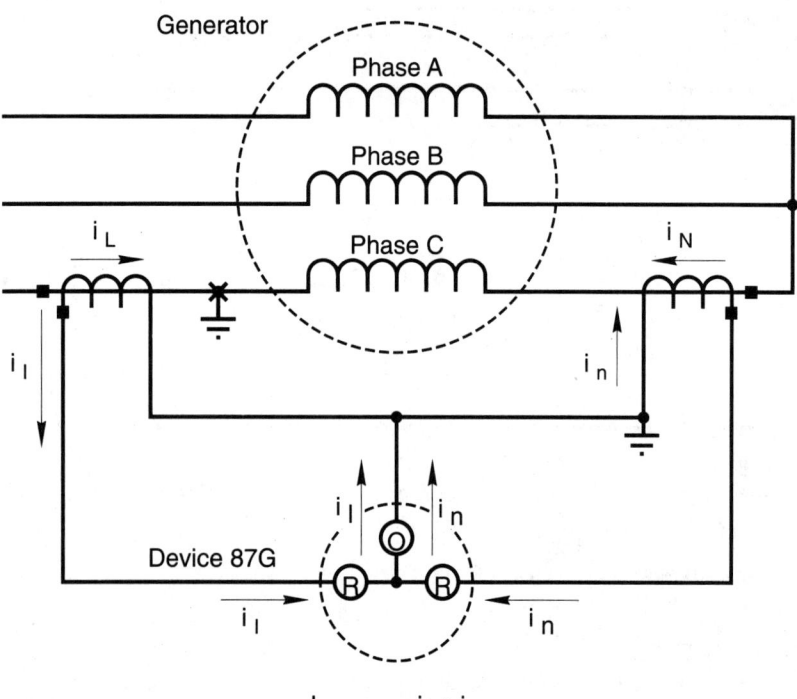

FIGURE 3–8 Differential relay system with an internal short circuit

$I_{operate} = i_l + i_n$

AC Generator Protection

operating current, Figure 3–9. Such relay systems are extremely fast and sensitive and found on all but the smallest of generators.

The differential scheme shown in Figure 3–7 uses a special type of differential relay called a percent slope differential relay. Such a relay uses restraint current windings which work against the operating winding. The coils are wound and connected in such a way that the operating current must exceed a certain percentage of the through fault current. This minimum ratio is called the *relays percent slope* and is calculated from the formula:

$$\%\text{Slope} = \left.\frac{I_o}{I_{sr}}\right|\text{balance} \tag{1}$$

This means that the percent slope is calculated by dividing the operating current, I_o by the smaller restraining current, I_{sr}, when the relay is just balanced. Such a characteristic is valuable when one of the two current transformers saturates during an external short circuit. Under such circumstances, the amount of unbalance created must exceed the relay percent slope before an improper operation may occur.

Another way of protecting for the same problem is to use the connection shown in Figure 3–10. In this connection, a single current transformer is

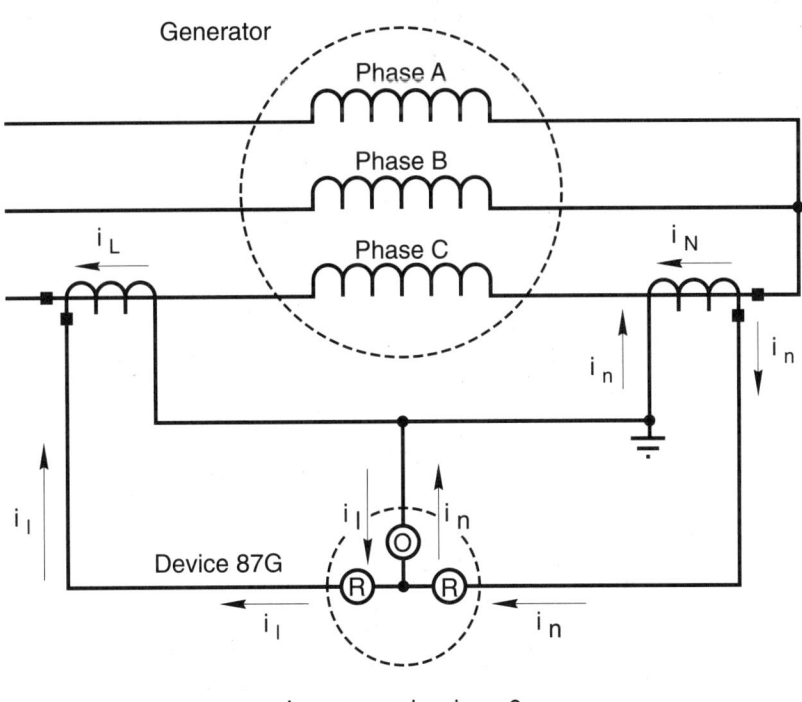

FIGURE 3–9
Differential relay system with an external short circuit

used in each phase and is connected to a simple, instantaneous overcurrent relay. The generator input and output leads for each phase are both passed through the current transformer. As long as there are no internal short circuits, the currents in the leads will be equal in magnitude, and the magnetic fields in the current transformers (CTs) will be zero. An internal fault will cause a magnetic unbalance and create an output current in the 87 relay(s) of the faulted phase(s). Since only one current transformer is used in each phase, no saturation problems will exist. Such a system is somewhat less sophisticated, but substantially less expensive than the percent slope differential scheme described previously. This system is used on many smaller generators.

FIGURE 3–10
Self-balancing generator differential scheme
(Courtesy of IEEE Std 242-1986, IEEE Recommended Practices for Protection and Coordination of Industrial and Commercial Power Systems)

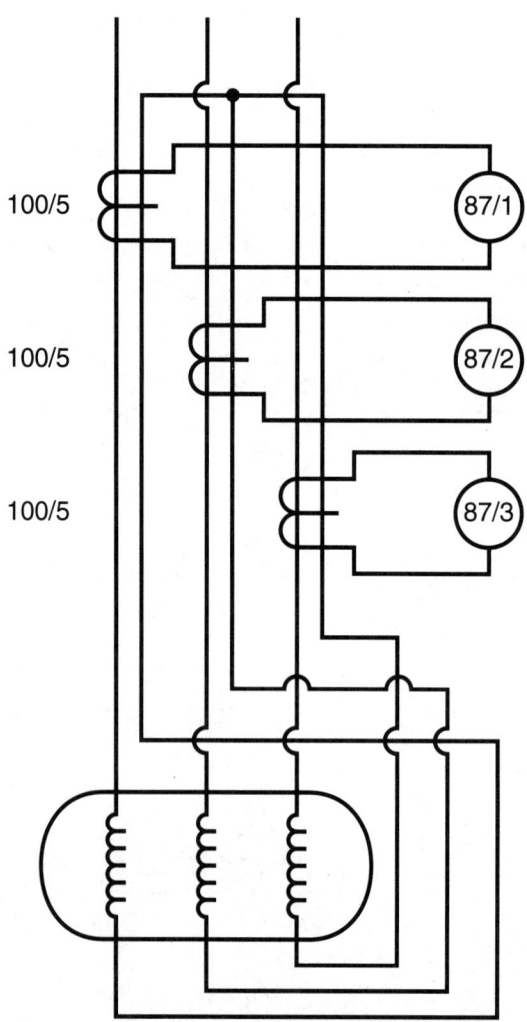

Ground Fault Protection

In Chapter 2 you learned that generators are usually grounded through a resistance. Such a connection limits the amount of fault current to a low level that is not usually damaging to the generator for short time periods. Ground fault protection is often applied as shown in Figure 3–11.

Figure 3–11a has a single current transformer on the generator neutral. Ground fault current, which is limited by the resistor, will flow in the primary of the CT. The CT transforms the current and applies it to the 51G relay. Since the resistor limits the ground fault current, the operating time of the relay can be set to fairly long values, thus allowing time for operator intervention and/or other external relays to clear the short circuit. Figure

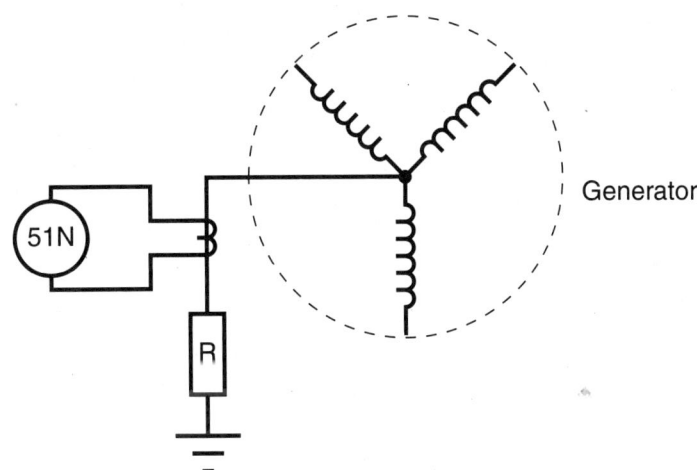

FIGURE 3–11
Generator ground relay connections (Courtesy of Cadick Professional Services)

a. Without Distribution Transformer

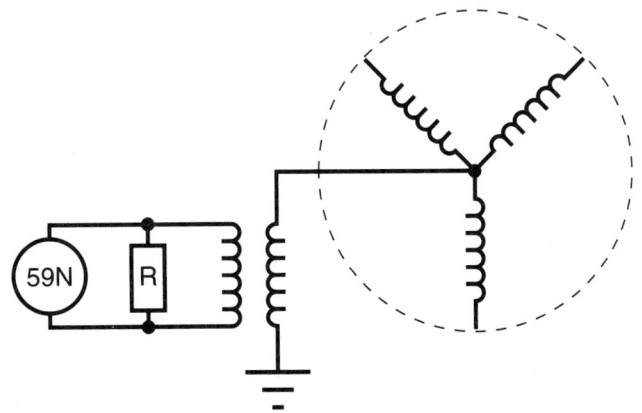

b. With Distribution Transformer

3–11b is the same type of protection with the ground relay located in the secondary of the grounding distribution transformer. The relay in this system is actually a voltage relay. Since the transformer's secondary voltage is proportional to the primary ground fault current, the relay can be set to respond in the same way as the overcurrent relay in Figure 3–11b.

Figure 3–12 is a generator ground differential scheme. The ground differential relay, 87GN, will operate only for internal ground faults. Such a scheme can be set very sensitively and yet will not operate for external ground faults. This allows the generator ground fault protection to operate independently from other downstream ground relays.

Loss of Excitation

This type of protection, type 40, senses when a generator loses its excitation system. Such protection is especially important when a generator is operating in parallel with other generators or an electric utility supply system.

When a synchronous generator loses its excitation system, it starts to behave like an induction generator. In such a configuration, the generator draws inductive excitation current from the other generators on the power system. Thus, to the power system, the generator looks like an inductive load.

The loss of excitation relay is capable of detecting such a condition by measuring the impedance of the system as referenced from the generator terminals. When it measures a capacitive impedance, a loss of excitation condition is indicated and the relay operates. A capacitive impedance measured at the generator terminals implies that the generator is drawing inductive current.

Loss of excitation is not normally considered to be an immediate tripping requirement. Because of this, many loss of excitation schemes do not trip instantaneously. Rather, they alarm and/or trip on a time delay basis.

Negative Sequence Protection

When the power system currents are unbalanced, a unique type of current called *negative sequence* appears. The negative sequence currents in each phase are equal in magnitude, 120° in phase angle, and have reverse phase rotation, Figure 3–13. These currents cause greatly increased stator heating in a generator. Negative sequence relays, such as that shown in Figure 3–14, are designed to sense the negative sequence currents and operate when they become excessively high.

The negative sequence relay has a sequence filter that is used to pass only the negative sequence currents to a standard overcurrent relay coil. The overcurrent relay coil operates when the negative sequence currents

AC Generator Protection

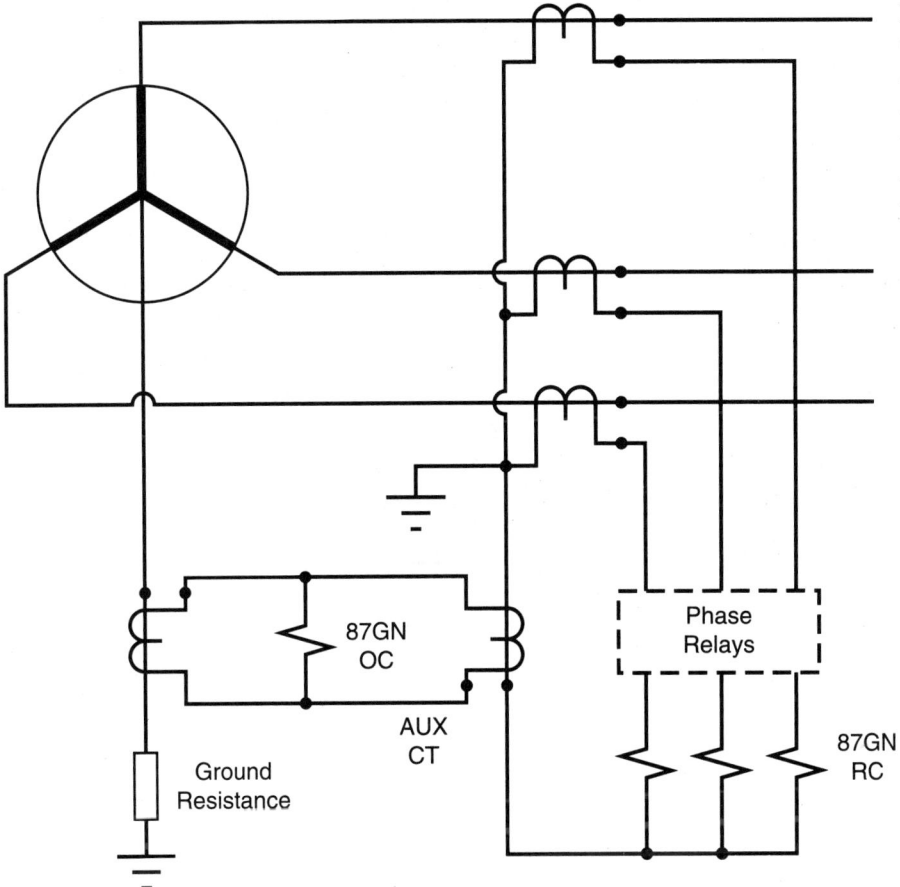

**FIGURE 3–12
Generator ground differential relay connection**
(Courtesy of IEEE Std 242-1986, IEEE Recommended Practices for Protection and Coordination of Industrial and Commercial Power Systems)

rise above a preset, allowable value. As with a standard overcurrent relay, the negative sequence overcurrent operates faster for higher current values. Negative sequence relays are assigned ANSI/IEEE type 46.

Reverse Power

Normally a generator supplies power to the electrical load that is connected to its terminals. If the generator prime mover fails, the generator becomes a motor and starts to draw power from the system. Such a condition is called *reverse power* or *motoring*. Reverse power relays (ANSI/IEEE type 32) are used to recognize this condition, and to trip circuit breakers to clear the generator.

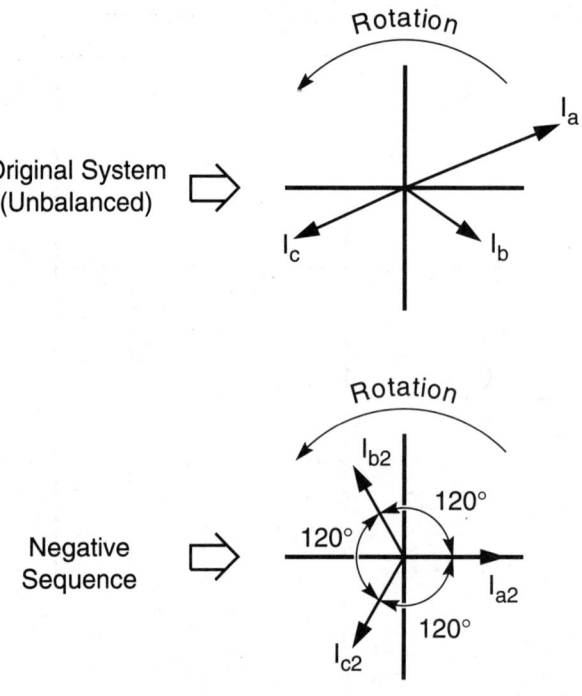

FIGURE 3–13 Negative sequence currents for an unbalanced system (Courtesy of Cadick Professional Services)

Note: The negative sequence is only part of the total current. Total current also includes the positive and zero sequence components.

A reverse power relay, such as that shown in Figure 3–15, is similar in construction to a watthour meter. The voltage and current coil are connected in such a way that when the generator is feeding power to the system, contact opening torque is developed. When power is feeding from the system, contact closing torque is developed. The amount of power fed to the generator during a motoring condition may be quite small—as little as a few **tenths** of a percent of the total generator capability. Because of this, the reverse power relay needs to be very sensitive.

EXTREMELY SMALL GENERATORS

Extremely small generators, such as those used for portable service, have relatively unsophisticated protection systems. Overload devices, such as fuses or circuit breakers, and simple bimetallic temperature sensors are commonly employed. Figure 3–16 shows a typical connection diagram.

AC Generator Protection

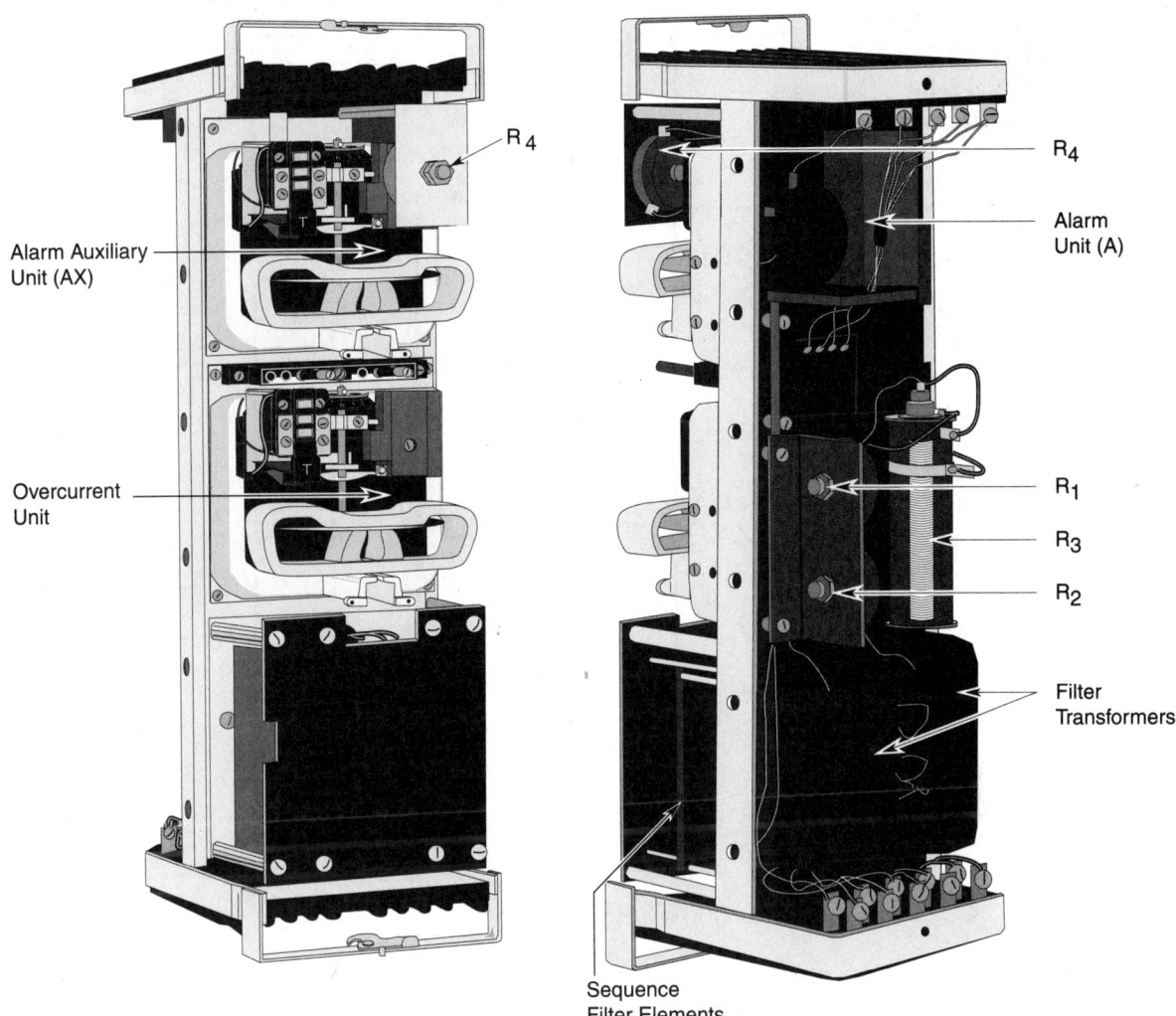

FIGURE 3–14
General Electric INC77 negative sequence relay front & rear view
(Courtesy of GE Protection and Control, Malvern, PA)

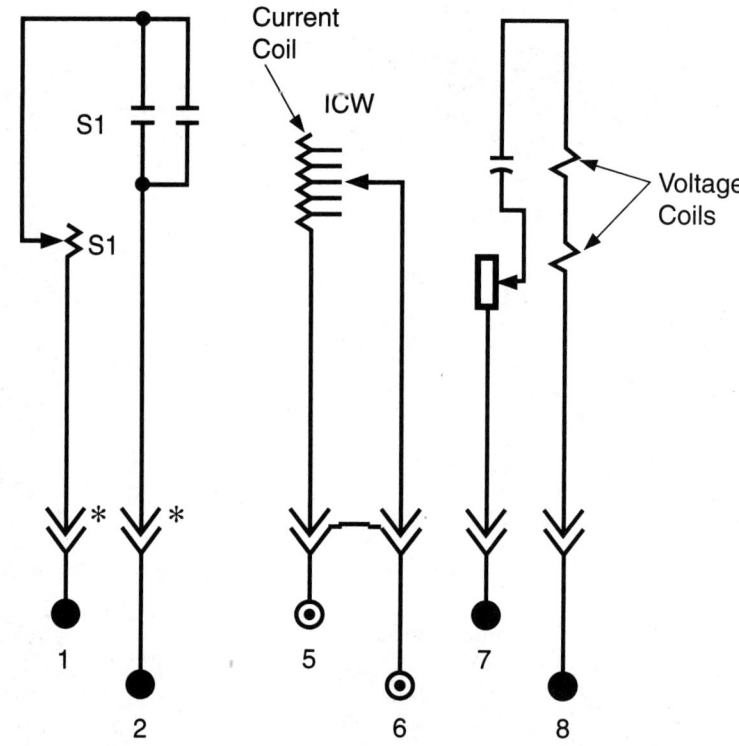

FIGURE 3–15 General Electric ICW power directional relay schematic diagram (Courtesy of GE Protection and Control, Malvern, PA)

* = SHORT FINGERS

SMALL GENERATORS

Small generator classification depends upon the voltage rating. Low voltage generators (those under 600 V) are classified as small in sizes up to 1000 kVA, and those above 600 V are classified as small in sizes up to 500 kVA.

The protection for small, isolated generators, Figure 3–17, includes:

- 1 device 51V, backup overcurrent relay, voltage restraint, or voltage controlled type
- 1 device 51G, backup ground time-overcurrent relay

If the generator is not isolated, that is, if it is being run in parallel with other generators in the system, the following additional protection should be included, Figure 3–18:

- 1 device 32, reverse power relay for anti-motoring protection
- 1 device 40, reverse VAR relay for loss of field protection

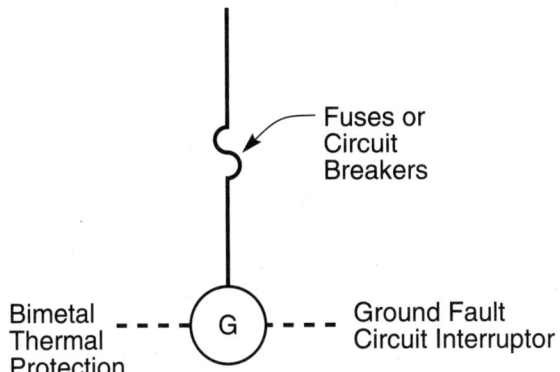

FIGURE 3–16
Extremely small generator protection (Courtesy of Cadick Professional Services)

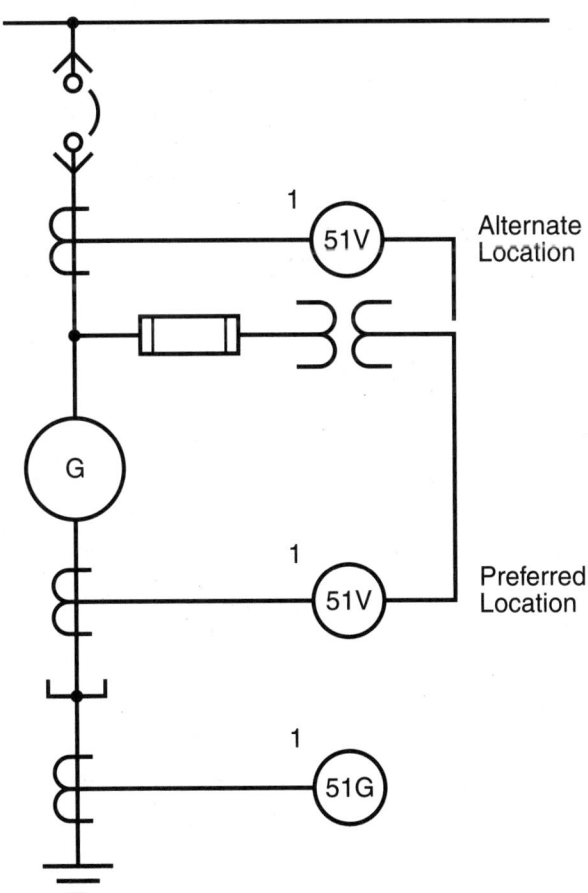

FIGURE 3–17
Typical protection scheme for a small, isolated generator
(Courtesy of IEEE Std 242-1986, IEEE Recommended Practices for Protection and Coordination of Industrial and Commercial Power Systems)

- 3 device 87, instantaneous overcurrent relays providing self-balancing differential protection, Figure 3–10.

MEDIUM-SIZED GENERATORS

Medium-sized generators are larger than small sizes and go up to 12,500 kVA. Typical protection is shown in Figure 3–19 and includes:

- 3 device 51V, backup overcurrent relays, voltage-restraint, or voltage-controlled type
- 1 device 51G, backup ground time-overcurrent relay
- 3 device 87, differential relays, fixed or variable percentage type, either standard speed, high speed, or the self-balance type
- 1 device 32, reverse power relay for anti-motoring protection

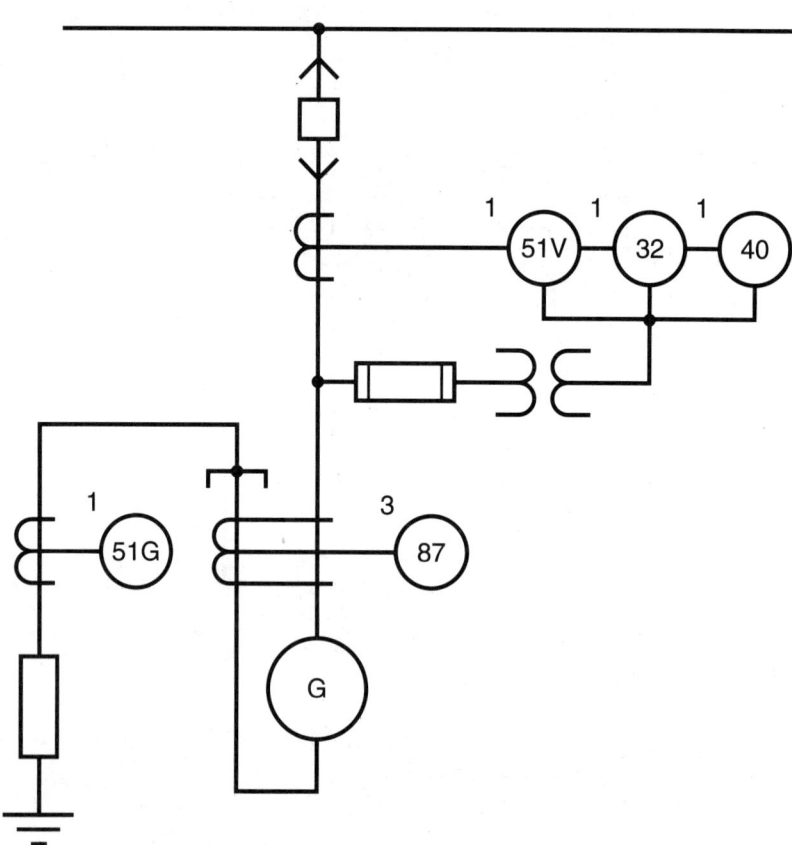

FIGURE 3–18
Typical protection scheme for a small non-isolated generator
(Courtesy of IEEE Std 242-1986, IEEE Recommended Practices for Protection and Coordination of Industrial and Commercial Power Systems)

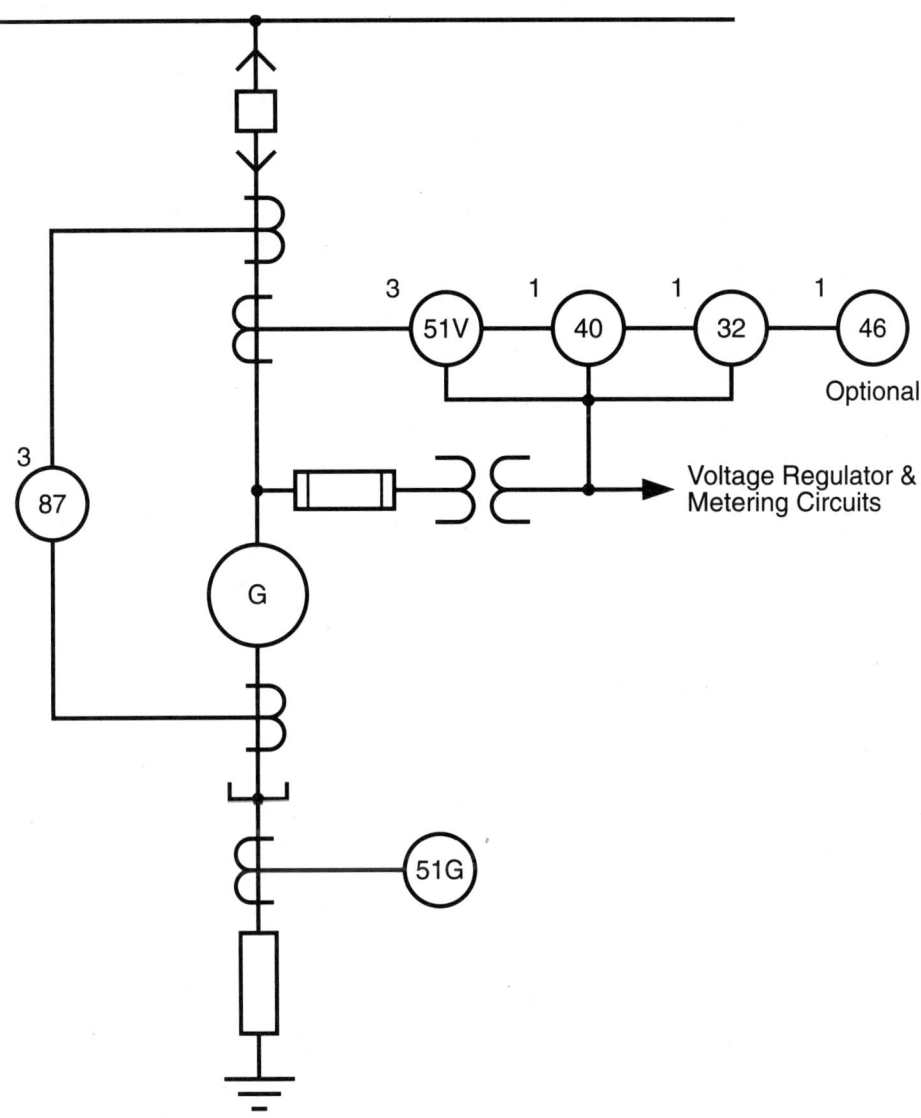

FIGURE 3–19
Typical protection scheme for a medium-sized generator
(Courtesy of IEEE Std 242-1986, IEEE Recommended Practices
for Protection and Coordination of Industrial and Commercial Power Systems)

- 1 device 40, impedance relay, offset mho type for loss of field protection
- 1 device 46, negative phase sequence overcurrent relay for protection against unbalanced conditions. This device is optional for all but the largest, medium-sized units.

LARGE GENERATORS

Large generators include units above 12.5 kVA. The following protection is suggested (Figure 3–20):

- 3 device 51V, backup overcurrent relays, voltage-restraint, or voltage-controlled types
- 1 device 51G, backup ground time-overcurrent relay
- 3 device 87, differential relays, high speed differential types
- 1 device 87G, ground differential relay, directional product type
- 1 device 40, impedance relay, offset mho type for loss of field protection, two-element type if recommended for greater sensitivity
- 1 device 46, negative phase sequence overcurrent relay for protection against unbalanced conditions
- 1 device 49, temperature relay to monitor stator winding temperature
- 1 device 64F, generator field ground relay, applicable only on machines having field supply slip rings. This is a special type of relay that detects double grounds on the generator field winding.
- 1 device 60, voltage balance relay. The 51V, 40, 32, and 46 protection devices depend on the availability of a correct, balanced voltage supply. The 60 relay monitors the voltage supply and disconnects 51V, 40, 32, and 46 if the voltage becomes improper.

SUMMARY

Generators may be plagued by a variety of problems, including overload, current unbalance, overheating, motoring, and short circuits. These faults, as they are called, may be caused by problems internal or external to the generator.

If the problem is internal, the generator must be disconnected as quickly as possible to mitigate the damage. If the problem is external, the generator protection must coordinate with external protection, that is, it must allow the external protection time to trip and minimize the amount of outage that occurs.

Generator protection schemes include overload/overcurrent protection, differential protection, ground fault protection, loss of excitation protection, negative sequence current protection, and reverse power (motoring) protection.

AC Generator Protection

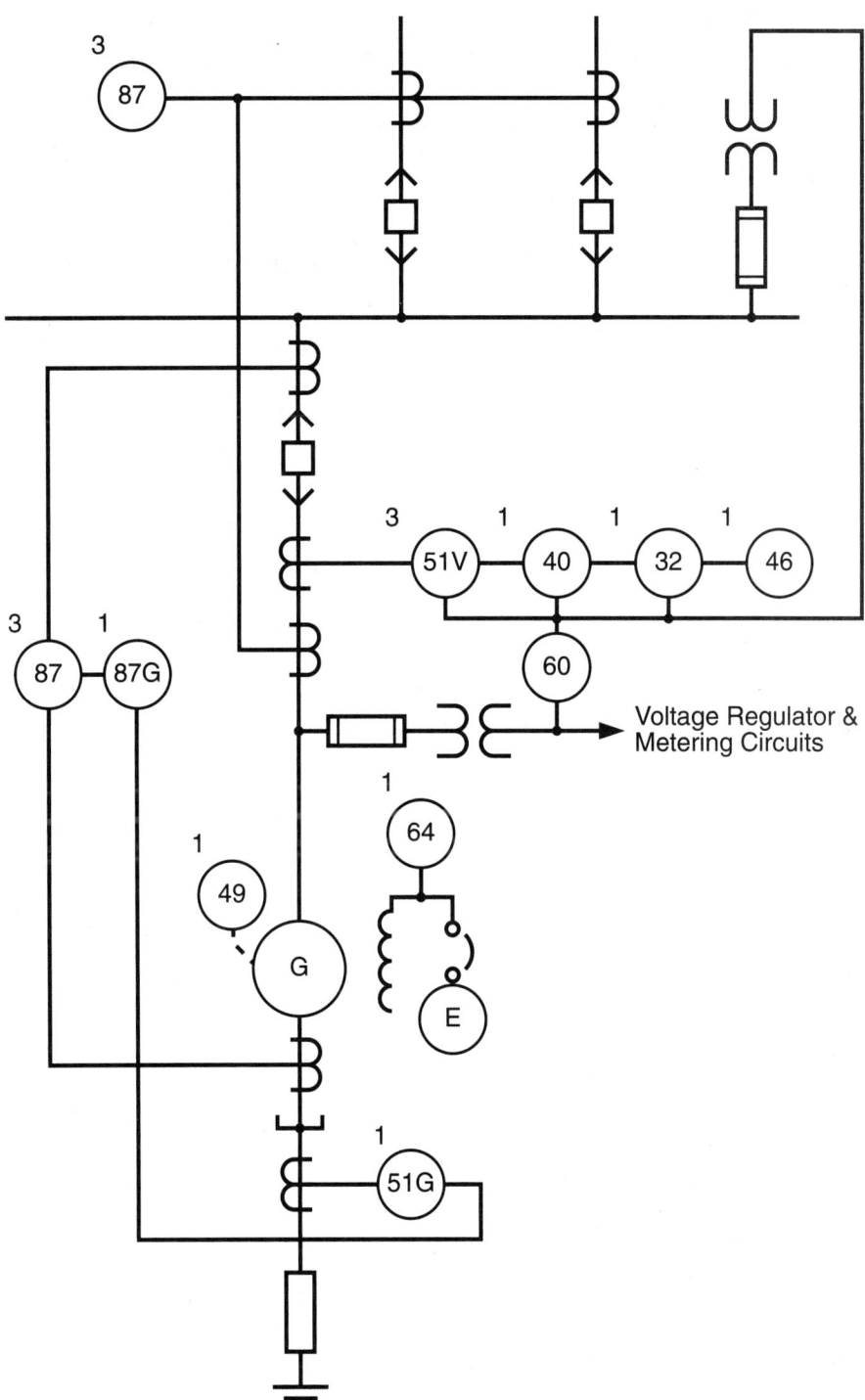

FIGURE 3–20
Typical protection scheme for a large generator
(Courtesy of IEEE Std 242-1986, IEEE Recommended Practices for Protection and Coordination of Industrial and Commercial Power Systems)

The type and number of protective devices that are used on a generator are usually based on the size of the generator, with the larger generators using more sophisticated protection. Extremely small generators may have simple fuses and/or temperature protection with larger machines using a full complement of schemes, including negative sequence and loss of excitation.

REVIEW QUESTIONS

1. Internal faults are those that occur _____ the generator.
   a. outside
   b. inside

2. List three types of generator faults that can occur.

   _____

   _____

   _____

3. What happens to the voltage on a generator's output when a short circuit occurs?

   _____

4. The overcurrent relay with voltage restraint has a voltage activated restraining coil. This means it will be _____ sensitive to overcurrent when the voltage is low.
   a. more
   b. less

5. The voltage/resistance curve for a 10-Ω RTD is shown in Figure 3–5. Such an RTD has a resistance of 10 Ω when its temperature is 25°C. Assume that you wish to calibrate an IRT temperature relay to operate just when the temperature is 65°C. What resistance should you apply to the relay? (Note: You may use the graph; however, a more accurate formula is given by the formula: $R_T = 9.04 + 0.0385T$.)

6. Write a brief description of the operation of a current differential scheme. You may use Figures 3–7, 3–8, and 3–9 for reference.

7. Draw simple diagrams that illustrate how a generator may be protected for ground fault currents. One of the diagrams should use a distribution transformer to isolate the relay from the high voltage levels on the generator neutral.

8. How can the negative sequence currents be used to protect a generator?

9. Draw typical protective schemes for extremely small, small, medium-sized, and large generators. You may use Figures 3–17 through 3–20 if you need to. When you have completed the drawings, write a brief description of the type of protection that each element provides.

AC Generator Testing 4

OBJECTIVES

After studying this chapter, the student should be able to:

- List the types of electrical tests that should be performed on generators.
- Describe the methods used and results expected for the electrical tests performed on generators.

INTRODUCTION

Before it is placed into service, a generator should be tested to make certain that it meets the manufacturer's specifications. Also, as equipment ages, it deteriorates. Temperature, vibration, chemicals, corona, voltage, and magnetic stresses all combine to age insulation, conductors, and structural members.

This chapter describes a variety of tests that may be performed to evaluate new equipment and/or determine the continued serviceability of service-aged equipment. All of these tests may be performed on electric motors. Their application to motors will be covered in later chapters.

> *WARNING!!!*
> *Many of the procedures described in this chapter employ lethal voltages. This material is intended for training purposes and should not be used as the basis for the development of step-by-step field procedures. For more details, refer to manufacturer's technical literature and other industry references.*

FIGURE 4–1
Insulation equivalent circuit (Courtesy of Cadick Professional Services)

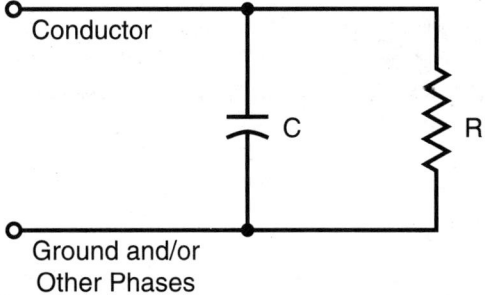

INSULATION TESTING

The Nature of Insulation

Physical Construction An electrical equivalent circuit of insulation is shown in Figure 4–1. The center conductor of the conductor is one terminal and the ground and/or other phase conductors are the other terminal. The layer of insulation between the two terminals creates a capacitor. While we normally think of insulation as allowing no current to pass, in reality a little current does flow through and over the surface of the insulation. This is represented by the resistor.

Insulation is tested by applying either AC or DC voltage to it and evaluating the current that flows. To understand this current, we will first describe the nature of DC versus AC current flow through an insulation system.

Direct Current When a DC voltage is applied to an insulation system, Figure 4–2, three current components flow, Figure 4–3. The magnitude and the duration of each of these currents will depend upon the condition of the insulation.

FIGURE 4–2
Direct current flow through insulation (Courtesy of Cadick Professional Services)

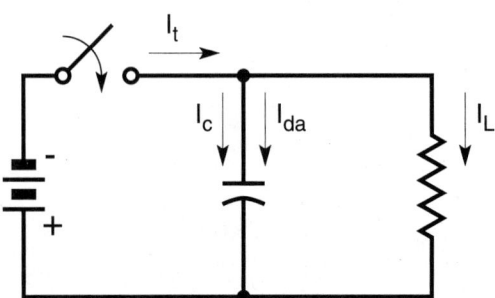

AC Generator Testing

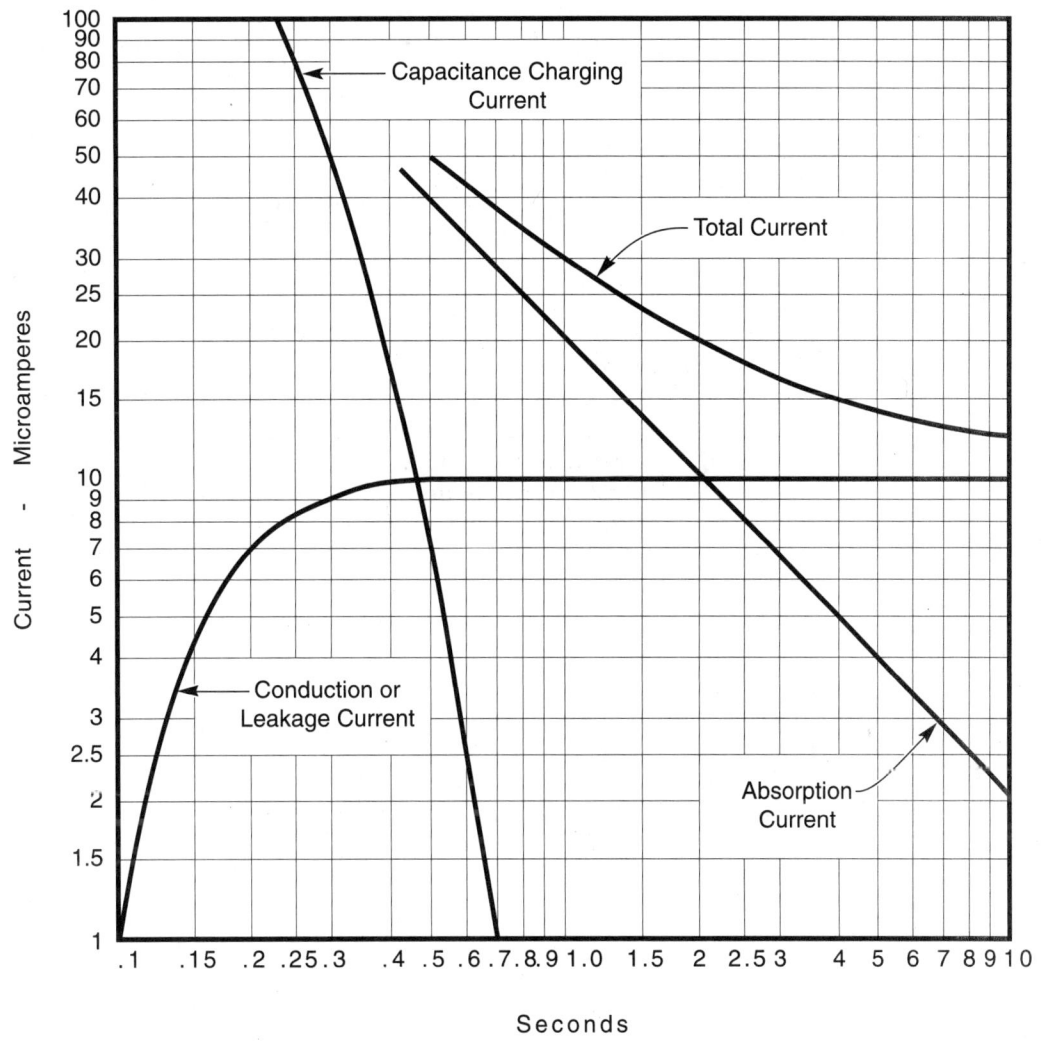

FIGURE 4–3
Components of DC current through insulation

- Capacitive Charging Current (I_C)
  When the switch in Figure 4–2 closes, electrons flow from the battery's negative plate to the capacitor's top plate. At the same time, electrons leave the capacitor's bottom plate and move to the battery's positive terminal. (Remember that the "plates" we discuss here are actually the electrical conductors and/or ground.)

Initially, when there is no excess or deficiency of electrons, the amount of current flow is very high. Gradually, a voltage builds up on the capacitor and the current reduces, eventually approaching zero. The capacitive current is shown in Figure 4–3.
- (Dielectric) Absorption Current (I_{da})
The insulation material has positive and negative charges in it. Normally, these charges are scattered randomly through the material leaving a net charge of zero. However, as the charges build up on the conductors (plates), the charges in the insulation are attracted or repelled, depending on whether they are positive or negative. The negative charges in the insulation are attracted to the bottom plate, and the positive charges are attracted to the top plate.

The net result of this action is a current flow in the circuit. The initial charge motion in the insulation is relatively large, but it gradually tapers as the charges reorient close to the plates with the opposite charge. The absorption current does not start until some charge builds up on the plates; but, it lasts for a much longer time.
- Leakage Current (I_L)
Even the best insulation has some leakage current. The leakage in good insulation will rise to a steady value and stabilize at that point, while the current in bad insulation will continue to rise and/or stabilize at an excessively high value.

The total direct current flow (I_t) will be the sum of the three components, and will be as shown in Figure 4–3. Notice that the rate at which the total current decays is a function of the type of insulation and the size of the insulation system. Larger pieces of equipment will take longer to decay. DC current in some small equipment may decay so quickly that the change is barely noticeable.

Alternating Current If an alternating voltage is applied to an insulation system, Figure 4–4, the current flow will be composed of a capacitive current and a resistive current. Figure 4–5 shows the two currents in a time relationship, and Figure 4–6 shows a vector diagram relationship.

The two points that define the relationship between the AC leakage and the AC capacitive current are:

1. The capacitive current in good insulation will be on the order of 100 times greater than the resistive current.
2. Since capacitive current leads resistive current by 90°, the total current will lead the resistive current by very close to 90°.

Temperature and Humidity Temperature and humidity both play a role in the amount of current that flows through insulation.

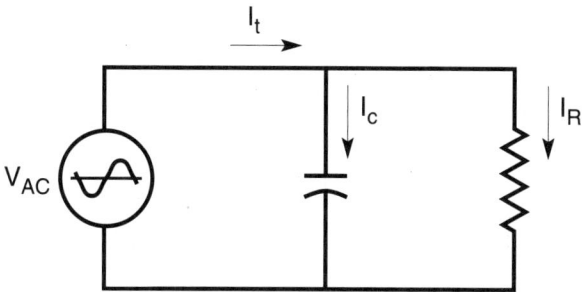

FIGURE 4–4
Alternating current flow through insulation (Courtesy of Cadick Professional Services)

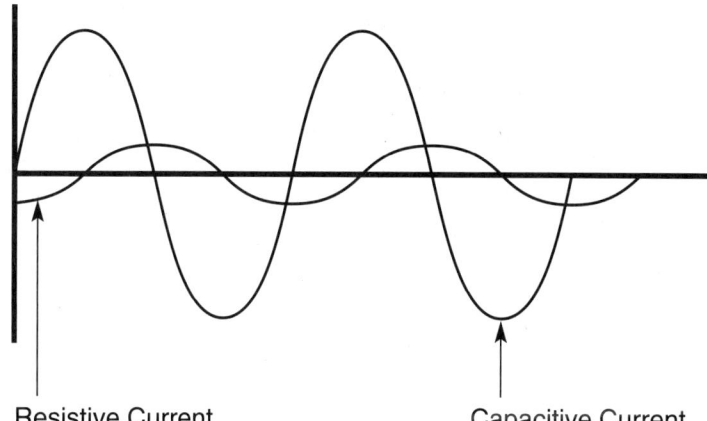

Resistive Current   Capacitive Current

FIGURE 4–5
Capacitive and resistive AC current flow in insulation (Courtesy of Cadick Professional Services)

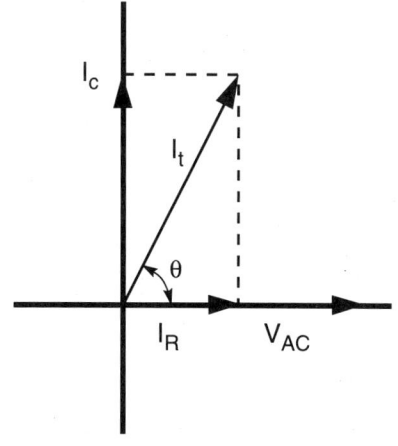

FIGURE 4–6
Vector relationship between capacitive and resistive alternating current flow

- Temperature
  Temperature and insulation resistance have an inverse relationship. That is, as the temperature of the insulation rises, the insulation resistance decreases. This means that hotter insulation will have a higher current flow for the same amount of voltage applied. Because of this fact, temperature correction tables must be applied. Table 4–1 is a temperature correction table for various types of insulation when tested using direct voltage instruments. The use of this table will be discussed later. Note that the correction factors are intended for direct voltage testing. Insulation resistance is not measured with alternating voltage because the high capacitive current masks the resistive current.
- Humidity
  Humidity has little effect on direct current flow through insulation unless the humidity is high enough to moisten the insulation material itself. Moisture *in* the insulation will cause a current increase, as well as other effects which will be described later.

  Very high humidity, in excess of 70% relative humidity, will affect the relationship between the alternating capacitive current and the alternating resistive current. Tests that use alternating current should be closely monitored in relative humidities above approximately 70%.

Insulation Resistance Testing

Test Equipment Insulation resistance testing uses direct voltage to create the test current. The most commonly used type of insulation resistance tester is the megohmmeter, Figure 4–7. The megohmmeter is commonly called a *Megger*®[1], even though *Megger*® is a registered trademark. The megohmmeter is a direct voltage instrument, and is available in a variety of outputs from 100 V to 15,000 V.

Megohmmeters have either analog or digital meters, and read out directly in ohms or megohms. The instrument usually has three terminal connections: line, earth, and guard.

1. Line—The line terminal is the so-called hot lead. It is connected directly to the specimen under test.
2. Earth—The earth terminal is the return terminal of the instrument. Current that leaves the line terminal passes through the insulation, and then returns to the instrument via the earth terminal. The earth terminal is usually the one that is in series with the meter.

---

[1]*Megger* is a registered trademark of AVO International.

AC Generator Testing

Table 4–1 Temperature Correction Factors for Insulation Resistance Tests

Temp.		Rotating Equipment		Transformers Oil-Filled	Cables							
°C	°F	Class A	Class B		Code Natural	Code GR-S	Performance Natural	Heat Resistance Natural	Heat Resist. & Perform. GR-S	Ozone Resist. Natural GR-S	Varnished Cambric	Impregnated Paper
0	32	0.21	0.40	0.25	0.25	0.12	0.47	0.42	0.22	0.14	0.10	0.28
5	41	0.31	0.50	0.36	0.40	0.23	0.60	0.56	0.37	0.26	0.20	0.43
10	50	0.45	0.63	0.50	0.61	0.46	0.76	0.73	0.58	0.49	0.43	0.64
15.6	60	0.71	0.81	0.74	1.00	1.00	1.00	1.00	1.00	1.00	1.00	1.00
20	68	1.00	1.00	1.00	1.47	1.83	1.24	1.28	1.53	1.75	1.94	1.43
25	77	1.48	1.25	1.40	2.27	3.67	1.58	1.68	2.48	3.29	4.08	2.17
30	86	2.20	1.58	1.98	3.52	7.32	2.00	2.24	4.03	6.20	8.62	3.20
35	95	3.24	2.00	2.80	5.45	14.60	2.55	2.93	6.53	11.65	18.2	4.77
40	104	4.80	2.50	3.95	8.45	29.20	3.26	3.85	10.70	25.00	38.5	7.15
45	113	7.10	3.15	5.60	13.10	54.00	4.15	5.08	17.10	41.40	81.0	10.70
50	122	10.45	3.98	7.85	20.00	116.00	5.29	6.72	27.85	78.00	170.00	16.00
55	131	15.50	5.00	11.20			6.72	8.83	45.00		345.00	24.00
60	140	22.80	6.30	15.85			8.58	11.62	73.00		775.00	36.00
65	149	34.00	7.90	22.40				15.40	118.00			
70	158	50.00	10.00	31.75				20.30	193.00			
75	167	74.00	12.60	44.70				26.60	313.00			

*Corrected to 20°C for rotating equipment and transformers; 15.6°C for cable.

FIGURE 4–7
Typical megohmmeter
(Biddle™ 15 kV
Megohmmeter)

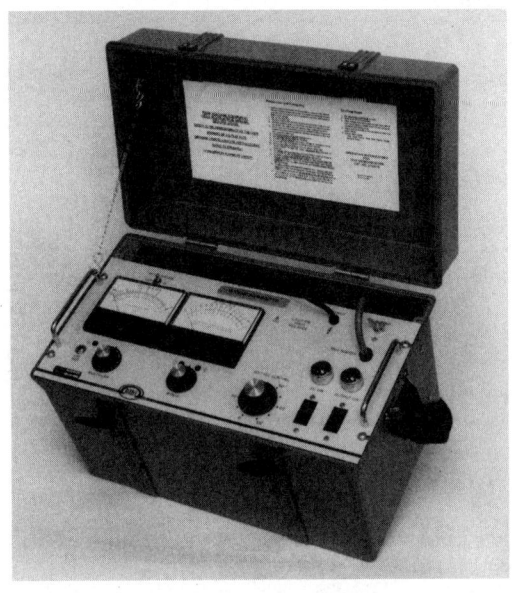

3. Guard—The guard terminal is also a return terminal, except it bypasses the instrument. Therefore, if the guard terminal is connected to some part of the insulation circuit, that part is "guarded" or bypassed.

Connections The general connection for the megohmmeter is shown in Figure 4–8. The line terminal is connected to the conductor, and the earth lead is connected to the insulation system. To test a generator, the earth lead will be connected to the ground and the metal frame of the stator.

FIGURE 4–8
Typical megohmmeter
connections

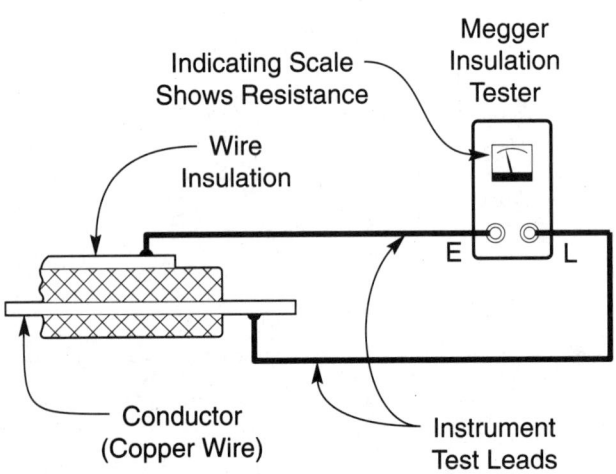

AC Generator Testing

Figure 4–9 shows the connections for measuring the insulation of a generator stator. If the generator is three-phase generator, all three phases should be jumpered together before the megohmmeter voltage is applied.

The generator shown in Figure 4–9 is a rotating field type. To measure the insulation of the field, the connections must be made to the exciter windings. This can be done at the slip rings if the generator is so equipped, and at the rectifier connection points for a brushless exciter. The line lead is connected to the windings and the earth lead is connected to the rotor shaft.

The connections for rotating armature generators are similar. The armature is tested by connecting the line lead to the slip rings and the earth lead to the shaft. If the generator is a three-phase machine, the three slip rings should be jumpered together. The stator is tested by first removing the exciter connections, shorting the field windings together, connecting the line lead to the field windings, and the earth lead to ground.

The connections are the same for the insulation resistance test, the polarization index test, and the step voltage test.

Insulation Resistance   The most commonly performed insulation test is the insulation resistance test. After the test connections are made as previously described, the megohmmeter is energized for one minute. The insulation resistance value is read from the meter and recorded. The one-minute wait is an industry rule of thumb, and ensures that every insulation reading is taken at exactly the same time. This way, standard readings are taken each and every time.

If the insulation is at some temperature other than 20°C, a corrected value must be obtained.

> *CAUTION:*
> *Be certain to disconnect the rectifier and any solid-state regulation components before the test is performed.*

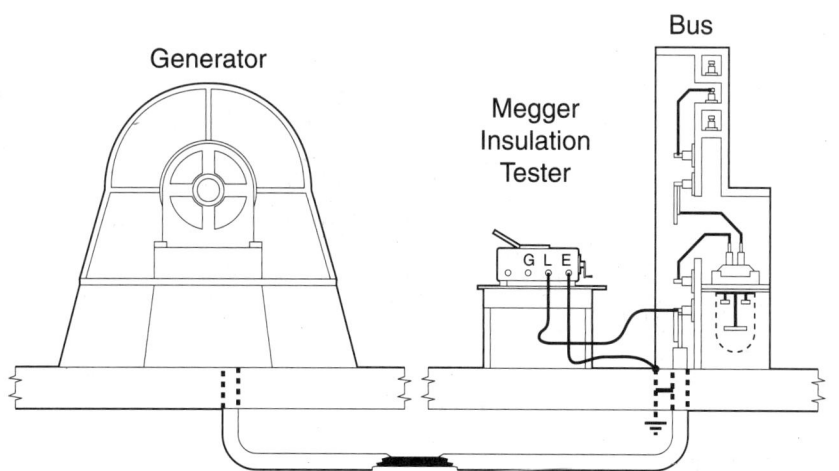

FIGURE 4–9 Megohmmeter connections for the stator of a generator

> **CAUTION:**
> *After the completion of any insulation test, the conductor should be grounded for a time period at least equal to the time that the test voltage was applied. This procedure drains the capacitive charge that has built up, and allows the worker to touch the conductor safely.*

Example:
Assume the following:

Measured Resistance: 75 megohms
Insulation Temperature: 30°C
Insulation Class: Class A

From the temperature correction table, Table 4–1, a multiplier of 2.2 is determined. Therefore, the corrected insulation resistance is

$$R_{corr} = 75 \times 2.2 = 165 \text{ megohms}$$

Equipment record cards should be kept for all insulation measurement. Figure 4–10 shows front and back views of a typical record card. The *A* curve shows the measured values and the *B* curve is the temperature corrected information. The back of the card is used to record the data as it is taken.

Dielectric Absorption Tests As described previously, good insulation will show a gradually increasing insulation resistance after the megohmmeter is applied to the circuit. This fact can be used to advantage, because good insulation will show increasing resistance in a very predictable manner.

After the connections are made as described previously, the megohmmeter is energized, and the insulation resistance is read at two different times—usually 30 seconds and 60 seconds, or 60 seconds and 10 minutes.

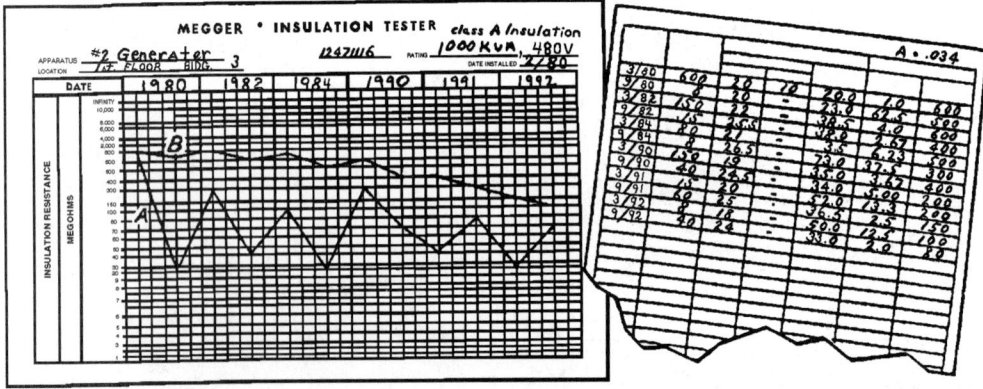

FIGURE 4–10
Typical insulation resistance record cards, front and back view

AC Generator Testing

The later reading is divided by the earlier to obtain a ratio. The 60-second reading divided by the 30-second reading is called the *dielectric absorption ratio*. The 10-minute reading divided by the 60-second reading is called the *polarization index*. Notice that both ratios require a resistance reading be taken at 60 seconds. This means that the standard insulation resistance reading can be obtained during the dielectric absorption test.

Example:

A megohmmeter is applied to a generator and the following insulation resistances are read:

30 seconds: $R_{30} = 50$ megohms
60 seconds: $R_{60} = 75$ megohms
10 minutes: $R_{10\,min} = 180$ megohms

The various ratios are calculated as:

dielectric absorption ratio (DAR) = 75/50 = 1.5
polarization index (PI) = 180/75 = 2.4

The DAR and PI can be evaluated using the information shown in Table 4–2.

Step Voltage Test   Moisture and contamination will normally be indicated by the insulation resistance and/or polarization index tests. Mechanical damage, however, may not show up at low voltages. To test for mechanical damage, the step voltage test is employed.

A dual voltage megohmmeter, connected as previously described, is used for this test. A low voltage, typically 500 V, is selected, and a standard insulation resistance test is performed. After the charge has been drained from the insulation by grounding the conductor, the voltage is increased to 2500 V, and the test is performed again. If more than a 25% difference exists between the two resistance readings, age deteriorated or damaged insulation is suspected. Figure 4–11 shows typical time-resistance curves for age deteriorated insulation.

High-Potential Testing

Overview   High-potential tests, also called overvoltage or overpotential, are used to obtain assurance that winding insulation will withstand the stresses expected in operation without breakdown. Because the voltage applied is higher than the rated operating voltage, the possibility exists for

Table 4–2 Condition of Insulation Indicated by Dielectric Absorption Tests

Insulation Condition	60/30-Second Ratio	10/1-Minute Ratio (Polarization Index)
Dangerous	—	Less than 1
Questionable	1.0 to 1.25	1.0 to 2*
Good	1.4 to 1.6	2 to 4
Excellent	Above 1.6†	Above 4†

These values must be considered tentative and relative — subject to experience with the time-resistance method over a period of time.

* These results would be satisfactory for equipment with very low capacitance such as short runs of house wiring.

† In some cases, with motors, values approximately 20% higher than shown here indicate a dry brittle winding which will fail under shock conditions or during starts. For preventive maintenance, the motor winding should be cleaned, treated, and dried to restore winding flexibility.

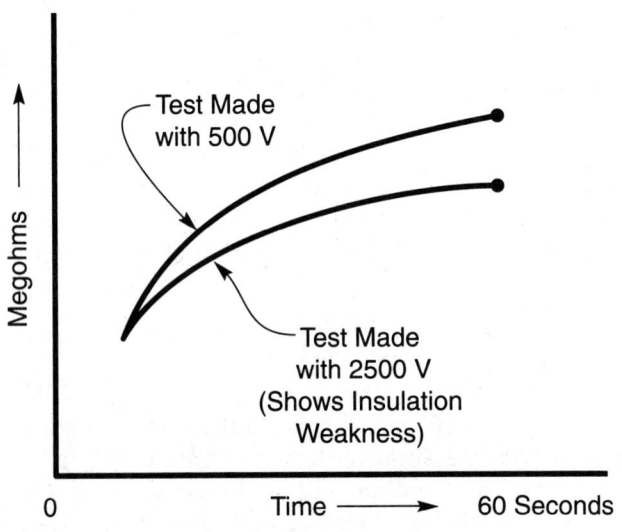

FIGURE 4–11 Step voltage time resistance curves for age deteriorated insulation

AC Generator Testing

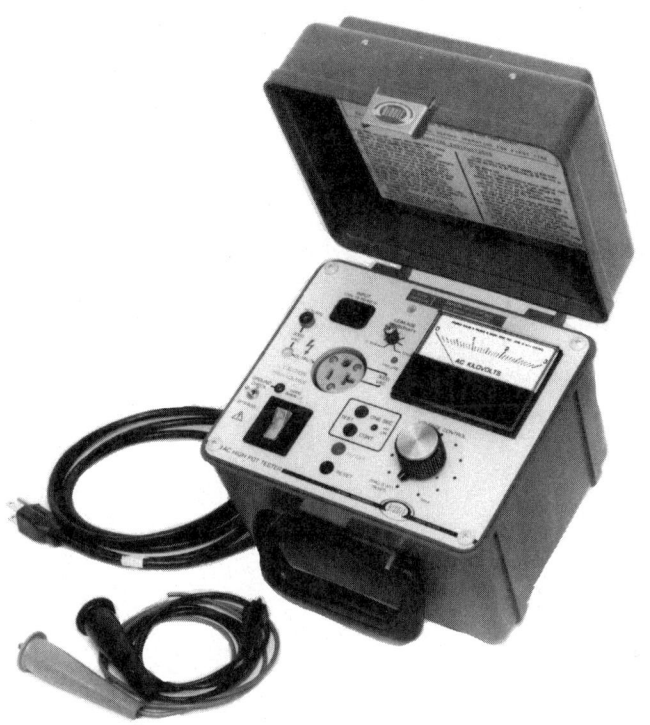

FIGURE 4–12
AC high-potential test set (Biddle™ AC High-Pot Tester)

causing severe damage to the windings under test. The manufacturer's instructions must be consulted to determine the proper voltage level to be applied in the test. The test should be conducted by experienced personnel, and all recommended safety precautions should be strictly adhered to.

High-potential testing should not be conducted until a complete visual inspection of the winding has been performed, and the insulation resistance test (covered previously) indicates that the winding is clean and dry.

Precautions must be taken to prevent the hazards of electrical shock. All windings not under test must be solidly grounded. The winding under test must be grounded upon completion of the test, before it is touched by personnel.

AC Testing When testing AC generator armatures or other alternating voltage windings, the AC test is preferred because it more closely approximates the stresses of operation. An AC high-potential test set, Figure 4–12, must be capable of providing a voltage that is relatively free from harmonics, and that is stable and controlled over the entire range required. During the test, the voltage should be slowly built up to the required level, held for one minute, and then slowly reduced to zero. Full voltage should not be applied or removed suddenly.

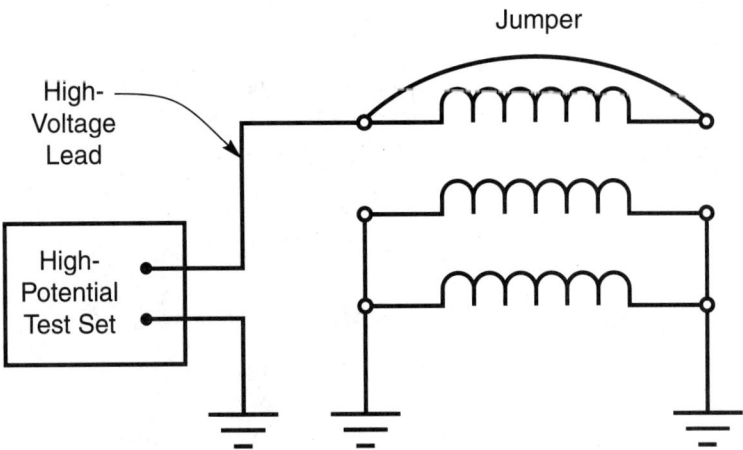

FIGURE 4–13 High-potential test connections (Courtesy of Cadick Professional Services)

To test a winding, the connections of Figure 4–13 are used. Notice that the neutral point is opened so that each winding is independent of the other. The winding to be tested is short-circuited, and connected to the high-voltage output lead of the high-potential test set. The windings that are not being tested are jumpered together and grounded. The test is then repeated for each of the other two windings.

The test voltage to be applied in high-potential tests must be determined from manufacturer's recommendations. The recommended or required values should not be exceeded.

Test voltage values vary. The following are fairly typical; however, **under no circumstances should high-potential tests be applied to a machine without first checking with the manufacturer. The following voltages are given for example only, Table 4–3.**

Note also that AC high-potential tests are classified as "go–no go" tests. This means that no quantitative results can be derived from an AC high-potential test. Upon completion, if the insulation did not fail, the test is deemed successful. No actual readings are taken.

DC Testing DC high-potential testing is not as commonly used for the AC windings of rotating equipment because AC testing more closely duplicates the actual service conditions. However, DC test equipment is generally smaller, lighter, and less expensive; therefore, many maintenance facilities prefer the use of DC. The following are the key features of a DC high-potential test when applied to a rotating machine:

Table 4–3 High-Potential Test Voltages

Acceptance Tests (New Equipment)	*Synchronous Machine Field Winding*	*10 × Rated Voltage, Not Less than 1500 V*
	All other AC or DC windings	2 × rated voltage + 1,000 V
Maintenance Tests (Service Aged Equipment)	75% of the values given for Acceptance Tests	

1. Connections are the same as those described for the AC high-potential test.
2. Test voltages are generally equal to 1.6 times the equivalent rms AC test voltage. As always, **check with the manufacturer before applying high voltage to any machine.**
3. Since DC is being used, the same principles apply as described previously under DC insulation resistance testing. Although the meters on high-potential test sets usually read out in microamperes (μA) or milliamperes (mA), the insulation resistance, dielectric absorption tests, and step voltage tests can all be performed with DC high-potential test sets. Figure 4–14 is a typical DC high-potential test set.

A direct voltage will generally leave a much higher voltage charge on insulation after it has been tested. Because of this charge, the requirement to drain the charge from the insulation after a test is even more stringent when using DC. To minimize stress on the insulation, use a resistor with a value from 1 to 6 Ω per volt of test voltage. For example, if a 25,000-V test voltage was employed, a resistor in the range of 25,000 Ω to 150,000 Ω would be applied to ground the conductor. Monitor the voltage periodically using a voltmeter rated for the voltage under test.

After the capacitive charge has dropped to 0 V, apply a solid copper ground wire to the conductor, and leave it in place for a time period at least equal to the length of time that the test voltage was present.

Insulation Power Factor Testing

Overview The basic concepts that underlie power factor testing can be determined from Figure 4–4. When an alternating voltage is applied to an insulation system, both capacitive and resistive current flow. The source

FIGURE 4–14
DC high-potential test set (Biddle™ DC High-Pot Tester)

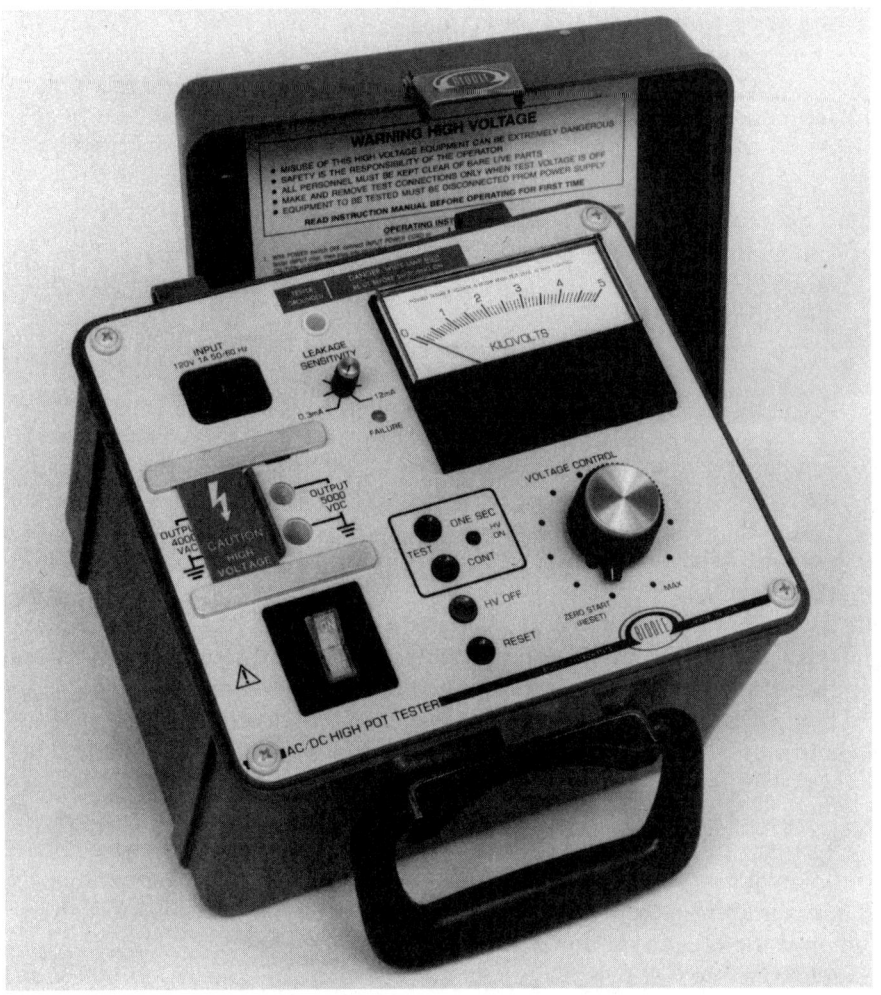

will, therefore, supply VAR to the capacitor and watts to the resistor. Power factor is defined as the ratio of watts to total volt-amperes or

$$\text{Power Factor (\%)} = \frac{\text{watts}}{\text{volt-amperes}} \times 100 \tag{1}$$

From Figure 4–4 this simplifies to:

$$\text{Power Factor} = \frac{V_{ac} \times I_R}{V_{ac} \times I_t} \times 100 = \frac{I_R}{I_t} \times 100 \tag{2}$$

If insulation is perfect, $I_R = 0$, and the power factor is zero. (Note the power factor discussed in this section is *not* the power factor of the system load. System load power factor is ideal at 100%.) This means that the power factor for good insulation will be very low. Moreover, two generators with the same structure and insulation type should have very similar insulation power factors.

Procedures Figure 4–15 shows a modern, computer-controlled insulation power factor test set. Power factor test sets have three output leads—a high-voltage lead, a low-voltage lead, and a ground lead. The high-voltage lead corresponds to the megohmmeter's line lead; however, the connection of the other two leads varies. The low-voltage lead is a current return lead, but it may also be guarded so that its current is not involved in the measurement. The ground lead always connects to ground; however, the current in the ground lead may also be measured at the discretion of the user. Figure 4–16 shows one type of connection that may be used for testing a three-phase generator stator.

With this connection, the connections can be switched to evaluate the insulation from A phase to ground, A phase to B and C phases, or A, B, and C phases to ground. After the test is performed, the power factor results are

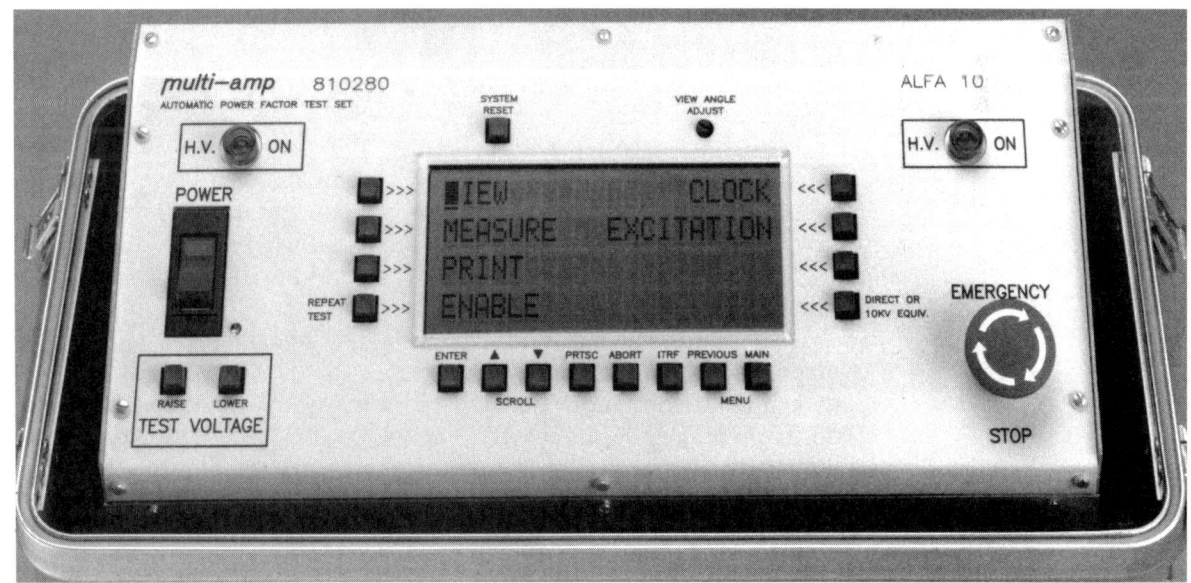

FIGURE 4–15
Computer-controlled insulation power factor test set (Multi-Amp® ALFA-10™)

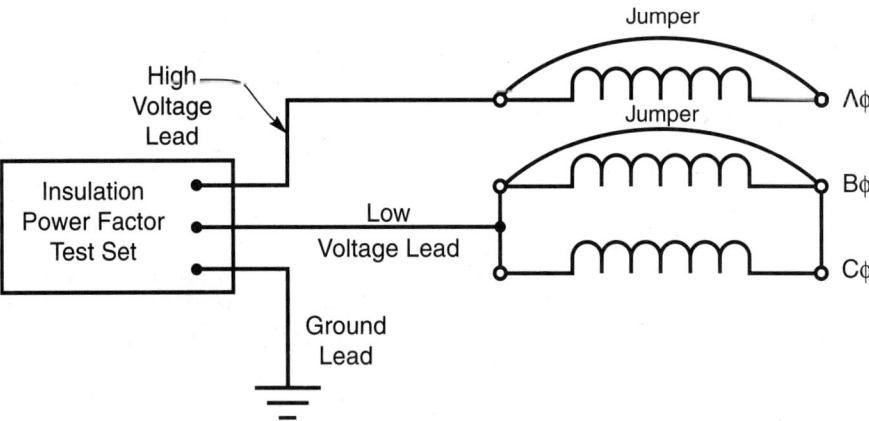

FIGURE 4–16 Sample connection diagram for insulation power factor testing (Courtesy at Cadick Professional Services)

compared to previous tests and/or tests on similar generators. For each piece of equipment a range of acceptable values is available. If the generator insulation falls outside of this range, the insulation is deemed either questionable or bad, depending on how far out it falls.

CORE LOSS TESTING

Core Loss and Its Effects

Core loss in the stator comes from two principal sources—*hysteresis loss* and *eddy current loss*. As the generator rotates, the magnetic flux shifts rapidly, constantly reorienting the polarity of the iron molecules in the core. This is called hysteresis loss. Eddy currents are the result of voltages generated in the core iron that result in a flow of current through the core's laminations. Eddy currents and hysteresis loss generate heat and thereby waste power.

An AC generator stator is composed of many thin, metallic structures called laminations, Figure 4–17. Thousands of these laminations are sandwiched together to form the long, cylindrical core, Figure 4–18. After the windings are placed into the slots, wedges are put into the slots to lock the windings in place, Figure 4–19.

Insulated laminations are used to reduce the eddy currents. If the insulation should weaken or fail, the eddy currents will greatly increase, Figure 4–20. This will cause increased loss and severe reduction in life for the generator.

AC Generator Testing

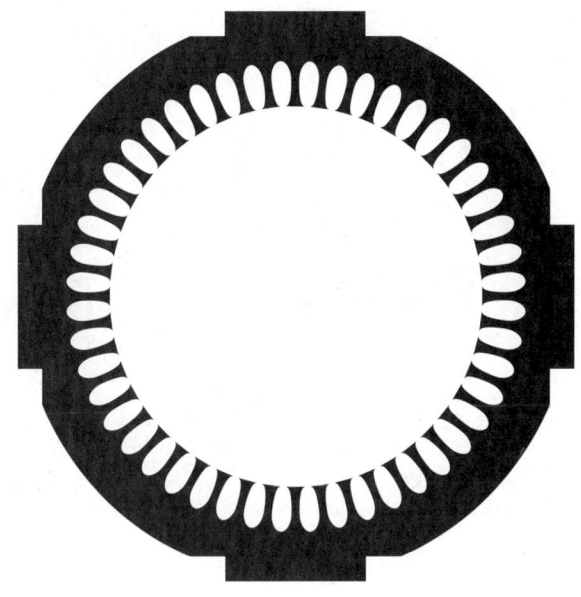

FIGURE 4–17
Typical lamination from a generator stator (Courtesy of Onan Corporation)

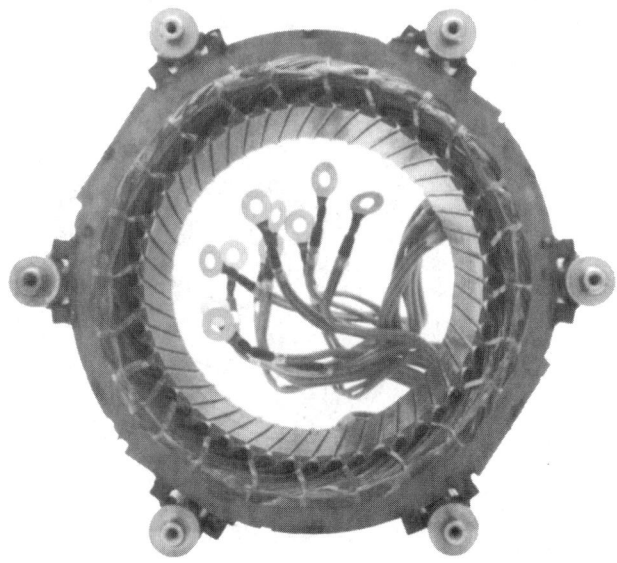

FIGURE 4–18
Generator stator (Courtesy of Onan Corporation)

FIGURE 4–19 Close-up view of lamination faces and coil wedges. Pencil is pointing to the laminations. (Courtesy of General Electric Company)

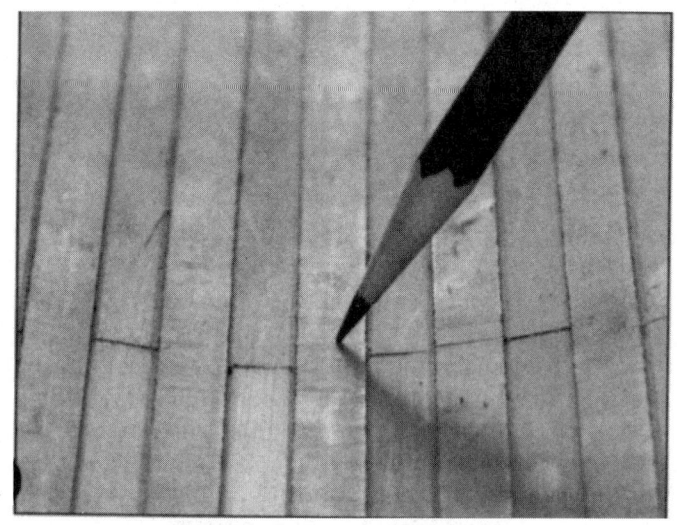

FIGURE 4–20 Eddy current flow caused by interlaminar insulation failure

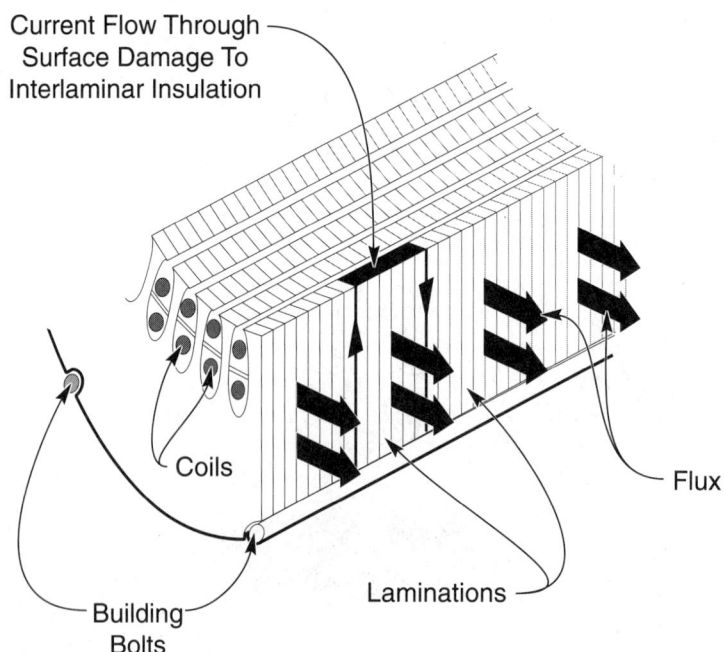

Core Loss Testing

Core loss testing is designed to detect interlaminar failures. The basic connection scheme is shown in Figure 4–23. One or more loops of wire are passed longitudinally through the core, and a current is passed through them. The current is adjusted to induce some percentage of the normal operating flux in the core. The amount of flux induced is determined by the method that will be used to identify the lamination problems.

If lamination problems exist, the eddy currents created by the induced flux will cause heating which can be detected by an infrared scan sensor. Trouble spots will show up as hot spots in the core. Of course, if the trouble spots are located far from the surface of the core, the heat will not appear at the surface for a long time. Because of this, other sensing methods have been developed.

Figure 4–21 is a motor core tester that uses a special type of sensor called a Chattock potentiometer, Figure 4–22. Figure 4–23 shows the general setup for testing the generator stator. While the exciting winding induces a magnetic field equal to approximately 5% of the rated flux density, the Chattock potentiometer, called the *sensing head* in Figure 4–23, is passed along the slots as shown. Excessive eddy currents are picked up by the Chattock potentiometer and fed to the instrumentation. The reference coil is placed on a slot, parallel to the excitation winding, and provides a baseline current value to which the output of the Chattock potentiometer is compared.

When properly set up, eddy fault currents of up to 100 mA are considered significant. Faults in excess of 200 to 300 mA should be repaired. Deep-seated faults can be repaired only by removing the windings; therefore, repair of such faults is generally not performed unless the eddy currents exceed 500 mA. These current values assume that the magnetic flux density is 5% of full-rated flux. If a lower flux is used, proportionally lower current levels are obtained.

Whatever method is used, the core damage test can identify problems which, if left uncorrected, can greatly reduce the life of the generator and possibly cause short term failure.

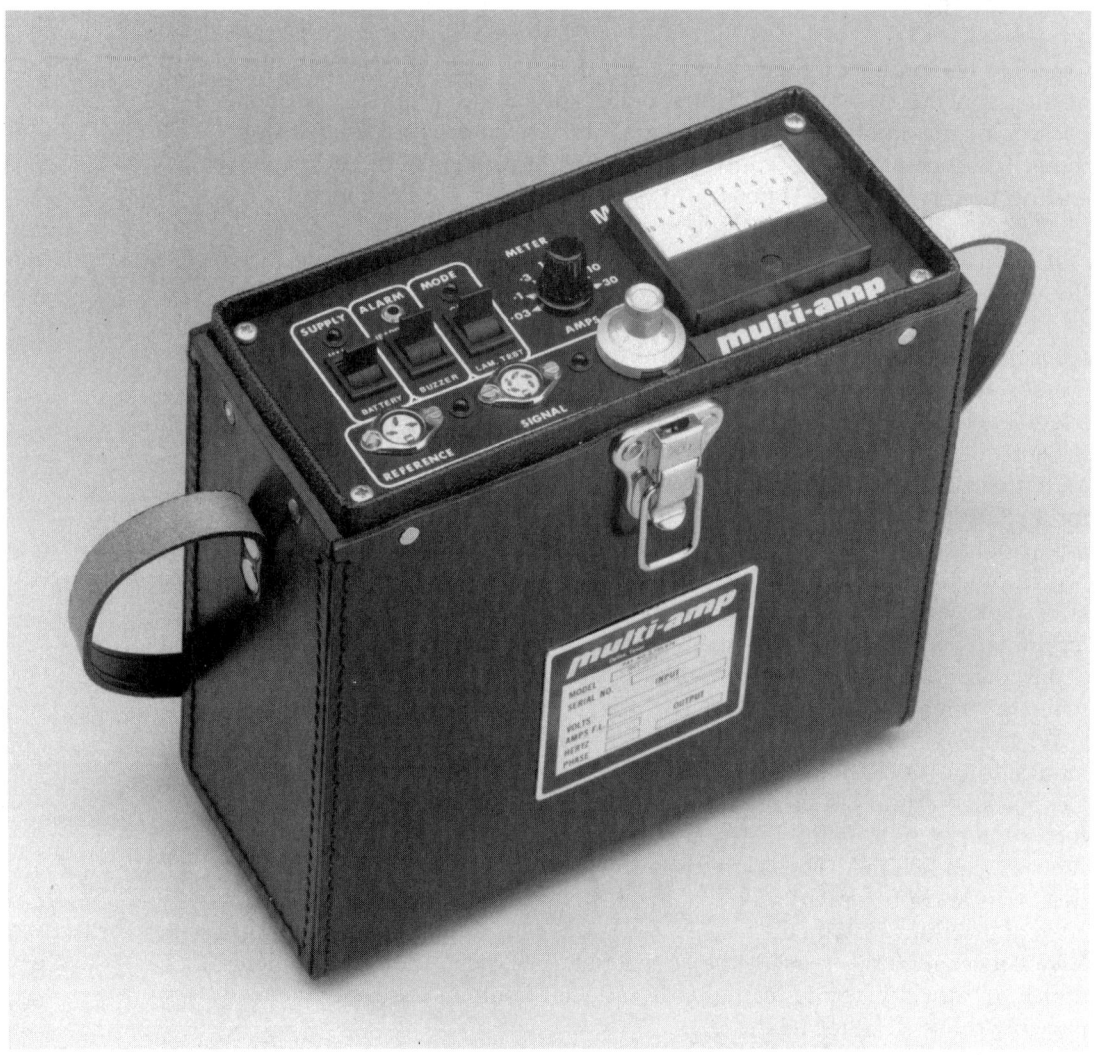

FIGURE 4–21
Motor core loss tester (Multi-Amp®)

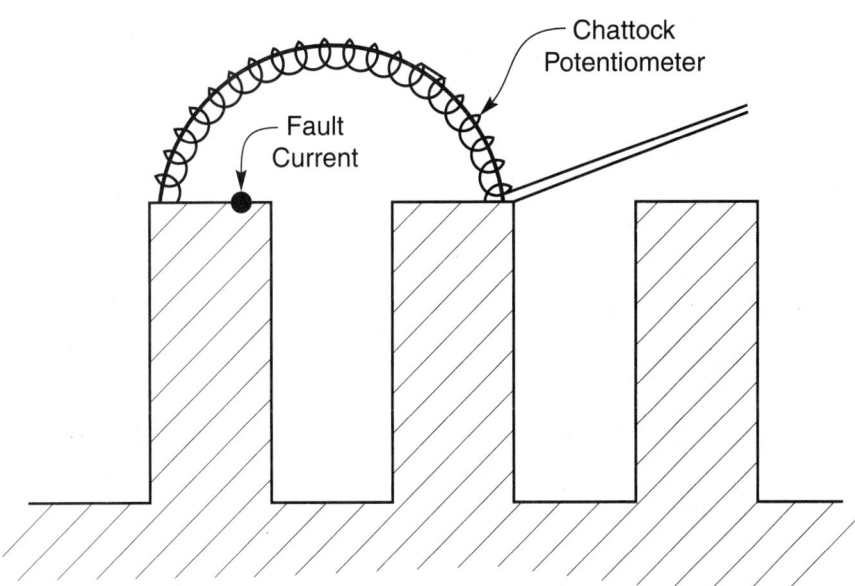

FIGURE 4–22
Chattock potentiometer

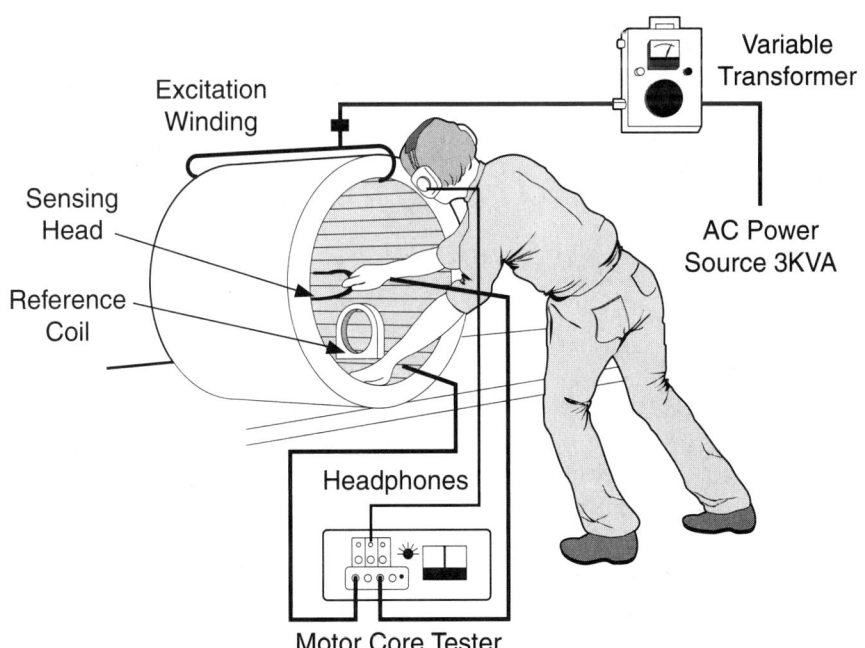

FIGURE 4–23
Setup for testing generator core using the test set of Figure 4–21

SURGE COMPARISON TESTING

Principles of Surge Comparison Testing

If a voltage pulse is applied to a capacitor and inductor in parallel, they will oscillate as follows:

- The voltage will charge the capacitor.
- The capacitor will discharge through the inductor, causing a magnetic field around the inductor.
- When the capacitor is completely discharged, the inductor magnetic field will start to collapse and recharge the capacitor.
- This process continues with the capacitor alternately charging and discharging, until all of the initial pulse energy is dissipated.
- The current that causes the charging/discharging will be at a frequency, called the ringing frequency, determined by the formula:

$$\text{Frequency} = \frac{1}{2\pi\sqrt{LC}} \qquad (3)$$

From this principle, you can see that if two identical coils have two identical pulses applied to them through identical capacitors, the ringing frequencies will be identical. Figure 4–24 shows a simplified diagram of a surge comparison tester.

A half-wave rectified AC signal is applied to two windings of a generator through two equal resistors. The current flow is established through the third winding of the stator. During the negative half of the AC waveform, the two silicon-controlled rectifiers are triggered, and the stator magnetic fields start to discharge through the capacitors. This creates the ringing effect that was described previously. If one of the stator windings is different, it will ring at a different frequency.

Methods and Results

A typical surge comparison tester is shown in Figure 4–25. This device applies the ringing waveforms to a cathode ray oscilloscope on the same axis. If the two windings are identical, only a single trace will appear as shown in Figure 4–26. If one of the windings is bad, however, a double trace will appear such as that shown in Figure 4–27.

In the field, the test set leads are connected to the generator stator leads, and the surge voltage is increased gradually from zero. In no case should the peak surge voltage exceed the following formula:

$$V\text{test}_{max} = (2 \times V_{rated}) + 1000 \qquad (4)$$

AC Generator Testing

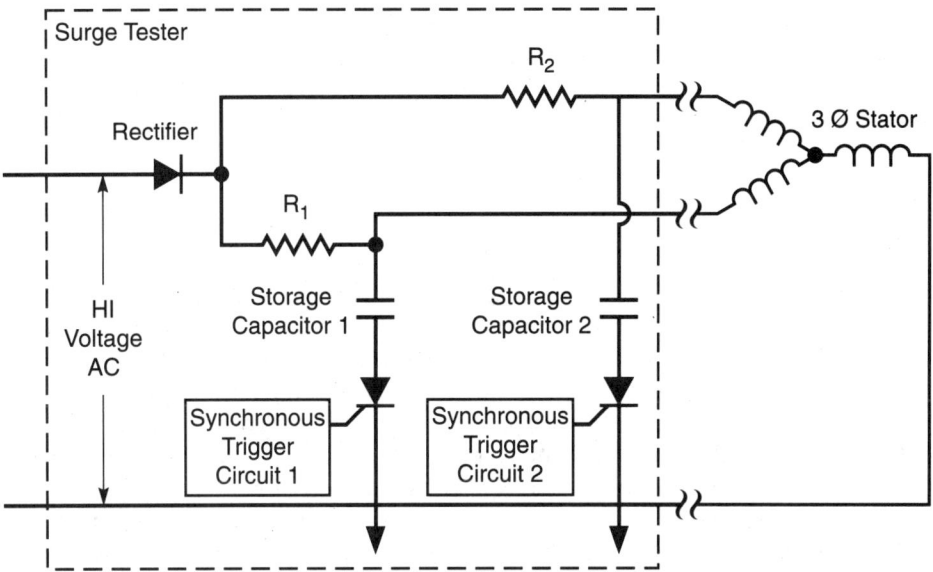

FIGURE 4–24
Schematic diagram of a surge comparison tester

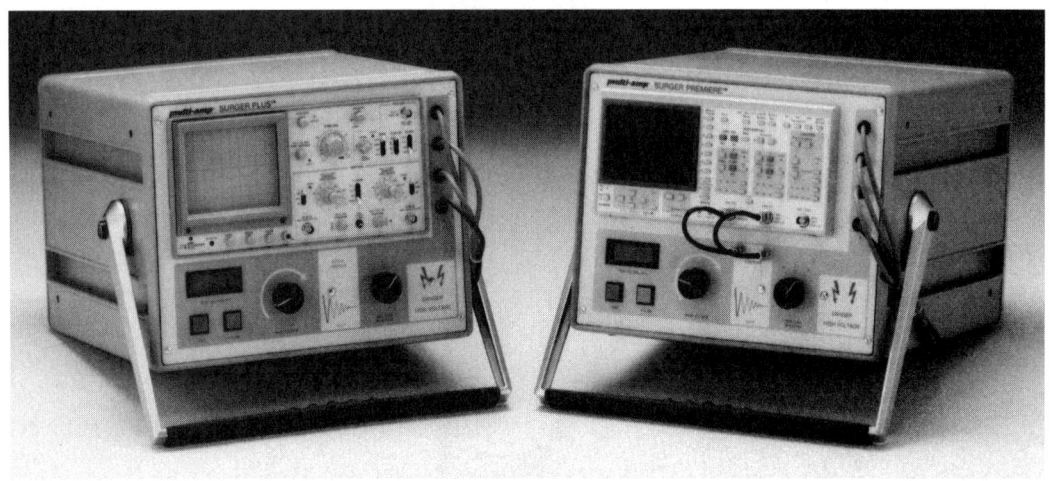

FIGURE 4–25
Surge comparison tester set (Mutli-Amp® Surger Plus)

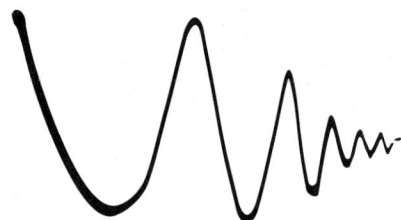

FIGURE 4–26
Good stator windings will produce a single trace on the CRT

If one of the windings is bad, a second ringing waveform will appear on the oscilloscope as the voltage is increased. The severity of the problem determines how much voltage is required to cause the problem to appear, and also determines the severity of the separation of the voltage waves. Figure 4–28 shows a variety of waveforms for various types of stator faults.

Note that the surge comparison test is capable of identifying a variety of problems including insulation failure, turn-to-turn shorts, phase-to-phase faults, and phase-to-ground faults, Figure 4–28. The surge comparison test is considered to be one of the most comprehensive of all generator tests. It can be used as a "go–no go" test, as well as a predictive test. The voltage at which the failure occurs can be correlated to other results to predict the amount of life left in the generator.

BEARING INSULATION

Slight variations in the magnetic circuit may cause a variation in the amount of flux passing through the shaft. This variation can cause current to flow in the circuit formed by the shaft, the bearings, and the base or frame. In operation there is no direct metal-to-metal contact at the bearings because the shaft and the bearings are separated by a film of oil. The cur-

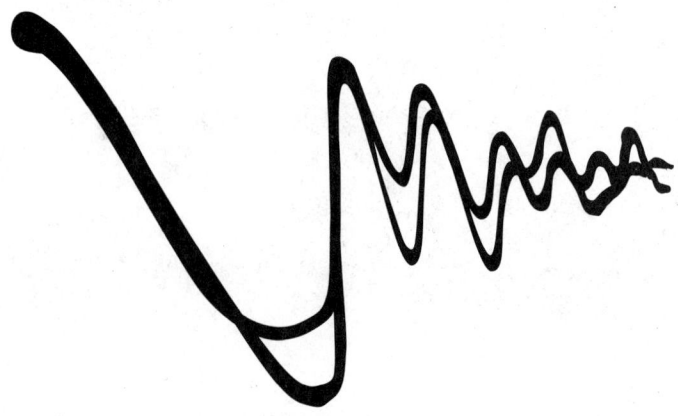

FIGURE 4–27
Bad stator windings will show two separate traces on the CRT

rents induced in the shaft can arc across the thin oil film, and cause pitting of the journal and the bearing surfaces. In order to prevent the flow of circulating currents, it is necessary to open the circuit at one point. Usually this point is chosen as the bearing farthest from the prime mover of a horizontal generator or the bearing above the rotor of a vertical machine.

When pedestal bearings are used, one or more of the pedestals may be insulated from the ground by thin shims of insulating material. Insulated sleeves are used around dowels or foundation bolts that mount the pedestal to the base plate or foundation. Bearings that are mounted in end shields or brackets may be insulated by placing a ring of insulating

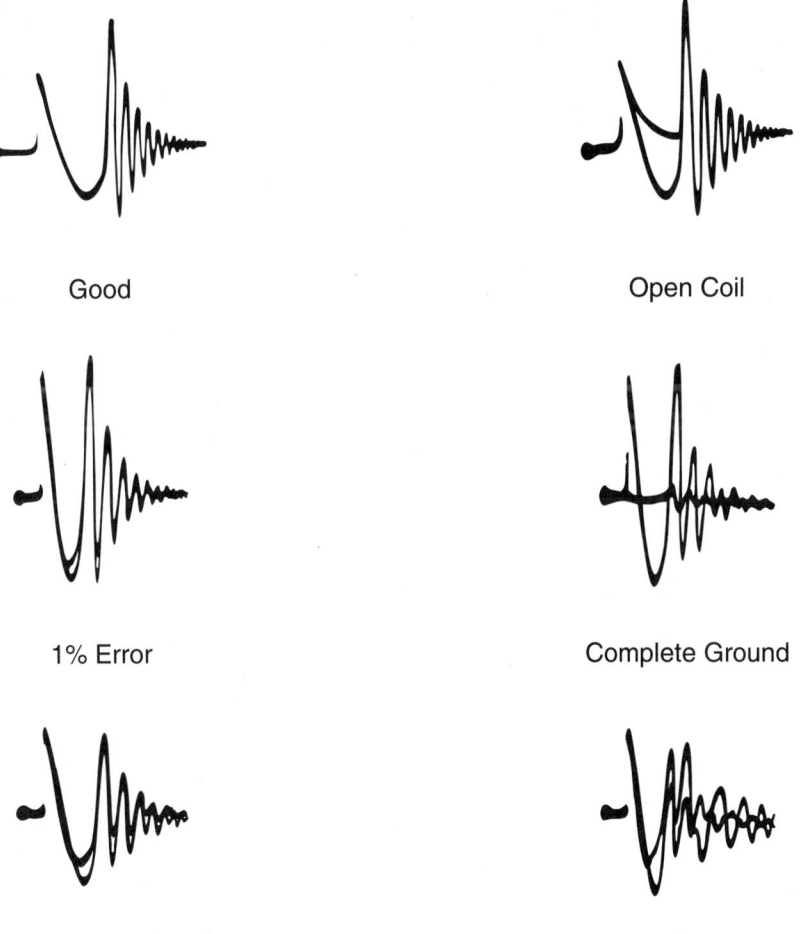

FIGURE 4–28
Surge comparison waveforms for various types of stator faults

material around the bearing shell. Insulated studs or dowels are used to locate the bearings and prevent movement.

In some machines, two layers of insulation are placed under or around the insulated bearing, Figure 4–29. The insulation layers are separated by a conductor layer called the high point. Each insulating layer can then be tested individually by connecting one test lead to the high point, and the other test lead to the bearing shell or machine frame.

Bearing insulation is tested using a 500-V megohmmeter. The insulation resistance should be *at least* 20,000 Ω. The general procedures used depend on the type of installation.

1. Machines with a high point are tested in place. The line lead on the megohmmeter is connected to the high point, the earth lead is connected to the shaft, and a one-minute insulation resistance reading is taken. The test is then repeated with the earth lead connected to the frame of the motor.

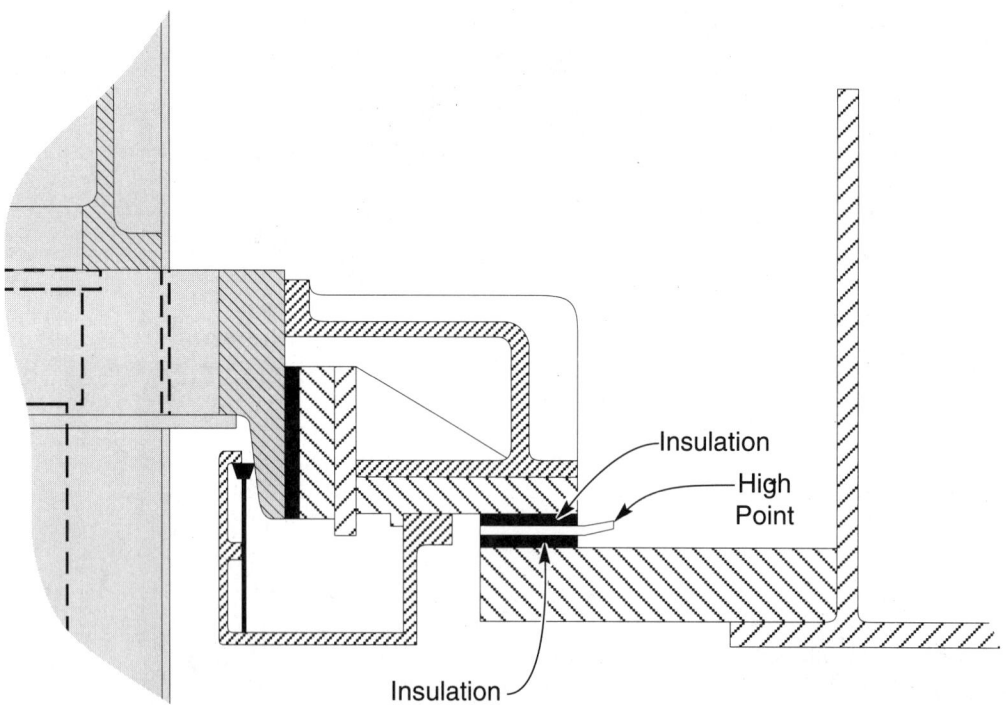

FIGURE 4–29
Bearing insulation with a metallic high point

2. Machines with a single layer of insulation at all bearings can also be tested in place. The line lead is connected to the shaft, the earth lead is connected to the frame of the motor, and a one-minute insulation resistance test is performed.
3. If the machine has insulation at only one bearing, and no high point, the uninsulated bearing must be separated from the shaft. This separation can be accomplished by placing paper or other insulating material between the shaft journals and the bearings. These materials must be removed after testing is completed.

In general, bearings should be tested after the installation of any lubrication piping, coolant piping, instrumentation, or other wiring that can contact the shaft or the insulated bearing. If only one bearing is insulated, external piping and wiring must be insulated to prevent bypassing the bearing insulation.

SUMMARY

Over the years, a variety of test procedures have been developed to evaluate both new and service-aged generators. The majority of these tests involve some form of insulation testing. Tests such as insulation resistance, dielectric absorption, step-voltage, and bearing insulation are commonly performed using simple, lightweight direct voltage test equipment. The megohmmeter has become almost as common as the digital multimeter.

Insulation power factor testing takes advantage of the fact that the AC capacitive current will be 100 or more times the AC resistive current in good insulation. Special instruments are used to measure the insulation power factor of the generators. The results are then compared to previous readings and/or test results from other similar equipment.

Generator core damage is costly in terms of efficiency, and may greatly reduce the life of the generator. When laminations are shorted together due to insulation failure, eddy currents can flow and create a significant amount of heating in the core. When the core loss test is performed, a magnetic flux is applied to the core, and core damage is sensed using an infrared camera or other sensing mechanism such as a Chattock potentiometer.

Surge testing is arguably the most comprehensive of all types of generator tests. Two controlled pulses are applied to a generator stator, and the resulting "ringing" signals are displayed on a cathode ray oscilloscope. Differences in the windings as small as a single turn short can be identified.

REVIEW QUESTIONS

1. Draw a simple equivalent circuit of an insulation system and label the branch currents that will flow when the insulation is subjected to a DC voltage.

2. In a good insulation system, how will the AC capacitive current compare to the AC resistive (leakage) current?

3. You have performed an insulation resistance test on a generator with class B insulation. The reading you got was 95 megohms, and the insulation temperature was 40°C. What is the 20°C corrected insulation resistance? (Refer to Table 4–1)

4. The following results were obtained on a generator stator:
   R = 150 megohms @ 500 V
   R = 105 megohms @ 2500 V
   Are these good results? Why or why not?

5. Describe how to perform an insulation resistance test on a three-phase, revolving armature generator.

6. The following insulation resistance readings are obtained from a generator field.
   R_{30} = 75 megohms
   R_{60} = 115 megohms
   $R_{1\,min}$ = 241 megohms
   Calculate the polarization index and the dielectric absorption ratio.

7. What is the maximum AC voltage to be applied to a *new* 250-VDC field winding?

8. A power factor test set gives the following results for a particular piece of insulation:
   milliwatts loss = 25
   millivolt-amperes applied = 5000
   What is the percent power factor of the insulation under test?

9. An eddy current of 250 mA is found during a core motor damage test. What condition might prevent you from repairing the insulation breakdown that is allowing the current?

10. You are surge-testing a 480-V generator winding. What is the maximum surge voltage that you should use?

Part III

AC Motors

AC Motor Theory 5

OBJECTIVES

After studying this chapter, the student should be able to:

- *Describe the operation of a three-phase AC motor stator, with emphasis on how it creates a rotating magnetic field.*
- *Describe the general operating principles and construction of synchronous and induction motors.*

INTRODUCTION

AC electric motors constitute a significant percentage of the world's total electrical load, and electric motors pervade every facet of residential, commercial, and industrial electric power. The basic operating principles of AC motors were developed by Nikola Tesla in the late 1800s and early 1900s. These principles are still used as the basis for the operation of the AC electric motor.

This chapter describes those operating principles and shows, in a general sense, how they are applied to working electric motors. The material that you will learn in this chapter will serve as a foundation for the more detailed information presented in later chapters.

THE ROTATING MAGNETIC FIELD

The underlying principle for all AC electric motors is the *rotating magnetic field*. The rotating magnetic field is created by allowing two or more magnetic fields to add in such a way that their net magnetic north and south poles seem to rotate. That is, the field strength is not stationary, but rather moves around in a circle. This process can be demonstrated using Figure 5–1.

Figure 5–1 shows a three-phase, two-pole motor stator. As with a generator, the motor stator is the part that does not move and the rotor is the part that rotates two poles because there will only be one net north pole and one net south pole. Each phase current is passed through two windings on opposite sides of the stator. For example, *C phase* current passes into the top winding and leaves through the bottom winding; *A phase* current enters

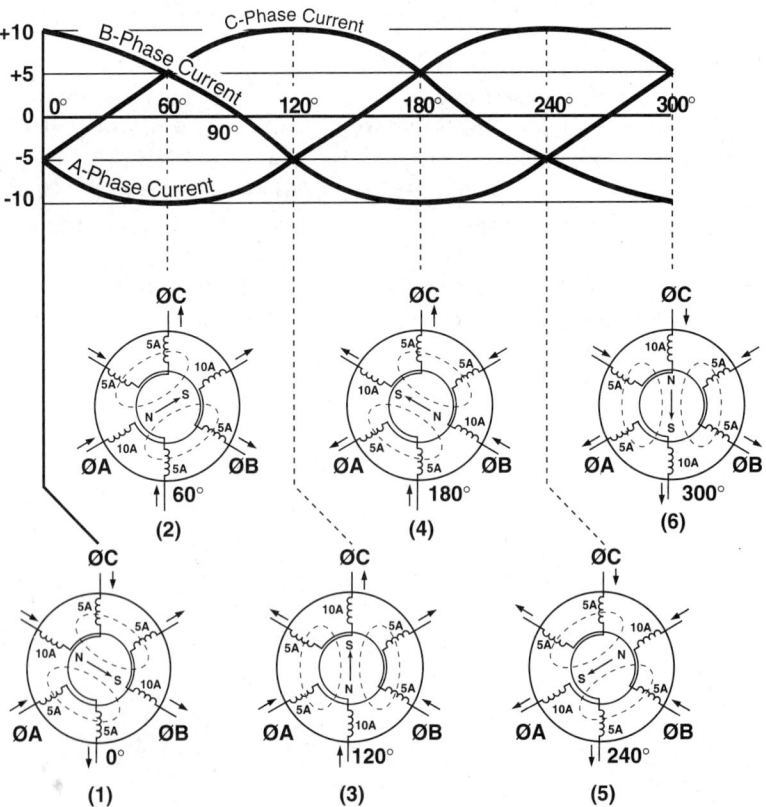

FIGURE 5–1
Development of a rotating magnetic field

through the lower left-hand winding and exits through the upper right winding; and *B phase* enters through the lower right winding and leaves through the upper left winding. The windings are configured in such a way that when the current is negative into a winding, a north magnetic pole is created in the interior of the stator.

Figure 5–1 shows the magnetic field condition at six instants during a cycle. Starting with instant (1), *B phase* current is 10 A, leaving its input terminal and entering its exit terminal. This means that the *B phase* windings will have a strong (10 A worth) north magnetic pole next to its exit winding and a strong south pole next to its entry winding. *A phase* is 5 A entering its entry winding and leaving its exit winding. This means that the *A phase* windings will have a moderate north pole at its entry winding and a moderate south pole near its exit winding. By similar reasoning, *C phase* will have a moderate north pole near its entry winding and a moderate south pole near its exit winding. The result of these fields is a net magnetic north pole in the upper left-hand half of the stator and a net magnetic south pole in the lower right-hand half. The strongest parts of the two poles are adjacent to the *B phase* windings.

One-sixth of a cycle (60°) later, the currents have changed to the values indicated by instant (2). Using the logic discussed previously, at instant (2) the net magnetic field has rotated counterclockwise by 60°. The strongest parts of the magnetic poles are located adjacent to the *A phase* coils, with the north pole being located to the lower left. By evaluating any instant between (1) and (2), you can show that the strong part of the north and south poles literally rotated continuously from (1) to (2).

Evaluating the magnetic field strength at any of the remaining four instants (3, 4, 5, and 6) will show that the magnetic field has relative strengths, as shown in Figure 5–1. If the diagram were extended onto the 360°-position, the magnetic field would return to the same condition as instant (1). This means that the magnetic field has made one full rotation in one electrical cycle. If the stator is wound in such a way that twice as many poles are developed, the field will rotate at half the speed. From this information the formula for the speed of rotation of the magnetic field can be developed as:

$$S = \frac{120 \times f}{P} \tag{1}$$

Where:
    S is the revolutions per minute of the field
    f is the frequency of the applied voltage in Hertz
    P is the number of poles produced by the three-phase windings.

Example:

In a 6-pole, 50-Hz motor, how fast will the magnetic field rotate?

$$S = \frac{120 \times 50}{6} = 1000 \text{ rpm} \qquad (2)$$

The speed at which an induction motor field rotates is referred to as its *synchronous* speed, because it is synchronized to the frequency of the power supply at all times. Notice, however, that the synchronous speed is not necessarily the speed at which the rotor turns. Only one type of motor, the synchronous motor, rotates at synchronous speed. Formula (1) is of course the same as the formula for generator frequency that was discussed in Chapter 2.

From formula (1) you can see that decreasing or increasing the number of poles will increase or decrease the synchronous speed respectively, and decreasing or increasing the applied frequency will decrease or increase the synchronous speed. Note also that the synchronous speed is independent of any load that is attached to the rotor shaft.

TYPES OF AC MOTORS

Synchronous Motors

Both of the general categories of motors use the same type of stator, one that creates a rotating magnetic field. The synchronous motor is very similar to a revolving field synchronous generator. The rotor is connected to a DC supply through slip rings or from rotating exciter system. The fundamental construction of a synchronous motor is shown in Figure 5–2.

When the synchronous motor is running, the rotor south pole locks into the stator north pole and the rotor north pole locks into the stator south pole. Because the stator poles are rotating at synchronous speed, they pull the rotor poles with them at synchronous speed. Thus, the synchronous motor rotates at constant, synchronous speed. Notice that the rotor must have the same number of poles as the stator. Thus, a rotor for a four-pole stator will have two north poles and two south poles.

Synchronous motors are frequently used for slow, constant speed applications. Because of this, their stator windings are frequently multiple poles like a vertically-mounted AC generator. Figure 5–3 shows a typical synchronous pole rotor and stator.

A synchronous motor cannot be started as such. Recall that the rotating magnetic field rotates at synchronous speed regardless of the load. If the

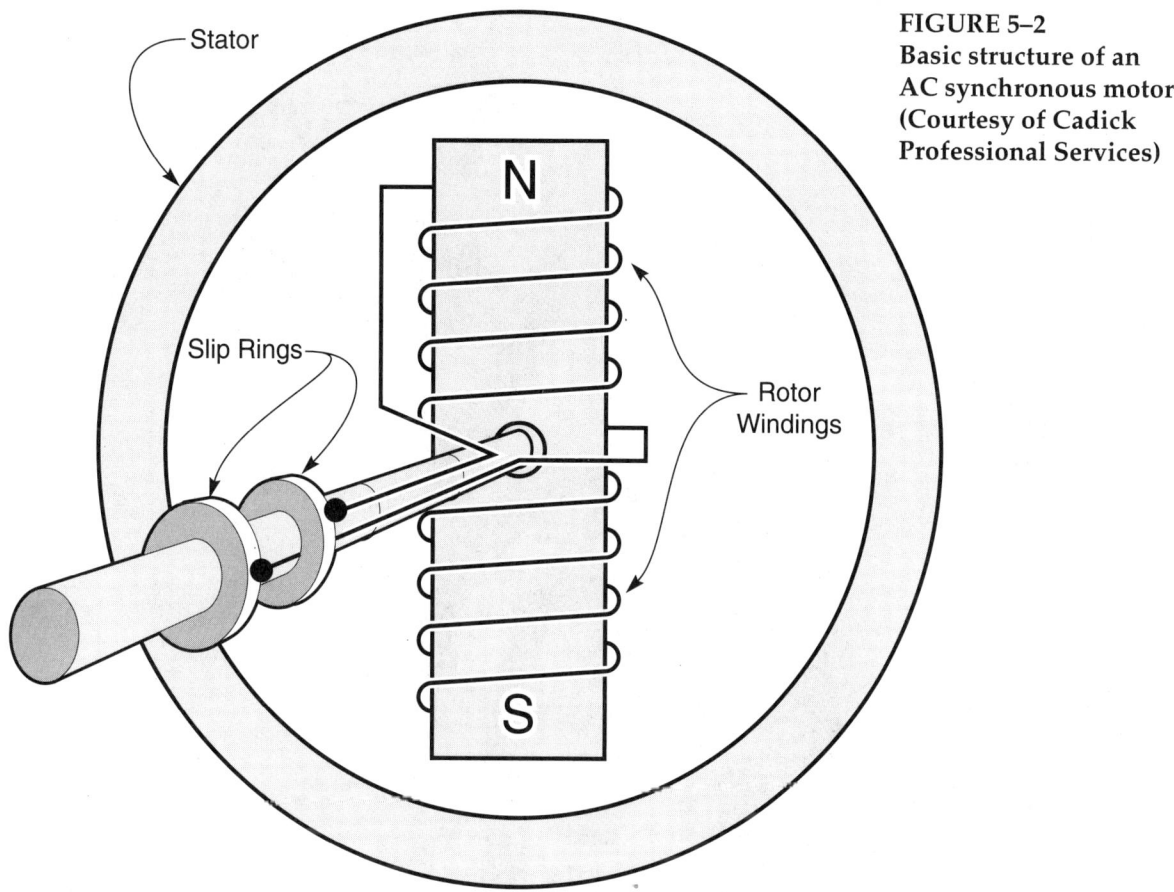

FIGURE 5–2
Basic structure of an AC synchronous motor (Courtesy of Cadick Professional Services)

rotor were energized and the stator voltage applied, the rotating magnetic field would literally smash into the rotor field at every half rotation. This would be extremely damaging to the motor. Because of this problem, a synchronous motor is equipped with induction motor windings, as well as synchronous windings on the rotor. As you will learn in the next section, an induction motor does not require a DC field and, therefore, will start smoothly when the stator voltage is applied. After the rotor is close to synchronous speed, the field current is applied and the rotor locks into synchronous speed. These starting windings are called *damper* or *amortisseur* windings. They also help to regulate speed when sudden step loads are applied or removed from the synchronous motor shaft.

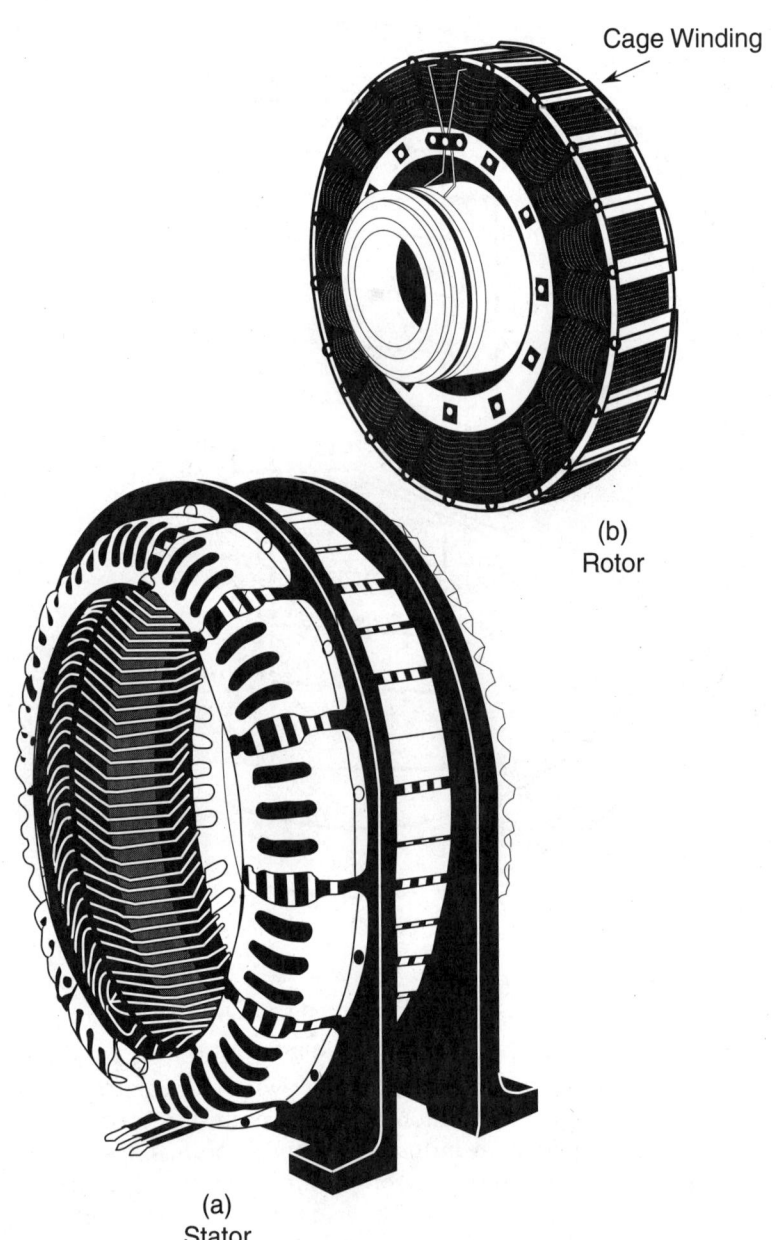

FIGURE 5-3
Typical synchronous motor stator and rotor

(a) Stator

(b) Rotor

Cage Winding

AC Motor Theory

Induction Motors

The stator of an AC induction motor is virtually identical to the stator of an AC synchronous motor. Since induction motors are usually operated at much higher speeds than synchronous motors, the induction motor stator looks more like a horizontal, AC generator, similar to Figure 4–20.

Squirrel Cage Rotors The induction motor rotors are available in two basic types—squirrel cage and wound rotor. The squirrel cage motor is by far the more common. The basic operating element of the squirrel cage rotor is an assembly of rotor bars and circular end pieces welded together in a configuration that looks somewhat like the small wheel used to amuse pet hamsters, Figure 5–4.

In this type of induction motor, the rotating torque is produced by Faraday's law. As the stator magnetic field rotates, it cuts through the rotor bars, creating a current flow in them. By Lenz's law, the current that flows creates an opposing magnetic field. Thus the rotor magnetic field creates an opposite pole to the stator magnetic field. The rotor field is attracted to the stator field and is "pulled" around. This creates rotation. The slower the rotor turns, the greater the number of lines of flux are cut per second; thus, a greater torque is created.

A squirrel cage built in air, as shown in Figure 5–4, would work as a motor; however, it would be extremely inefficient because of the very low permeability of air. Because of this, the squirrel cage is mounted in a structure of laminated steel plates. The steel plates are pressed on the rotor shaft and are insulated from each other every place except the shaft. Figure 5–5 shows a generalized diagram of how a squirrel cage rotor looks without the squirrel cage.

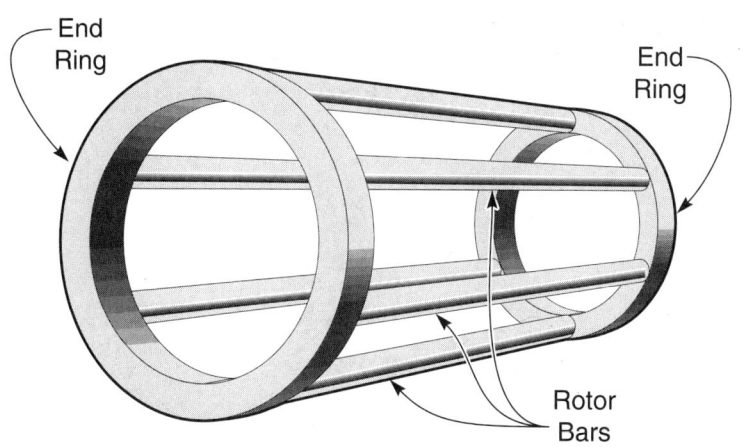

FIGURE 5–4
Squirrel cage rotor without laminations (Courtesy of Cadick Professional Services)

FIGURE 5–5
Conceptual drawing of rotor without squirrel cage

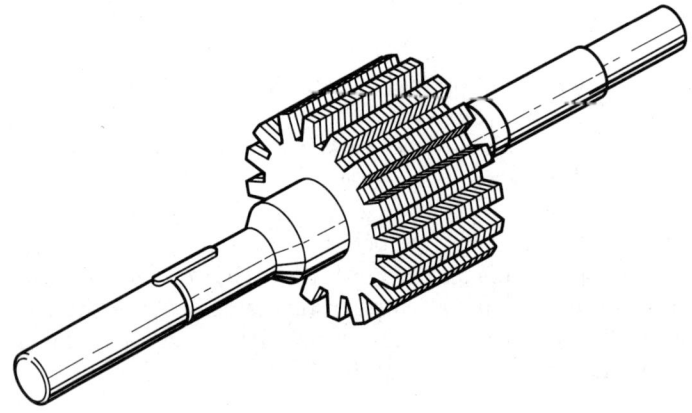

The squirrel cage is placed, or sometimes poured, into the laminations. A diagram of an assembled squirrel cage rotor is shown in Figure 5–6, and a photograph of such an assembly is shown in Figure 5–7. Notice that the rotor bars are not parallel to the rotor shaft. Skewing the rotor bars in this fashion reduces noise and causes a much smoother acceleration.

Figure 5–8 shows a cutaway view of an assembled AC induction motor. If you look closely, you can see lighter areas on the cutaway end of the rotor. These light colored regions are the cross sections of the cut off rotor bars.

Wire Wound Rotors The name *wound rotor* explains the fundamental construction difference between the wound rotor induction motor and the squirrel cage induction motor. Figure 5–9 shows a diagram of the wound rotor motor and Figure 5–10 shows a schematic diagram.

Recall that the synchronous motor rotor must have the same number of poles as the stator. The squirrel cage rotor and the wound rotor motor must be constructed in such a way that they have the same number of poles as the stator.

FIGURE 5–6
Rotor for a squirrel cage motor (Courtesy of the Lincoln Electric Company)

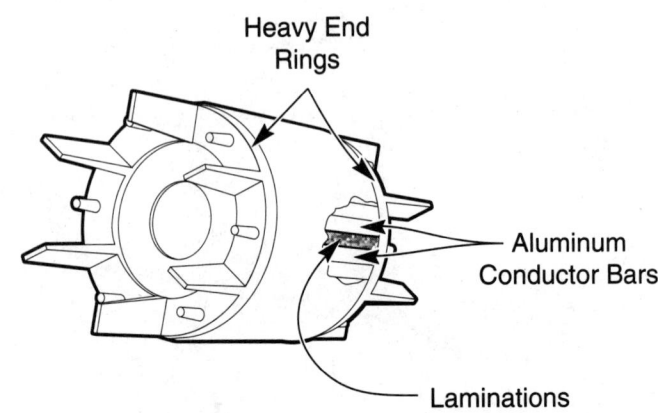

FIGURE 5–7
Assembled rotor for squirrel cage motor (Courtesy of the Lincoln Electric Company, Cleveland, OH 44117)

When a short circuit is placed across the windings of the rotor, it works in much the same way as a squirrel cage rotor. The key value of a wound rotor motor is to control speed of operation by placing a rheostat, or stepped resistors, in series with the windings, Figure 5–11. In this figure, three sets of switches—S_1, S_2, and S_3—can be closed and opened to control the motor speed.

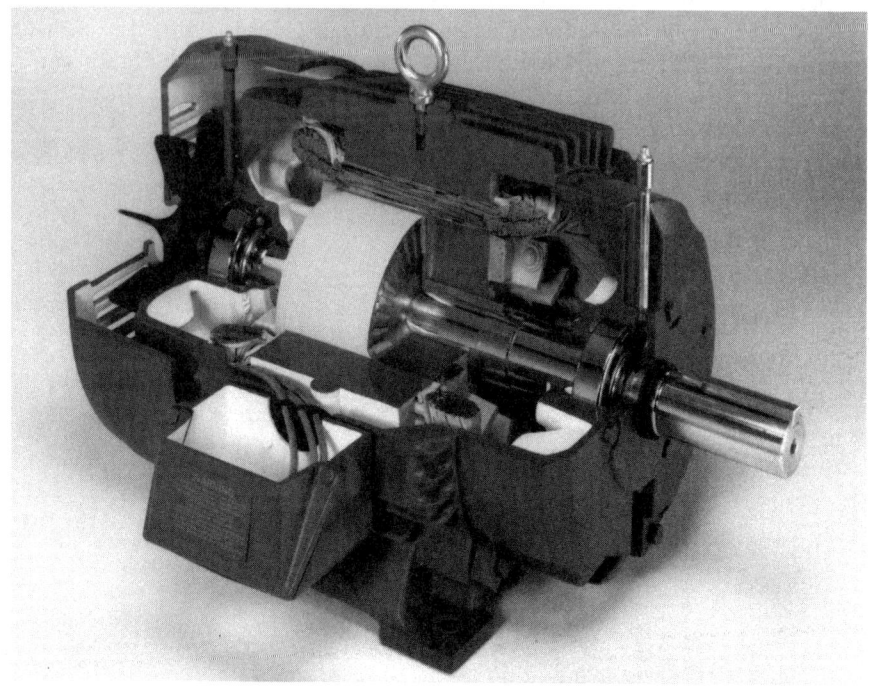

FIGURE 5–8
Energy efficient squirrel cage AC induction motor (Courtesy of Reliance Electric Company)

FIGURE 5–9
Induction motor wound rotor

With all switches open, the rotor has maximum resistance in series with it. This reduces rotor current and therefore, rotor speed. If the S_1 switches are closed, some of the resistance is removed and the motor speeds up. This process continues until all switches are closed and the rotor is moving at maximum speed. This method of speed control has been very common; however, more modern methods are replacing it.

Single-Phase AC Motors

Single-phase motors must generate a rotating magnetic field using only one phase. This can be done, but, a single-phase motor will not generate sufficient starting torque with only the single winding. To make up for this, single-phase motors employ a second winding called a starting winding. For normal torque requirements, the starting winding will be taken out of service after the motor reaches some percentage of its running speed.

In Figure 5–12, this single-phase motor is started with a main winding and a starting winding. The starting winding is supplied through a resistor

FIGURE 5–10
Schematic diagram—induction motor wound rotor (Courtesy of Cadick Professional Services)

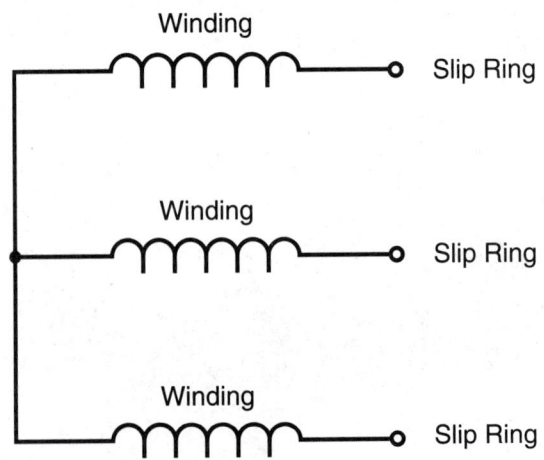

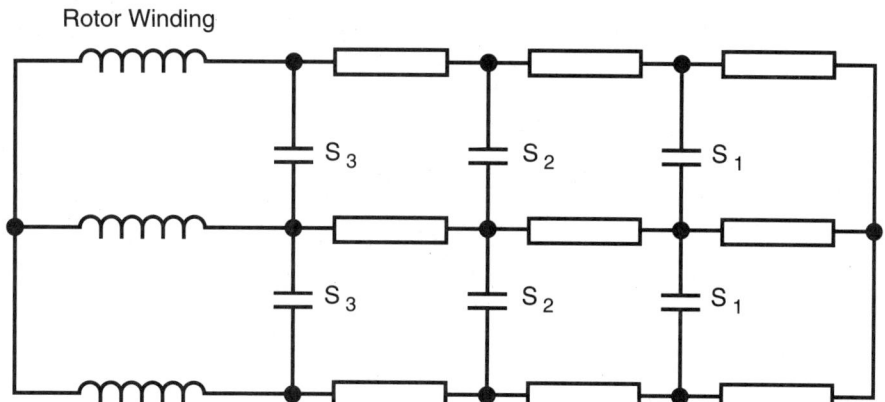

FIGURE 5–11
Wound rotor induction motor speed control (Courtesy of Cadick Professional Services)

to provide a phase shifted flux. The two windings together generate sufficient torque to start the motor. When the motor reaches a percentage of full speed, the centrifugal switch operates and opens the starting winding circuit. Other types of single-phase motors use different configurations. These will be discussed in the next chapter.

AC Motor Speed

The speed at which an AC motor rotates depends upon the applied frequency and the motor type. The following paragraphs describe the major considerations involved in the speed of rotation for induction and synchronous motors.

Synchronous Motors As previously discussed, an AC synchronous motor operates at synchronous speed. Thus, its speed of rotation will be given by formula (1). This will be true until the shaft load is increased beyond the capability of the motor. When this occurs, the synchronous motor will *slip a pole*. Slipping a pole implies that the rotor momentarily loses synchronism and slows until the next magnetic pole catches up. At that time, if the load has been reduced, the rotor will re-synchronize. Pole slippage is an extremely serious problem and can result in severe damage to the motor.

Induction Motors Since the induction motor develops torque when the stator field cuts through the rotor winding, the induction motor cannot operate at synchronous speed. It always operates at some percentage of synchronous speed called *slip*. Slip is defined as

$$\text{Slip} = \frac{N_s - N_a}{N_s} \times 100 \qquad (3)$$

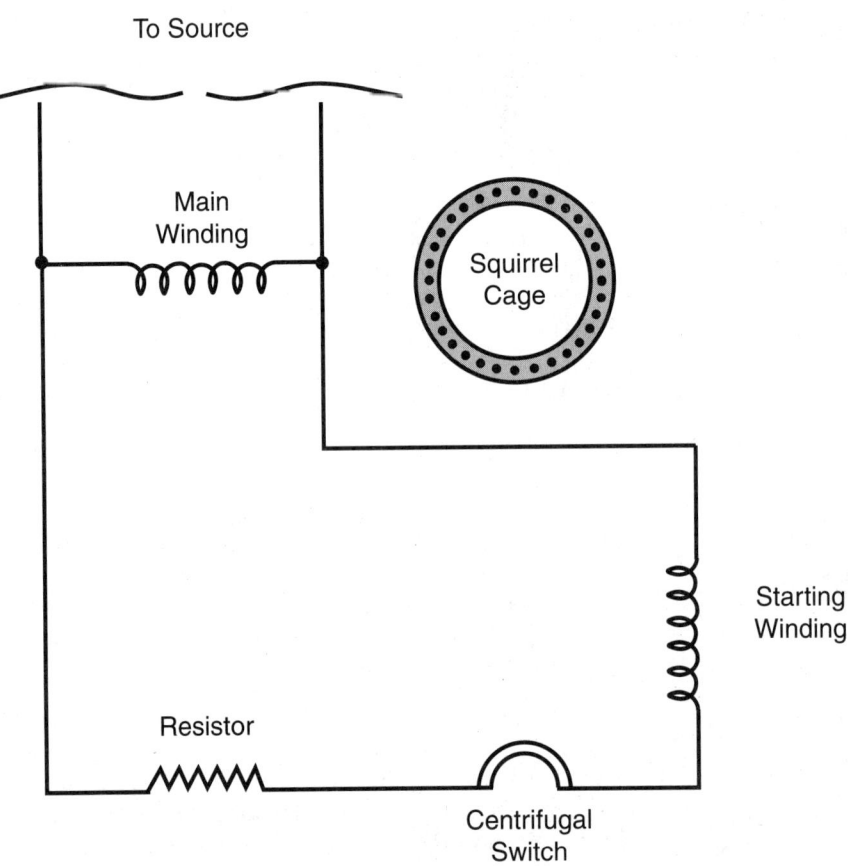

FIGURE 5–12
Resistance-start, split-phase motor (Courtesy of Cadick Professional Services)

Where:
N_s is the synchronous speed from formula (1)
N_a is the actual rotor speed

As the motor load increases, the rotor slows. With the rotor moving more slowly, the number of flux lines being cut per second increases. This increase results in the torque increasing to a value sufficient to rotate the load. Because of this factor, the lighter the load, the smaller the slip. At very light loads, induction motor will have slips on the order of 2%, depending on the type of motor design. At full load, the slip increases to values as high as 20% in some designs. Slip is one of the many motor design considerations.

SUMMARY

When two or more alternating magnetic fields are properly added in a motor stator, they create a rotating magnetic field. This rotating magnetic field literally pulls the rotor around in a circular motion. Like all magnets, the

AC Motor Theory

rotating magnetic field has at least one north and one south pole. The stator is referred to by the number of poles in the rotating magnetic field. Thus, if the stator is wound to create two north poles and two south poles, it is referred to as a four-pole stator.

The speed of rotation of the magnetic field is called the synchronous speed and is given by the formula:

$$S = \frac{120 \times f}{P} \tag{4}$$

From this formula you can see that the speed is proportional to the applied frequency and inversely proportional to the number of poles.

The synchronous motor has a rotor that creates a DC magnetic field which is pulled along by the rotating magnetic field. The induction motor rotor field is actually created by the stator field as it rotates and cuts through the rotor windings. The synchronous motor always rotates at synchronous speed while the induction motor always rotates slightly slower than synchronous speed. The greater the load on an induction motor, the greater the slip. Slip is defined by the equation:

$$\text{Slip} = \frac{N_s - N_a}{N_s} \times 100 \tag{5}$$

Single-phase motors operate in much the same way as three-phase rotors except they use only a single-phase voltage supply. The single-phase motor has a special winding called the starting winding, which helps create the rotating magnetic field. The starting winding is usually taken out of service after the starting cycle finishes.

AC motor stators are made much like AC generator stators. A sandwich of metallic, slotted laminations is created, and insulated windings are laid into the slots. The laminations are mounted inside a frame which holds them in place and provides electrical connection at the mounting point.

AC motor rotors are also made of a slotted lamination sandwich. The laminations are pressed onto the rotor and the windings (wound rotor motor), or rotor bars (squirrel cage motor) are installed into the slots.

REVIEW QUESTIONS

1. Draw a simple diagram of a four-pole stator. Show the entry and exit connections for eac phase. (Hint: Remember that entry connections for the three phases must be 120° apart, the poles for each phase must alternate [north - south - north - south], and the poles for each phase must be 90° apart.) From this diagram, draw the north and south pole configuration for current the same as those at instant 1 in Figure 5–1. Use an underlined capi-

tal letter to indicate a strong north or south pole (N or S).

2. What is the synchronous speed of an 8-pole, 60-Hz motor?

3. Is there any condition that will cause a synchronous motor to rotate at something other than synchronous speed?

4. The stator of an wound rotor induction motor is wound with 12 poles. How many poles should the rotor have?

5. A synchronous motor is running at half load. The mechanical process suddenly applies an additional 25% of full load to the motor shaft. Describe how the amortisseur windings on the motor will help to prevent pole slippage.

6. Describe, in your own words, how a rotating magnetic field, and a squirrel cage rotor combine to create torque.

7. What advantage does a wound rotor induction motor offer?

8. Why is a starting winding used in a single-phase induction motor?

9. The nameplate for a 6-pole, 60-Hz motor says that the motor rotates at 1150 rpm at full load. What is the motor slip at full load?

10. In a squirrel cage induction motor, what prevents current from flowing in the metal laminations, parallel to the shaft?

AC Motor Construction and Application 6

OBJECTIVES

After studying this chapter, the student should be able to:

- Describe the general construction features of AC induction and synchronous motors.
- Describe the types of electrical connections for three-phase motors.
- Describe the various types of single-phase motors.
- Describe the type of information found on a typical motor nameplate.
- Describe a procedure that may be used to select a motor for typical applications.

INTRODUCTION

AC electric motors are available in a wide variety of sizes, types, and styles ranging from tiny fractional horsepower units to huge machines of 20,000 hp and more. This chapter discusses many of the significant construction features of single- and three-phase motors.

Much of the material in this chapter is taken from the National Electrical Manufacturer's Association publication MG-1. MG-1 is the definitive reference for motor and generator specifications and construction standards. Unless otherwise noted, the specifications and standards listed in this chapter apply to all types of rotating machinery—generators and motors, AC and DC.

This chapter does not attempt to provide all of the material covered in NEMA standard MG-1. Rather, a sampling is given to familiarize you with the general construction characteristics of AC motors. For more detail on this subject, you should refer to NEMA standard MG-1.

GENERAL CONSTRUCTION

Machine Classifications

Size NEMA frame numbers are assigned as either two- or four-digit numbers according to their size. A two-digit frame number is created by multiplying the motor's *D dimension* by 16. The *D dimension* is the distance, in inches, from the shaft center line to the bottom of the mounting feet.

The first two digits in a four-digit frame number are obtained by multiplying the D dimension by 4. The second two digits are taken from Table 6–1. The 2F dimension is the distance between center lines of mounting holes in the motor feet when viewed from the side.

Note that the frame number will also have a suffix letter or letters. The two-digit frame number will have one or more of the letters B, C, G, H, J, K, M, N, Y at the end, while four-digit frames will have A, C, CH, D, E, HP, HPH, JM, JP, LP, LPH, P, PH, R, S, T, U, V, VP, X, Y, and/or Z. These suffix letters imply variations in motor application or construction. You should refer to section 11 of MG-1 for the specific meanings of each of these letters.

NEMA classifies machines as either small, medium, or large. A medium-sized AC machine is assigned a four-digit frame number, and has ratings up to and including those listed in Table 6–2.

A medium-sized DC machine is also built in a four-digit frame number, and has a continuous rating up to and including 1.25 hp per rpm for motors, or 1.0 kw per rpm for generators.

Small motors (AC or DC) are those that are either built with a two-digit frame number, therefore physically smaller than four-digit frame number sizes, or are built in smaller frames than medium size, with open construction ratings at 1700–1800 rpm of 1 hp for motors, or 0.75 kw for generators.

Large motors are those that are larger than the definitions given for medium-sized motors.

Application Motors are also classified according to their application. The following are including in NEMA MG-1:

- General-Purpose Alternating Current Motor
  Less than 200 horsepower
  Open construction
  Rated continuous duty
  Service factor in accordance with MG 1–12.47
  Class A insulation system with a temperature rise as specified in MG 1–12.43 for small motors, or Class B insulation for a temperature rise as specified in MG 1–12.43 for medium motors
- General-Purpose Direct Current Small Motor
  A small motor as previously defined with construction suitable for general use

Table 6–1   Method for Determining 3rd and 4th Digits in a 4-Digit Frame Number
(Table MG-1-1993)

Frame Number Series	D	\multicolumn{7}{c}{Third/Fourth Digit in Frame Number}
		1	2	3	4	5	6	7
		\multicolumn{7}{c}{2F Dimensions}
140	3.50	3.00	3.50	4.00	4.50	5.00	5.50	6.25
160	4.00	3.50	4.00	4.50	5.00	5.50	6.25	7.00
180	4.50	4.00	4.50	5.00	5.50	6.25	7.00	8.00
200	5.00	4.50	5.00	5.50	6.50	7.00	8.00	9.00
210	5.25	4.50	5.00	5.50	6.25	7.00	8.00	9.00
220	5.50	5.00	5.50	6.25	6.75	7.50	9.00	10.00
250	6.25	5.50	6.25	7.00	8.25	9.00	10.00	11.00
280	7.00	6.25	7.00	8.00	9.50	10.00	11.00	12.50
320	8.00	7.00	8.00	9.00	10.50	11.00	12.00	14.00
360	9.00	8.00	9.00	10.00	11.25	12.25	14.00	16.00
400	10.00	9.00	10.00	11.00	12.25	13.75	16.00	18.00
440	11.00	10.00	11.00	12.50	14.50	16.50	18.00	20.00
500	12.50	11.00	12.50	14.00	16.00	18.00	20.00	22.00
580	14.50	12.50	14.00	16.00	18.00	20.00	22.00	25.00
680	17.00	16.00	18.00	20.00	22.00	25.00	28.00	32.00

Frame Number Series	D	\multicolumn{8}{c}{Third/Fourth Digit in Frame Number}
		8	9	10	11	12	13	14	15
		\multicolumn{8}{c}{2F Dimensions}
140	3.50	7.00	8.00	9.00	10.00	11.00	12.50	14.00	16.00
160	4.00	8.00	9.00	10.00	11.00	12.50	14.00	16.00	18.00
180	4.50	9.00	10.00	11.00	12.50	14.00	16.00	18.00	20.00
200	5.00	10.00	11.00	...	...	...	...	...	...
210	5.25	10.00	11.00	12.50	14.00	16.00	18.00	20.00	22.00
220	5.50	11.00	12.50	...	...	...	...	...	...
250	6.25	12.50	14.00	16.00	18.00	20.00	22.00	25.00	28.00
280	7.00	14.00	16.00	18.00	20.00	22.00	25.00	28.00	32.00
320	8.00	16.00	18.00	20.00	22.00	25.00	28.00	32.00	36.00
360	9.00	18.00	20.00	22.00	25.00	28.00	32.00	36.00	40.00
400	10.00	20.00	22.00	25.00	28.00	32.00	36.00	40.00	45.00
440	11.00	22.00	25.00	28.00	32.00	36.00	40.00	45.00	50.00
500	12.50	25.00	28.00	32.00	36.00	40.00	45.00	50.00	56.00
580	14.50	28.00	32.00	36.00	40.00	45.00	50.00	56.00	63.00
680	17.00	36.00	40.00	45.00	50.00	56.00	63.00	71.00	80.00

All dimensions in inches

- Industrial Direct-Current Motor
  A medium-sized motor suitable for industrial use.
- Definite-Purpose Motor
  A motor that is designed to standard ratings and operating characteristics for use under unusual service conditions.
  Other application classifications may be found in MG-1.

Electrical Type   The electrical type classifications include all of the generally broad categories such as electrical motor, electrical generator, induction motor, asynchronous motor, and so on. Other type classification criteria are also used. The following information summarizes some of the more common design and construction features:

- Design A—Three-phase, squirrel cage, medium induction motors
  These motors are usually intended for applications where extremely high efficiency and high full-load speed are required. They will withstand full-voltage starting and develop locked rotor torque values that are similar to the Design B motor. Because of the high efficiency, low slip characteristics, Design A motors tend to be special application motors.
- Design B—Three-phase, squirrel cage, medium induction motors
  Design B motors are the general purpose work-horse designs of polyphase induction motors. They tend to have relatively low slips and moderate torque characteristics as compared to the other designs.
- Design C—Three-phase, squirrel cage, medium induction motors
  Design C motors feature low slip and high torque as compared to Design B.

Table 6–2   Maximum Horsepower Ratings for Medium-Size Motors
(Courtesy of NEMA)

Synchronous Speed, rpm	*Motors Hp*	*Generators Kilowatt at 0.8 Power Factor*
1201 – 3600	500	400
901 – 1200	350	300
721 – 900	250	200
601 – 720	200	150
515 – 600	150	125
451 – 514	125	100

AC Motor Construction and Application

- Design D—Three-phase, squirrel cage, medium induction motors
  Design D motors develop extremely high torque but do not regulate speed as well as the others and have a high slip. Table 6–3 summarizes Design B, C, and D motor characteristics.
- Wound Rotor Motors
  These are often used in speed control applications. The electrical characteristics of the wound rotor, polyphase induction motor are summarized in Table 6–4. Notice that the wound rotor motor develops very high torque values.
- Synchronous Motors
  Table 6–5 is a summary of the general characteristics of synchronous motors.
- Small Single-phase Induction Motors
  Small single-phase induction motors are designed for full voltage start with locked rotor currents as summarized in Table 6–6, p. 132. Note that general-purpose, single-phase motors are not allowed to exceed the values for Design N.

Environmental Protection and Cooling This type of classification refers to the way in which the machine is constructed to protect it from environmental effects and/or to protect the environment from the machine. NEMA classifies three major categories including the *open machine, the totally-enclosed machine, and the machine with encapsulated or sealed windings.*

1. Open Machines:
   Open machines have ventilated openings which permit the passage of external cooling air over and around the windings. Open machines come in several sub-categories including:
   Drip-proof machines: The ventilating openings are constructed so that drops of liquid or solids falling on the machine at any angle not greater than 15° from the vertical, cannot enter the machine.
   Guarded machines: All the ventilating openings are limited to specified size and shape to prevent insertion of fingers or rods to avoid accidental contact with rotating or electrical parts.
   Splash-proof machines: The ventilating openings are so constructed that drops of liquid or solid particles falling on the machine, or coming toward the machine in a straight line at any angle not greater than 100° from the vertical, cannot enter the machine.
2. Totally-Enclosed Machines:
   These motors are constructed to prevent the free exchange of air between the inside and outside of the case. Totally-enclosed machines are not, however, airtight. Common totally-enclosed sub-categories include:
   Totally-enclosed nonventilated is constructed so that no cooling occurs by any means external to the enclosing parts.

Table 6–3 Summary of Characteristics Polyphase Induction Motors—Designs B, C and D

Speed Regulation	Speed Control	Starting Torque	Breakdown Torque	Applications
General Purpose Squirrel (Design B)				
Drops about 3% for large to 5% for small sizes	Adjustable frequency and voltage, multi-speed (two to four constant speeds)	100% for 200 hp, 300% for 1-hp, four-pole unit	200% of full load	Constant-speed service where starting torque isn't excessive. Fans, blowers, rotary compressors, and centrifugal pumps
High-Torque Squirrel-Cage (Design C)				
Drops about 3% for large to 6% for small sizes	Adjustable frequency and voltage, multi-speed (two to four constant speeds)	250% of full load for high-speed to 200% for low-speed designs	200% of full load	Constant speed where fairly high starting torque is required infrequently with starting current about 550% of full load. Reciprocating pumps and compressors, crushers, etc.
General Purpose Squirrel (Design D)				
Drops about 10% to 15% from no load to full load	Adjustable frequency and voltage, multi-speed (two to four constant speeds)	225-300% of full load, depending on speed with rotor resistance	200%. Will usually not stall until loaded to maximum torque, which occurs at standstill	Constant speed and high starting torque, if starting is not too frequent, and for high-peak loads with or without flywheels. Punch presses, shears, elevators, etc.

Table 6–4 Summary of Polyphase Wound Rotor Induction Motor Characteristics

Speed Regulation	Speed Control	Starting Torque	Breakdown Torque	Applications
With rotor rings short circuited, drops about 3% for large to 5% for small sizes	Speed can be reduced to 50% by rotor resistance. Speed varies inversely with load.	To 300%, depending on external resistance and its distribution in rotor circuit.	300% when rotor slip rings are short circuited.	Where high starting torque with low start-current or where limited speed control is required. Fans, centrifugal and plunger pumps, compressors, conveyors, hoists, cranes, etc.

Totally-enclosed fan-cooled has a fan that blows cooling air across the external frame. This machine is very popular for use in dusty, dirty, and corrosive atmospheres.

Explosion-proof is designed to withstand the explosion of gas or vapor within it, and to prevent ignition of gas or vapor surrounding the machine by sparks, flashes, or explosions which may occur within the machine casing.

Table 6–5 Summary of Polyphase, Synchronous Motor Characteristics

Speed Regulation	Speed Control	Starting Torque	Breakdown Torque	Applications
Constant	None	40% for slow to 160% for medium	Unity-pf motors 225%. Specials up to 300%	For constant-speed service, direct connection to slow-speed machines, and where power-factor correction is required.

Table 6–6 Locked Rotor Currents for Design Letters O and N
(Courtesy of NEMA)

Hp	2-, 4-, 6-, and 8-Pole, 60-Hertz Motors, Single-Phase			
	Locked-rotor Current, Amperes			
	115 Volts		230 Volts	
	Design O	Design N	Design O	Design N
1/6 and smaller	50	20	25	12
1/4	50	26	25	15
1/3	50	31	25	18
1/2	50	45	25	25
3/4	...	61	...	35
1	...	80	...	45

3. Encapsulated or Sealed Windings Machines:
   Encapsulated windings are covered and/or impregnated with a heavy coating of resin to protect them from moisture, dirt, abrasion, and other such damage. These machine types are impervious to environmental conditions which will destroy other, non-encapsulated motors.

Variability of Speed   Motors are also constructed for various types of speed and/or speed control. NEMA Standard MG-1 defines the following:

- Constant Speed Motor:
  The speed of operation of this motor type is constant or very nearly constant. Synchronous motors are constant speed motors, but so are induction motors with small slip values and DC shunt-wound motors.
- Varying-Speed Motor:
  These types of motors have varying speeds as the load changes. Examples include the DC series motor and the repulsion motor.

- Adjustable-Speed Motor:
  This is a motor whose speed can be varied over a considerable range. Once the desired speed has been selected, the motor speed is constant with respect to load. The DC shunt motor with a field resistance speed control is an example of this type of motor. The adjustable speed motor always has a base speed which is defined as the lowest rated speed obtained at rated load and rated voltage at the temperature rise specified in the rating.
- Adjustable Varying-Speed Motor:
  This motor is similar to the adjustable speed motor, except that its speed will vary to a considerable degree with change in load. A DC compound wound motor with field control and a wound rotor induction motor are examples of adjustable varying-speed motors.
- Multi-Speed Motors:
  These motors can be operated at any one of two or more definite speeds. Induction motors with windings capable of various pole groupings are typical of multi-speed motors.

Since the mid-1980s, frequency speed control of AC motors has become increasingly common. Remember that the motor itself determines the speed classification. Thus, a high-slip induction motor used in a frequency speed control system would, in general, be categorized as an adjustable varying-speed motor. Speed control will be covered in more detail in the next chapter.

Insulation Excessive temperature has two effects on insulation. First, higher temperatures cause lower insulation resistance. Second, excessively high temperature can damage or prematurely age the insulation. Because of these facts, insulation systems are designed in four basic classes of temperature rating: A, B, F, and H. Do not confuse these categories with the NEMA design categories discussed previously in this chapter.

- Class A—This type of insulation is one which, by experience or accepted test, can be shown to have a suitable thermal endurance when operated at the limiting Class A temperature of 105°C. Typical materials used in Class A insulation include cotton, paper, cellulous acetate films, enamel-coated wire, and similar organic materials impregnated with suitable substances.
- Class B—The limiting Class B temperature is 130°C. Typical materials include mica, glass fiber, asbestos and other materials, not necessarily inorganic, with compatible bonding substances having suitable thermal stability.

- Class F—Class F has a limiting temperature of 155°C. Typical materials are similar to Class B except with higher temperature bonding materials.
- Class H—The Class H limiting temperature is 180°C. Typical materials for Class H include mica, glass fiber, asbestos, silicone elastomer, and other materials. The bonding substances include such materials as silicone resins.

Ratings

General To help standardize the industry, NEMA publishes standards for rotating machinery. As long as minimum requirements are met, manufacturers are free to build motors to values other than NEMA standards; however, such special built motors usually involve a price premium.

Voltage Rating Voltage rating is the voltage that must be maintained at the motor terminals to allow it to produce its rated horsepower and speed. The applied voltage should not vary from the rated voltage by more than ±10%. NEMA publishes voltage ratings for small, medium, and large size motors.

Voltage ratings for small and medium motors are as follows:

1. Universal motors – 115 and 230 V
2. Single-phase motors
    a. 60 Hz – 115 and 230 V
    b. 50 Hz – 110 and 220 V
3. Polyphase motors
    a. 60 Hz – 115, 208, 230, 460, 575, 2300, 4000, 4600, and 6600 V
    b. Three-phase 50 Hz – 220 and 380 V

Voltage ratings for large motors are a function of horsepower according to Table 6–7. All large motor voltage ratings are the phase-to-phase voltage of a three-phase system.

Frequencies All alternating current motors are rated for either 50 Hz or 60 Hz.

Horsepower and Speed The horsepower rating of a motor is the amount of shaft horsepower the motor will supply at rated voltage and frequency. For small and medium motors, NEMA publishes tables of standard horsepower and speed ratings in standard MG-1. Tables 6–8 and 6–9 are two such tables. The horsepower rating for small and medium motors is based on the breakdown torque value.

Table 6–7 Voltage Ratings for Large Polyphase Induction Motors
(Courtesy of NEMA)

Voltage (60 Hz)	Horsepower
460 or 575	100 – 600
2300	200-4000
4000 or 4600	400-7000
6600	1000-12,000
13,200	3500-25,000

Standard horsepower ratings for large polyphase induction motors are given incrementally as a function of size. Table 6–10, p. 138 lists the standard incremental horsepower ratings for large, polyphase induction motors.

Torque Torque is defined as twisting force. All motors are rated for four basic types of torque.

1. Full-load torque—Full load torque is the torque required to produce rated horsepower at full-load speed.
2. Locked-rotor torque—This is the minimum torque that the motor develops when the rotor is held stationary.
3. Pull-up torque—Pull-up torque is the minimum torque developed by a motor during the acceleration period from rest to the breakdown torque. If a motor does not have a breakdown torque, the pull-up torque is defined as the minimum torque developed by the motor as it accelerates to full speed.
4. Breakdown torque—The maximum torque developed by a motor at rated voltage and frequency without a sudden change in speed is called the breakdown torque. Torque requirements beyond the breakdown torque will cause the magnetic field to "slip" and the motor will change speed suddenly.

Other torques are defined in the NEMA standards; however, these four are the most commonly encountered.

Table 6–8 Horsepower and Speed Ratings for Small Induction Motors
(Courtesy of NEMA)

	All Motors Except Shaded-Pole and Permanent-split Capacitor		Permanent-split Capacitor Motors	All Motors Except Shaded-Pole and Permanent-split Capacitor		Permanent-split Capacitor Motors
	60-hertz Synchronous Rpm	Approximate Rpm at Rated Load		50-hertz Synchronous Rpm	Approximate Rpm at Rated Load	
1, 1.5, 2, 3, 5, 7.5, 10, 15, 25, and 35 millihorsepower	3600	3450	...	3000	2850	...
	1800	1725	...	1500	1425	...
	1200	1140	...	1000	950	...
	900	...	...			
¹⁄₂₀, ¹⁄₁₂, and ⅛ horsepower	3600	3450	...	3000	2850	...
	1800	1725	...	1500	1425	...
	1200	1140	...	1000	950	...
	900	850	...			
⅙, ¼ and ⅓ horsepower	3600	3450	...	3000	2850	...
	1800	1725	...	1500	1425	...
	1200	1140	...	1000	950	...
	900	850	...			
½ horsepower	3600	3450	3250	3000	2850	2700
	1800	1725	1625	1500	1425	1350
	1200	1140	1075	1000	950	900
¾ horsepower	3600	3450	3250	3000	2850	2700
	1800	1725	1625	1500	1425	1350
1 horsepower	3600	3450	3250	3000	2850	2700

AC Motor Construction and Application

Table 6–9   Horsepower and Speed Ratings, Polyphase Medium Induction Motors
(Courtesy of NEMA)

Hp	60-hertz Synchronous Rpm							50 Rpm hertz Synchronous Rpm			
½	...	...	...	900	720	600	514	...	...	...	750
¾	...	...	1200	900	720	600	514	...	...	1000	750
1	...	1800	1200	900	720	600	514	...	1500	1000	750
1½	3600**	1800	1200	900	720	600	514	3000	1500	1000	750
2	3600**	1800	1200	900	720	600	514	3000	1500	1000	750
3	3600**	1800	1200	900	720	600	514	3000	1500	1000	750
5	3600**	1800	1200	900	720	600	514	3000	1500	1000	750
7½	3600**	1800	1200	900	720	600	514	3000	1500	1000	750
10	3600**	1800	1200	900	720	600	514	3000	1500	1000	750
15	3600**	1800	1200	900	720	600	514	3000	1500	1000	750
20	3600**	1800	1200	900	720	600	514	3000	1500	1000	750
25	3600**	1800	1200	900	720	600	514	3000	1500	1000	750
30	3600**	1800	1200	900	720	600	514	3000	1500	1000	750
40	3600**	1800	1200	900	720	600	514	3000	1500	1000	750
50	3600**	1800	1200	900	720	600	514	3000	1500	1000	750
60	3600**	1800	1200	900	720	600	514	3000	1500	1000	750
75	3600**	1800	1200	900	720	600	514	3000	1500	1000	750
100	3600**	1800	1200	900	720	600	514	3000	1500	1000	750
125	3600**	1800	1200	900	720	600	514	3000	1500	1000	750
150	3600**	1800	1200	900	720	600	...	3000	1500	1000	750
200	3600**	1800	1200	900	720	...	...	3000	1500	1000	750
250	3600**	1800	1200	900	...	...	...	3000	1500	1000	750
300	3600**	1800	1200	...	...	...	...	3000	1500	1000	...
350	3600**	1800	1200	...	...	...	...	3000	1500	1000	...
400	3600**	1800	...	...	...	...	...	3000	1500	...	...
450	3600**	1800	...	...	...	...	...	3000	1500	...	...
500	3600**	1800	...	...	...	...	...	3000	1500	...	...

*For frame assignments, see MG-13.
**Applies to squirrel cage motors only.

Table 6–10   Standard Horsepower Ratings for Large Polyphase Induction Motors (Courtesy of Cadick Professional Services)

Horsepower Range	Horsepower Increments
100-200	25
250-500	50
500-1000	100
1000-2500	250
2500-6000	500
6000-20,000	1000
20,000-30,000	2500
30,000-50,000	5000
	10,000

Time Ratings   The time rating of a motor is the amount of time that a motor is expected to operate at full load. Standard intervals are 5, 15, 30 > 60 minutes and continuous. Most general purpose motors are designed for continuous operation.

Code Letters for Locked Rotor   When a motor is first started it draws an extremely large current until it reaches operating speed. This current may be 5, 6, or more times the full load current. NEMA has standardized the multiples of full load current that a given motor is allowed to draw when it starts. The amount of locked rotor current is identified by a code letter from the following table:

Table 6–11 NEMA Locked Rotor Letter Codes
(Courtesy of NEMA)

Letter Designation	kVA per Horsepower*	Letter Designation	kVA per Horsepower*
A	0 – 3.15	K	8.0 – 9.0
B	3.15 – 3.55	L	9.0 – 10.0
C	3.55 – 4.0	M	10.0 – 11.2
D	4.0 – 4.5	N	11.2 – 12.5
E	4.5 – 5.0	P	12.5 – 14.0
F	5.0 – 5.6	R	14.0 – 16.0
G	5.6 – 6.3	S	16.0 – 18.0
H	6.3 – 7.1	T	18.0 – 20.0
J	7.1 – 8.0	U	20.0 – 22.4
		V	22.4 and up

*Locked KVA per horsepower rance includes the lower figure up to, but not including, the higher figure. For example, 3.14 is designated by letter A and 3.15 by letter B.

The following example illustrates the use of Table 6–11.

Example problem:

A manufacturer is building a 200-hp, 480-V, three-phase motor that has a NEMA locked rotor letter code H. What is the maximum locked rotor current that the manufacturer's motor may draw?

Solution:

According to Table 6–11, the multiplier for a letter code H is 6.3 to 7.1. The <u>maximum</u> locked rotor kVA is then 7.1 × 200 = 1420 kVA. The locked rotor current can be calculated from

$$I_{LR} = \frac{1420}{(.48) \times \sqrt{3}} = 1708 \text{ Amperes}$$

Nameplate Temperature The temperature upon which the allowable temperature rise is based. For all general purpose motors, the nameplate temperature is 40°C.

Temperature Rise The allowable temperature rise is the maximum, observable temperature difference between the ambient, cooling air at the motor intake, and each of the motor parts. Temperature rise varies depending on the method that is used to take the measurement, as shown in Table 6–12.

Service Factor Many motors have the ability to operate at loads in excess of their rated value. The service factor defines the ability of the motor to operate at a higher than rated value without exceeding the specified temperature rise and other ratings. Note that to be operated at the higher value, the motor must be operated at correct terminal voltage and frequency.

Motor Nameplates

General Motor nameplates provide full information for a motor including many of the ratings that have already been discussed. The following paragraphs identify the major elements included on two typical nameplates.

Model Number The model number is the manufacturer's identification of motor design and construction. This information is not a NEMA standard item.

Serial Number This is the manufacturer's production sequence number.

HP (Horsepower) The rating of the motor when operated at the other parameters specified on the plate.

Service Factor This is the capability of the motor to be operated at higher than nameplate horsepower. A designation of 1.0 means that there is no such built-in capability. A designation of 1.15 means the motor can be used at 15% above its nameplate horsepower, but the motor will run warmer. As described previously, the motor must be operated at rated terminal voltage and frequency.

RPM This is the motor speed at full load in revolutions per horsepower. If the motor were a synchronous motor, the rated speed would be synchronous speed as previously calculated.

Time Rating Indicates the length of time the motor is designed to operate at full load. The CONT in Figures 6–1 and 6–2 means that continuous operation is permitted.

Table 6–12 Allowable Large Motor Temperature Rise
(Courtesy of NEMA)

Item	Machine Part	Method of Temperature Determination	Temperature Rise, Degrees C° Class of Insulation System			
			A	B	F	H
a.	Insulated windings					
	1. All horsepower (kw) ratings	Resistance	70	90	115	135
	2. 1500 horsepower and less	Embedded detector**	80	100	125	150
	3. Over 1500 horsepower (1120 kw)					
	(a) 7000 volts and less	Embedded detector**	75	95	120	145
	(b) Over 7000 volts	Embedded detector**	70	90	115	135
b.	The temperatures attained by cores, squirrel-cage windings, collector rings, and miscellaneous parts (such as brushholders and brushes, etc.) shall not injure the insulation or the machine in any respect.					

*For machines which operate under prevailing barometric pressure and which are designed not to exceed the specified temperature rise at altitudes from 3300 feet (1000 meters) to 13,000 feet (4000 meters), the temperature rises, as checked by tests at low altitudes, shall be less than those listed in the foregoing table by 1 percent of the specified temperature rise for each 330 feet (100 meters) of altitudes in excess of 3300 feet (1000 meters).

**Embedded detectors are located within the slot of the machine and can be either resistance elements or thermocouples. For machines equipped with embedded detectors, this method shall be used to demonstrate conformity with the standard. (See 20.63.)

NOTE: Temperature rises in the foregoing table are based upon a reference ambient temperature of 40°C. However, it is recognized that induction machines may be required to operate in an ambient temperature higher than 40°C. For successful operation of induction machines in ambient temperatures higher than 40°C, it is recommended that the temperature rises of the machine given in the foregoing table be reduced, as indicated below, for the ranges of ambient temperature given (Exception: totally-enclosed water-air-cooled machines)[1].

Ambient Temperature Degrees C	Values by which Temperature Rises in the Foregoing Table Should be Reduced, Degrees C
Above 40 up to and including 50	10
Above 50 up to and including 60	20

Authorized Engineering Information 5-24-1960, revised 11-8-1973.

1. For totally-enclosed water-air-cooled machines, the temperature of the cooling air is the temperature of the air leaving the coolers. Totally-enclosed water-air-cooled machines are normally designed for the maximum cooling water temperature encountered at the location where each machine is to be installed. With a cooling water temperature not exceeding that for which the machine is designed:
   a. On machines designed for cooling water temperature up to 30°C – the temperature of the air leaving the coolers shall not exceed 40°C.
   b. On machines designed for higher cooling water temperatures – the temperature of the air leaving the coolers shall be permitted to exceed 40°C provided the temperature rises of the machine parts are then limited to values less than those given in the table by the number of degrees that the temperature of the air leaving the coolers exceeds 40°C.

2. See Referenced Standards MG 1-1.01.

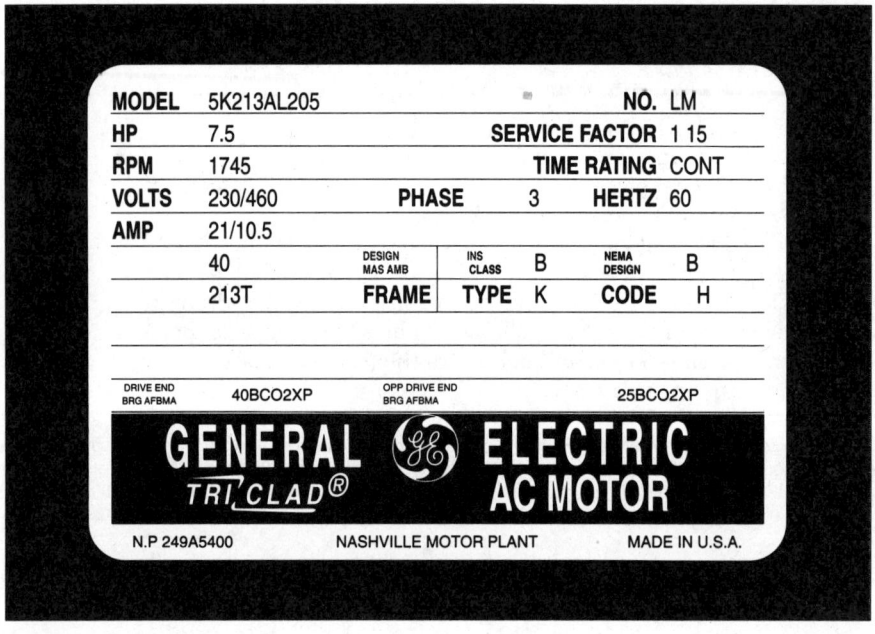

FIGURE 6–1
Typical motor nameplate (#1)
(Courtesy of General Electric Company)

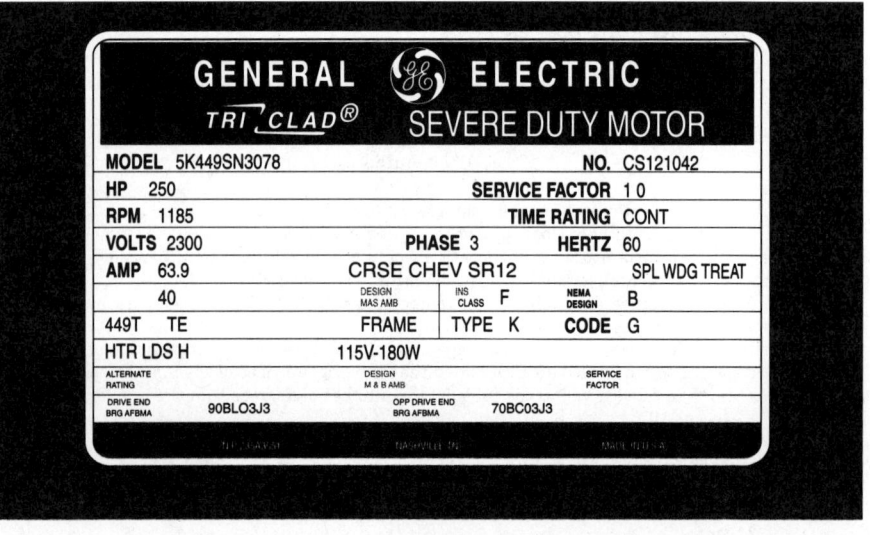

FIGURE 6–2
Typical motor nameplate (#2)
(Courtesy of General Electric Company)

AC Motor Construction and Application

Volts   This is the voltage level at which the motor is designed to operate. Normally a 10% variation in level is permitted. On Motor 1, two voltages are indicated and can be selected by changing the connection of the motor leads.

Phase   Indicates the number of electrical phases of power that the motor is designed for.

Hertz   The motor design frequency.

AMP   This is the full load line current of the motor as measured on any of the phase wires. Motor 1 shows the current for both of the two allowed voltages.

SPL WDG Treat   This rating indicates that a special winding treatment was used to allow for some unusual or extreme operating condition.

Deg C Max Amb   This is the maximum ambient temperature at which the motor can be operated at full load. It is the same as the nameplate temperature discussed previously.

INS Class   Indicates the temperature class of the motor insulation. Table 6–12 can then be used to determine the maximum temperature rise.

NEMA Design   This is motor design letter A, B, C, or D.

Frame   This is the NEMA frame number which was discussed previously.

Type   This is a manufacturer selected designation. It is not a NEMA standard.

Code   This is the NEMA locked rotor letter code.

Extra Lines   Most nameplates have extra lines so the manufacturer can convey additional operation or service information. In Motor 2, the plate indicates that the motor has heaters with Class H leads and the heaters are 180 W connected to 115 VAC.

Drive End AFBMA & OPP Drive End BRG AFBMA   This is a designation by AFBMA (Antifriction Bearing Manufacturer's Association) of the bearing type used in the motor. These numbers are all inclusive, defining the diameter of bore, the tolerance, internal clearance, cage, enclosure, type, and any other important characteristics that whoever is replacing it should know.

STATOR ELECTRICAL CONNECTIONS

Three-Phase

Three-phase stators can be connected in a variety of ways to obtain different voltage and speed characteristics. NEMA has established standards which all manufacturers follow so that users can interchange one manufacturer's motor for another without concern about the connections. The following section illustrates many of the most common connections. Note that the stator leads are designated by the letter T, and wound rotor leads are designated by the letter M.

- Figure 6–3 is the simplest of all connections—single voltage, single speed. The three phases (L_1, L_2, L_3) are connected to T_1, T_2, and T_3.
- Figures 6–4 and 6–5 are two of the most common dual voltage motors. Assume that the motor of Figure 6–4 is designed for use on either 240 or 480 V. To use the motor on a 240-V system, the connections would be made as indicated in the "low" voltage row of the table. This has the effect of putting the two sets of windings in parallel. For high voltage operation (480 V) the windings are placed in series so that the total voltage divides between each winding section. Figure 6–5 is a similar configuration except that the motor windings are connected in delta.

The circuit of Figure 6–6 can be used for a single voltage, wye-delta start. To use it in this manner, the windings are first connected to the

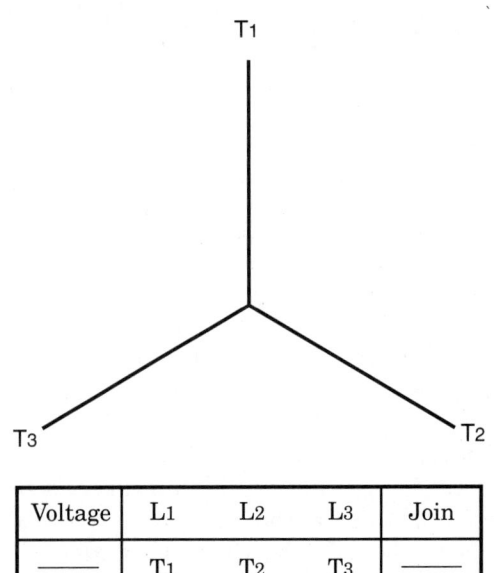

FIGURE 6–3 Three-phase, single-speed stator connection, wye connection

Voltage	L1	L2	L3	Join
——	T1	T2	T3	——

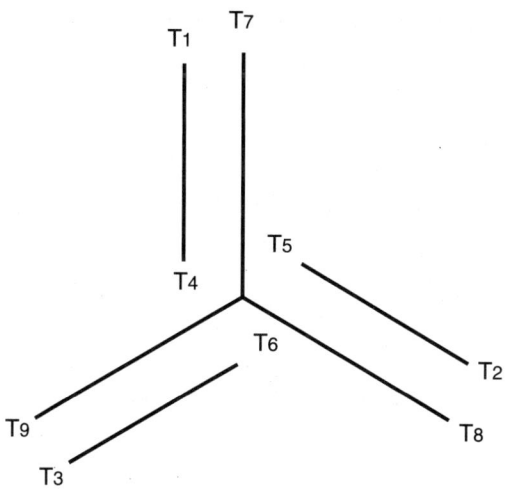

FIGURE 6–4
Wye-connected, dual voltage stator connection (Courtesy of NEMA)

Voltage	L1	L2	L3	Join
Low	(T1, T7)	(T2, T8)	(T3, T9)	... (T3, T5, T9) ...
High	T1	T2	T3	(T4, T7) (T5, T8) (T6,T9)

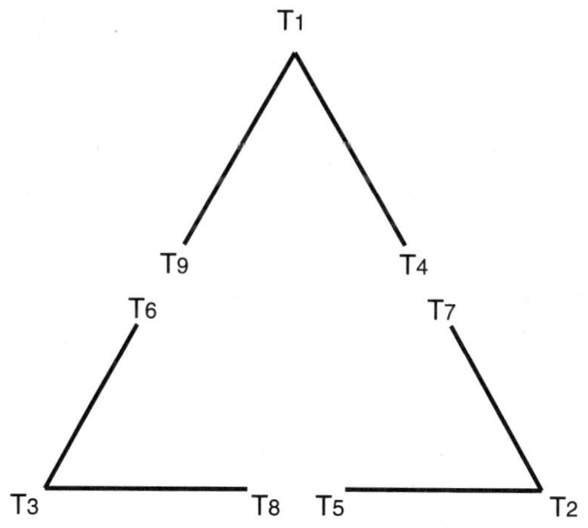

FIGURE 6–5
Delta-connected, dual voltage stator connection (Courtesy of NEMA)

Voltage	L1	L2	L3	Join
High	(T1, T6, T7)	(T2, T4, T8)	(T3, T5, T9)	... ...
Low	T1	T2	T3	(T4,T7) (T5,T8) (T6,T9)

supply voltage in a wye configuration as shown in the "start" row of the first truth table. This has the effect of placing line-to-neutral voltage on the windings, which produces a lower locked rotor current than phase-to-phase voltage. After the motor has reached close to full speed, an external switching network modifies the connections as shown in the "run" row. This applies the phase-to-phase voltage to the motor with the attendant increase in torque.

- Figure 6–6 can also be used as a dual voltage motor with a $\sqrt{3}$ to 1 voltage ratio. Assume that the motor windings are designed to operate at

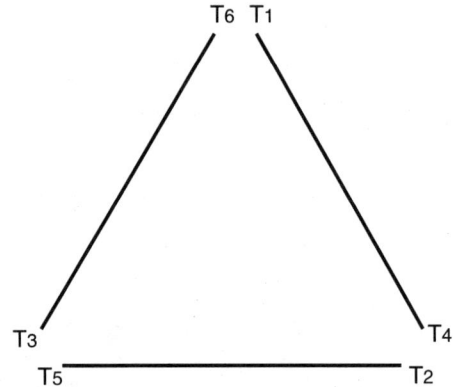

Y-Connected Start, Delta-Connected Run, Single Voltage

	L1	L2	L3	Join
Start	T1	T2	T3	(T4, T5, T6)
Run	(T1, T6)	(T2, T4)	(T3, T5)	...

Wye-Delta-Connected Dual Voltage (Voltage Ratio $\sqrt{3}$ to 1)

Voltage	L1	L2	L3	Join
High	T1	T2	T3	(T4, T5, T6)
Low	(T1, T6)	(T2, T4)	(T3, T5)	...

FIGURE 6–6
Wye-delta connected stator connections (Coutresty of NEMA)

AC Motor Construction and Application

2300 V, but the system has a 4160-V source. If the windings are connected as shown in the "high voltage" row of Figure 6–6, only 2300 V will be applied to the windings (2300 is 4160 divided by $\sqrt{3}$). If the system has a 2300-V supply, the windings can be connected in delta as indicated in the "low voltage" row of Figure 6–6. This will also apply 2300 V to the windings.

- Examples of multiple speed motors and their connections are shown in Figures 6–7 and 6–8. Recall that the number of poles is one of the determining factors for motor speed. Since all of the connections in Figures 6–7 and 6–8 are connected to the same frequency, each connection must create a different number of poles. For example, if the motor of Figure 6–7 is to have synchronous speeds of 3600, 1800, and 900 rpm, the T_1, T_2, T_3 windings will have to create magnetic poles. The other two windings (T_{11}, T_{12}, T_{13} and T_{21}, T_{22}, T_{23}) will have 4 poles and 2 poles respectively.
- Figure 6–9 shows the standard connections for a three-phase, wound rotor.

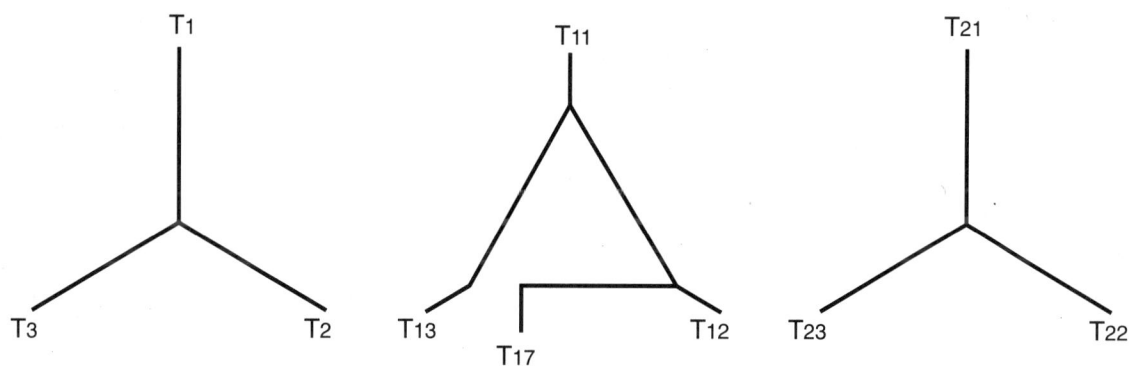

Three-Speed Motor Using Three Windings

Speed	L1	L2	L3	Insulate Separately	Join
Low	T1	T2	T3	T11- T12- T13- T17- T21- T22- T23	. . .
Second	T11	T12	(T13, T17)	T1- T2- T3- T21- T22- T23	. . .
High	T21	T22	T23	T1- T2- T3- T11- T12- T13- T17	. . .

FIGURE 6–7
Three-speed motor using three windings (Courtesy of NEMA)

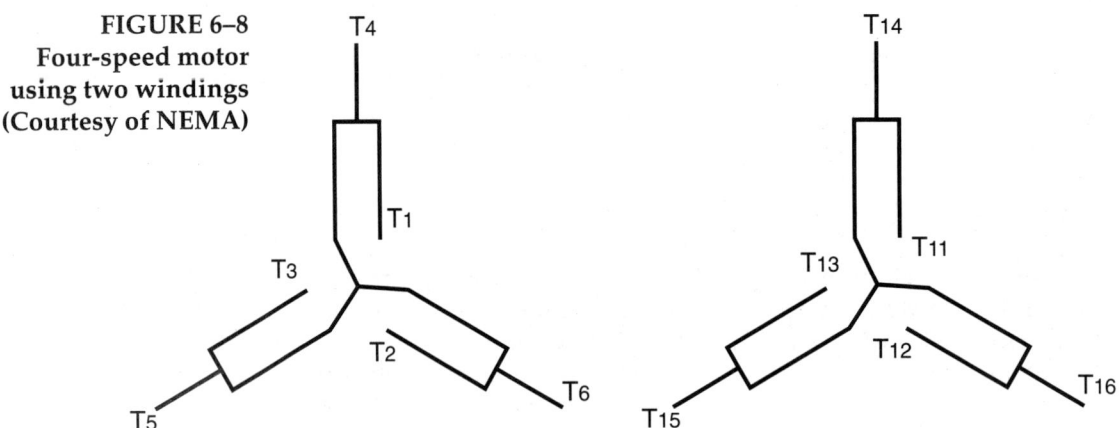

FIGURE 6–8 Four-speed motor using two windings (Courtesy of NEMA)

Four-Speed Motor Using Two Windings

Speed	L1	L2	L3	Insulate Separately	Join
Low	T1	T2	T3	T4-T5-T6-T11-T12-T13-T14-T15-T16	. . .
Second	T11	T12	T13	T1-T2-T3-T4-T5-T6-T14-T15-T16	. . .
Third	T6	T4	T5	T11-T12-T13-T14-T15-T16	(T1, T2, T3)
High	T16	T14	T15	T1-T2-T3-T4-T5-T6	(T11, T12, T13)

Single-Phase Motors

General The single-phase AC motor stator does not produce a rotating magnetic field. Instead, it produces a stationary, pulsating field. This pulsating field induces a current in the rotor which is also pulsating. Although the two fields pulsate, they would not interact sufficiently to create torque. A variety of methods are used to create torque so the motor will turn. The following sections describe the methods that are used to create this torque. Single-phase stators are also wound with different numbers of poles to create different synchronous speeds. Thus, a 4-pole single-phase motor will have a synchronous speed of 1800 rpm.

Note that the rotors of single-phase induction motors are usually of squirrel cage construction and virtually identical to three-phase induction

AC Motor Construction and Application

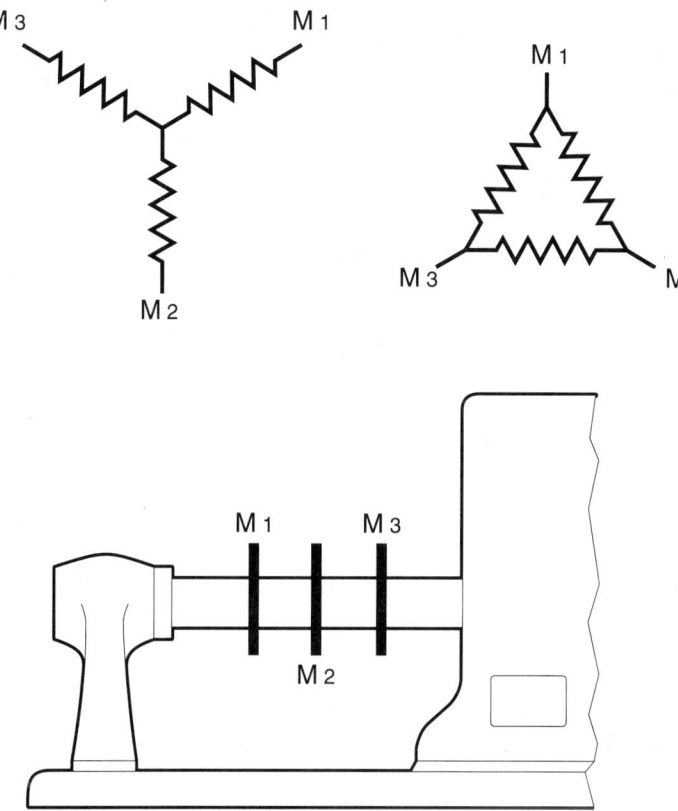

FIGURE 6–9
Standard connections for wound rotors
(Courtesy of NEMA)

motors. Note also that only the more common motors are discussed in this section.

Split-Phase Motor Figure 6–10 is a conceptual diagram of a split-phase, single-phase AC induction motor. To create torque, a second, starting winding is wound using high resistance, and displaced 90° from the main winding. The combination of the resistance and the physical displacement causes the starting winding and main magnetic fields to interact so that a rotating magnetic field is established. This rotating field creates sufficient torque to turn the rotor. After the rotor is close to full speed, the centrifugal switch opens and removes the starting winding. Thus, interaction between the rotor and stator fields is sufficient to maintain the motor rotation.

Notice that the windings in a single-phase motor encompass 360° of the stator. They are shown as they are in the figures to clarify that the starting and main windings are displaced by 90°.

FIGURE 6–10 Split-phase single-phase stator connection (Courtesy of Cadick Professional Services)

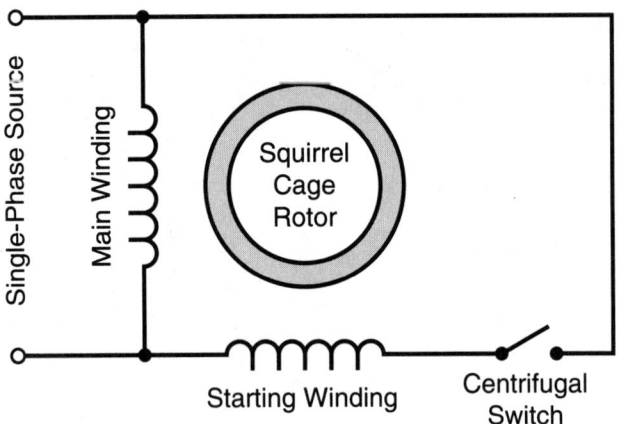

Capacitor Start The capacitor start motor, Figure 6–11, is very similar to the split-phase. A capacitor is placed in series with the starting winding to obtain an even greater phase shift, and a stronger rotating magnetic field. The capacitor start motor has a much stronger starting torque than the split-phase.

Capacitor Run A capacitor run motor, Figure 6–12, keeps the capacitor in the circuit at all times. A capacitor run motor has higher starting and running torques than a split-phase motor.

Shaded Pole Motor Figure 6–13 is a drawing of a shaded pole motor. The shading winding is short-circuited and positioned in such a way that it covers approximately one third of the main winding pole face. The current induced into the shading winding creates a magnetic field which opposes

FIGURE 6–11 Capacitor start single-phase induction motor (Courtesy of Cadick Professional Services)

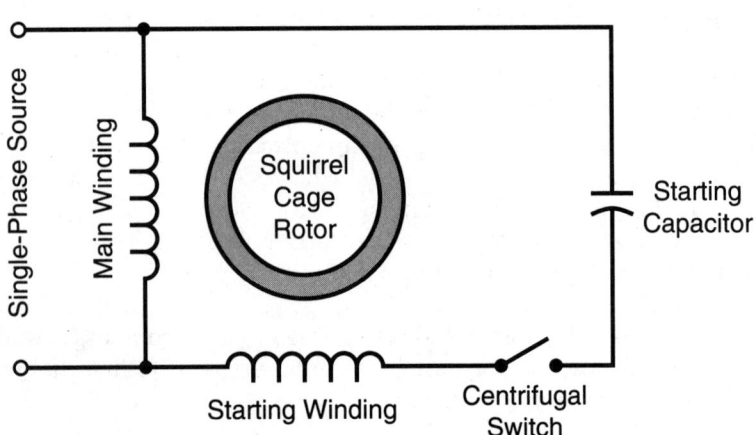

AC Motor Construction and Application

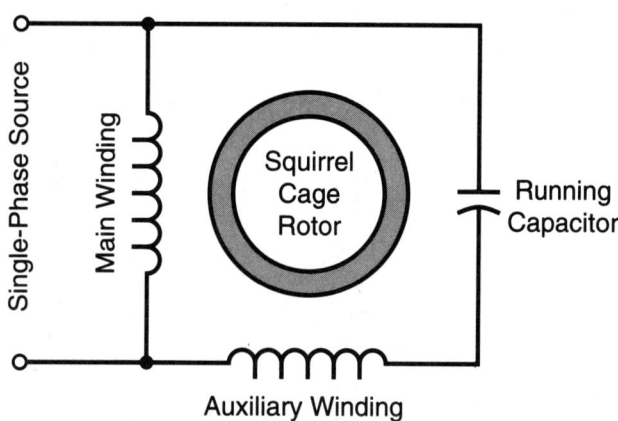

FIGURE 6–12
Single-phase capacitor run induction motor (Courtesy of Cadick Professional Services)

the main field. This opposition reduces the field over a portion of the pole face and creates an out-of-phase flux which has the effect of creating a rotating field. The shaded pole motor is extremely inefficient, has very high slip, and is used only for very small motors. Some speed control can be obtained by placing a variable reactor in series with the shading winding.

SELECTING THE RIGHT MOTOR

Selecting the proper AC motor for a given application is a relatively straightforward procedure. The following steps are derived in large part from the Lincoln Electric Company application data D1T entitled *How to Select a Motor.*

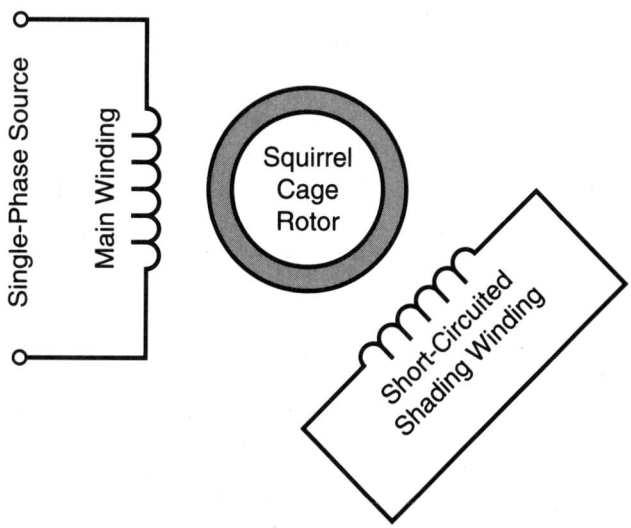

FIGURE 6–13
Shaded pole motor (Courtesy of Cadick Professional Services)

Determine the Type of Motor Required

If a constant speed or power factor control is required, a synchronous motor must be selected. Otherwise, an induction motor is satisfactory.

Determine the Power Characteristics Required

Establish the horsepower required to operate the driven machine. This is usually done by one of the following methods:

From specifications on the nameplate of the driven machine.
By actually testing the driven machine.
By comparison to similar machines with similar loads.
By calculating from known load data.
By operating the machine at rated load and voltage and measuring the input current and temperature rise of a motor you think will do the job.

Torque and Speed
The horsepower rating of a motor is determined from its rated load output characteristics of torque and speed. For a particular application, you must, of course, choose a motor with a rated load torque to drive the machine at the required speed. However, there are three other torque characteristics to be considered—locked rotor or starting torque, pull-up torque, and breakdown torque. The motor must have sufficient starting and pull-up torque to bring the driven machine to operating speeds, and it must be able to overcome peak loads without stalling.

Squirrel cage motors operate only at certain speeds. These speeds depend on the input power frequency and type of winding. The relationship between torque characteristics, horsepower, and speed for sizes of motors can be found in their manufacturer's literature.

Select a Standard Motor

After the horsepower or torque and speed required is determined, select the smallest motor that will produce all the output characteristics required. For special starting torque requirements, select a motor larger than would be needed for normal rated load horsepower.

Check the Motor Dimensions

The NEMA frame size of the motor selected depends on the horsepower and speed required. After choosing the motor, check the dimensions of that frame size to see if it fits in the space available. Check the manufacturer's literature for dimensions of standard motors.

AC Motor Construction and Application

If a standard motor does not fit, determine whether it is most economical to change the driven machine, use a different standard motor, or adapt the standard motor by using a mechanical modification or accessory.

Safety

When selecting a motor, carefully consider the application and provide proper safeguards for personnel to prevent contact with rotating or live parts.

Effect of External Conditions

Ambient Temperature Most standard motors are NEMA-rated for operation at 40°C maximum ambient temperature. If operated at rated load at higher ambient temperatures, the motor will tend to overheat, thus reducing insulation life. For such application, a motor of higher horsepower rating may have to be used. Because a larger motor operates at less than rated load, its temperature rise is less, and the motor, if properly selected, will not overheat. Manufacturers will usually supply derating/uprating tables for their motors to operate in conditions other than NEMA-specified.

Altitudes The temperature rise for NEMA motors is based on operation at altitudes of 3300 ft or less, and a maximum ambient temperature of 40°C. Motors having Class B or F insulation systems, and a temperature rise in accordance with NEMA, will operate satisfactorily at altitudes above 3300 ft, in those locations where the ambient temperature is not greater than shown in Table 6–13.

Table 6–13   Altitude versus Temperature for NEMA Standard Motors
(Courtesy of Lincoln Electric Company, Cleveland OH 44117)

Maximum Altitude	Ambient Temperature
3300 feet	40 degrees C
6600 feet	30 degrees C
9900 feet	20 degrees C

Table 6–14 Altitude versus Temperature Derating Factors
(Courtesy of Lincoln Electric Company, Cleveland, OH 44117)

Altitude Range	*Motors Having a 1.15 Service Factor*	*Motors Having a 1.0 Service Factor*
3300 to 9000 feet	100% of Rating	93% of Rating
9000 to 9900 feet	98% of Rating	91% of Rating
9900 to 13,200 feet	92% of Rating	86% of Rating
13,200 to 16,500 feet	85% of Rating	79% of Rating

When the motors are operated at altitudes above 3300 ft, in locations where an ambient temperature of 40°C may exist, they must be derated according to the manufacturer's specifications, Table 6–14.

Environment For extremely harsh environments, encapsulated or sealed windings may be required.

Check Available Input Power

The selected motor must operate on the power available from the power system. If rated voltage is not available, contact the manufacturer for suggestions.

Starter Sizing

Consult a specific starter manufacturer to determine the proper size starter for the motor application. Starter size for a particular motor is based on the rated load horsepower and the nameplate current and voltage of the motor.

SUMMARY

The classification of AC motors includes such characteristics as size, horsepower, application, electrical type, enclosures, and speed variability. Each motor classification is defined by the National Electrical Manufacturing Association in their standard MG-1.

AC Motor Construction and Application

Whether three-phase, single-phase, small, or large, all motors have certain common standards. Many of a motor's ratings and standards are shown on the nameplate. Items on the nameplate include such information as horsepower, speed, voltage, frequency, and full load current.

Proper selection of a motor for a given application is not difficult, as long as a standard procedure is used. This chapter has presented a procedure that may be used for the selection of most electric motors.

REVIEW QUESTIONS

1. Using Table 6–1, and the nameplate shown in Figure 6–1, determine the height of the center line of the shaft, and the distance between the center lines of the motor feed mounting holes.

2. What AC motor design should be employed for extremely high torque applications when slip speed is not important?

3. What AC motor type might be employed for high torque applications when speed control is required?

4. A 6600-V, 2000-hp motor with class H insulation is to be used in an environment where the ambient temperature may rise to 55°C. What is the maximum winding temperature that this motor may be subjected to, without exceeding the allowable rise?

5. An engineer has calculated the following requirements for a given motor:
   Minimum Horsepower: 54
   Maximum Slip: 5%
   Torque: Moderate
   Efficiency: Moderate
   What horsepower and design might be used for this application?

6. For the motor selected in question 5, what voltage rating would be best if 230 V and 460 V are available?

7. A 30-hp, 230-V motor has a NEMA locked rotor designation of E. What are the minimum and maximum locked rotor currents that may flow when this motor is started?

8. Look at Figure 6–1. Assuming that the motor terminal voltage stays within ±10% of 230 V and the frequency does not change, what is the maximum horsepower that this motor may deliver? Will the motor operate any differently when delivering this horsepower?

9. Assume that the motor of Figure 6–1 is a delta-connected, 9-lead motor. How should the leads be connected to set the motor up for 460-V operation?

10. A certain load must be driven at a constant speed of 900 rpm. What sort of motor would you choose for this application?

AC Motor Control and Starting 7

OBJECTIVES

After studying this chapter, the student should be able to:

- *Draw circuit connections for across-the-line, resistance, autotransformer, wye-delta, and part winding starting methods.*
- *Compare the various types of motor starting methods.*
- *Describe the construction and ratings of standard across-the-line starters.*
- *Describe a general method that may be used to troubleshoot standard across-the-line starters.*
- *Explain various methods used in speed control of AC motors.*

INTRODUCTION

This chapter describes several of the more common methods that are used to start and control AC motors. Each of these methods offers some advantage in cost, simplicity, efficiency, or precision. The final choice of which to use must lie with the user and the specific application.

Troubleshooting motor starters need not be a difficult and time-consuming task. This chapter presents a straightforward set of techniques that may be used to diagnose and troubleshoot standard NEMA starters.

Speed control has grown from crude, brute force techniques such as wound rotor resistors, to modern techniques that use efficient frequency control units to regulate speed efficiently and precisely.

MOTOR STARTING CIRCUITS

Across-the-Line

Across-the-line starting, also called full voltage starters, are the most common of all starting mechanisms. Figure 7–1 is a typical across-the-line circuit. When the *Start* pushbutton is pressed, the M coil is energized, closing contacts M_1 and M. When the M contacts energize the motor, it starts. The M_1 contact seals around the *Start* pushbutton, thus allowing the motor to continue running after the button is released. The M_1 contacts are called the *holding contacts*.

To stop the motor, the operator depresses the *Stop* pushbutton, which releases the M coil. The M and M_1 contacts open, thus de-energizing the motor and resetting the starting mechanism.

Across-the-line starting draws extremely high locked rotor currents for time periods of up to several tens of seconds. During this time, the system voltage drops to very low values and other equipment may suffer. Because of this, some power companies may restrict motor starting by:

- Specifying a maximum starting current or kVA.
- Specifying a maximum locked rotor kVA to hp ratio. (The NEMA letter code.)
- Requiring a reduced voltage start on motors above a certain size.

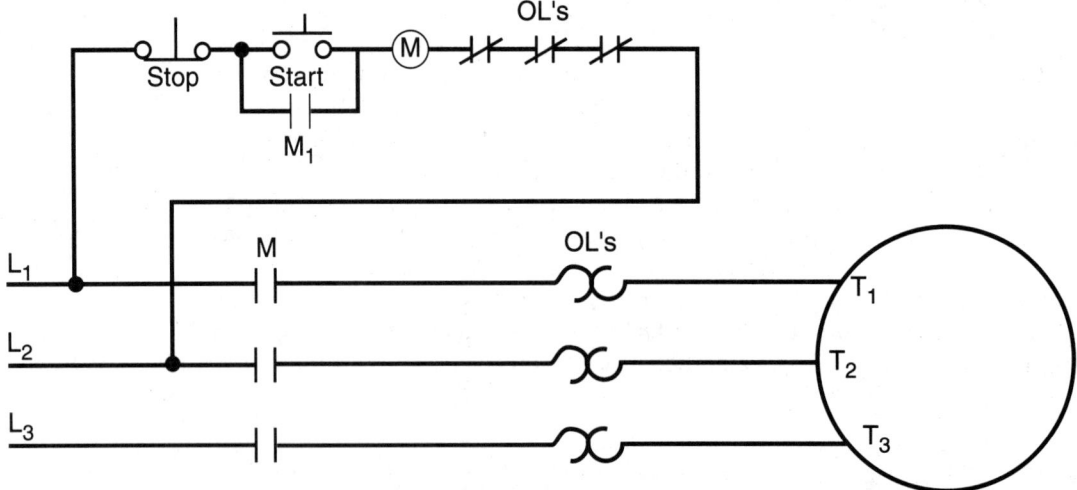

FIGURE 7–1
Simple across-the-line starter with start and stop pushbuttons (Courtesy of Cadick Professional Services)

AC Motor Control and Starting

Resistance Starters

Primary Resistance The primary resistance starter, Figure 7–2, reduces the starting voltage to the motor by placing resistors in series with the motor windings. When the *Start* pushbutton is depressed, the M coil is energized, closing the M and M_1 contacts. This energizes the motor windings through the starting resistors, R. The timing relay coil, TR, is also energized. The TR relay is a timing relay set to allow sufficient time for the motor to reach some percentage of running speed.

When the TR operates, it picks up the R relay which closes the R contacts, shorting out the resistors in the motor circuit, and applying full voltage to the motor. In some applications the TR relay may be replaced by a centrifugal switch, which applies full voltage after the motor reaches some reset percentage of full speed.

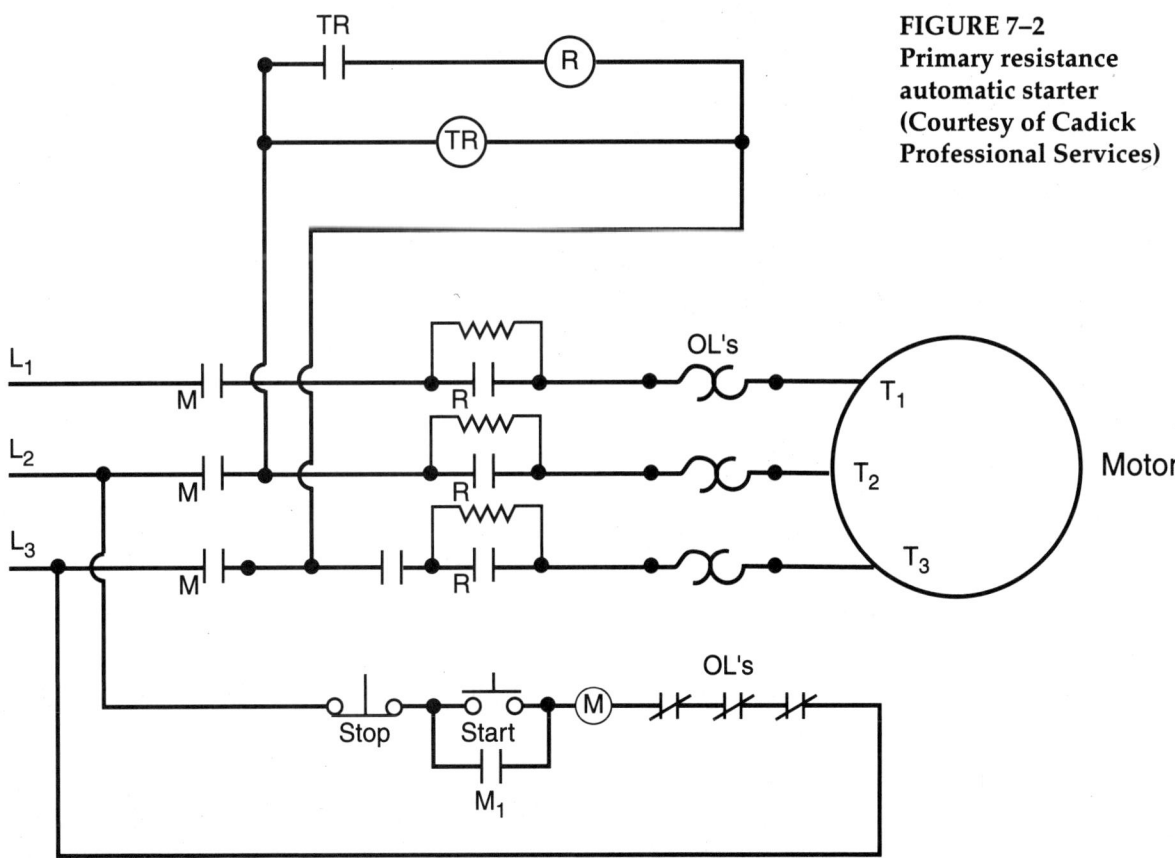

FIGURE 7–2
Primary resistance automatic starter (Courtesy of Cadick Professional Services)

Secondary Resistance A motor is a rotating transformer. A resistance placed in the rotor winding will be reflected into the stator winding—thus reducing the flow of current in the stator. Figure 7–3 is an example of such a circuit. When the M contactor is closed, the resistors are in the rotor circuit, and reduce the locked rotor current by virtue of the reflected resistance. After the time delay relay TR operates, the resistors are removed from the circuit, and the motor runs as a standard induction motor.

Autotransformer Reduced Voltage

Figure 7–4 shows how three autotransformers can be used to reduce the starting voltage for a dual voltage start. Depressing the start button energizes the M coil which seals around the *start* button (M_1) and energizes the

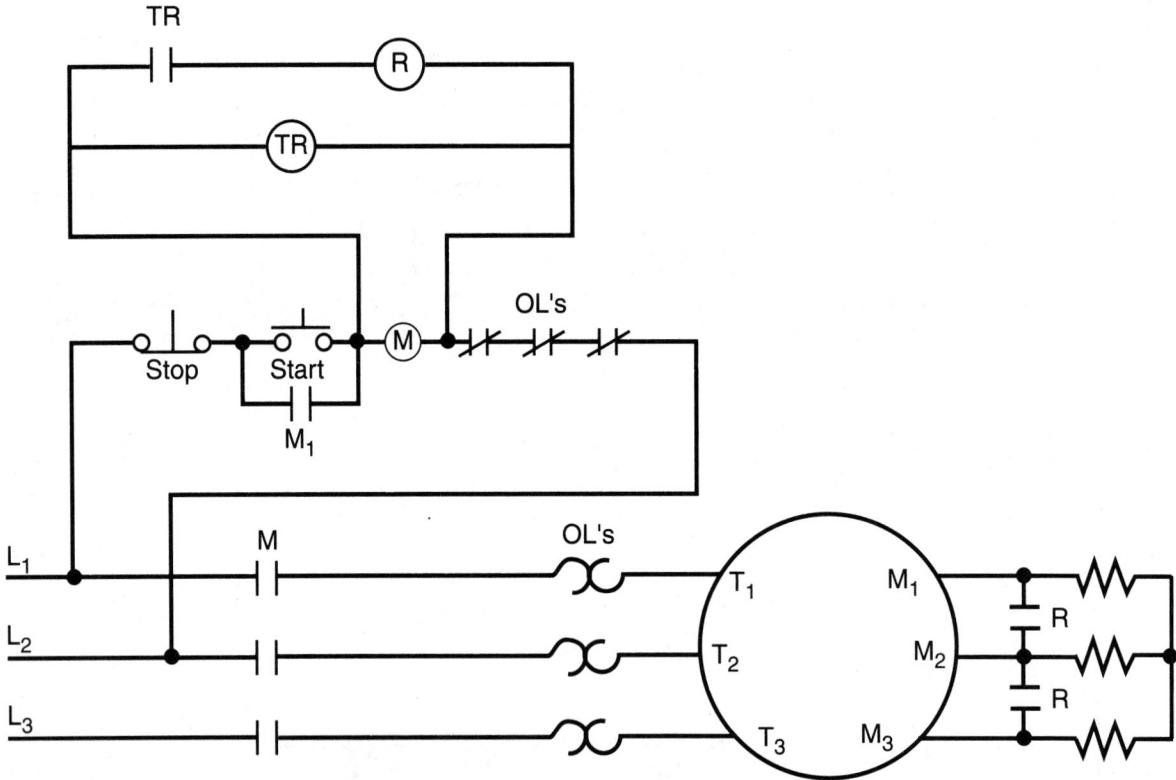

FIGURE 7–3
Secondary resistance starting circuit for a wound rotor induction motor (Courtesy of Cadick Professional Services)

AC Motor Control and Starting

TR coil. The M contacts all pickup and create a three-phase autotransformer circuit with the tap voltages being fed to T_1, T_2, and T_3. After TR picks up, it energizes the R and closes all of the R contacts. This bypasses the autotransformers and applies full voltage to the motor.

Part Winding Starters

Part winding starters (PWS) employ a two- or three-step control mechanism to connect the motor to the line. The two-step starter is the more common, Figure 7–5. The first stage of the two-step PWS connects a portion of the winding to the line. Then, after a time delay (usually 4 seconds maximum), the second-step contactor closes and connects the remaining windings. Since only a portion of the winding is initially connected to the line, the starting current is greatly reduced.

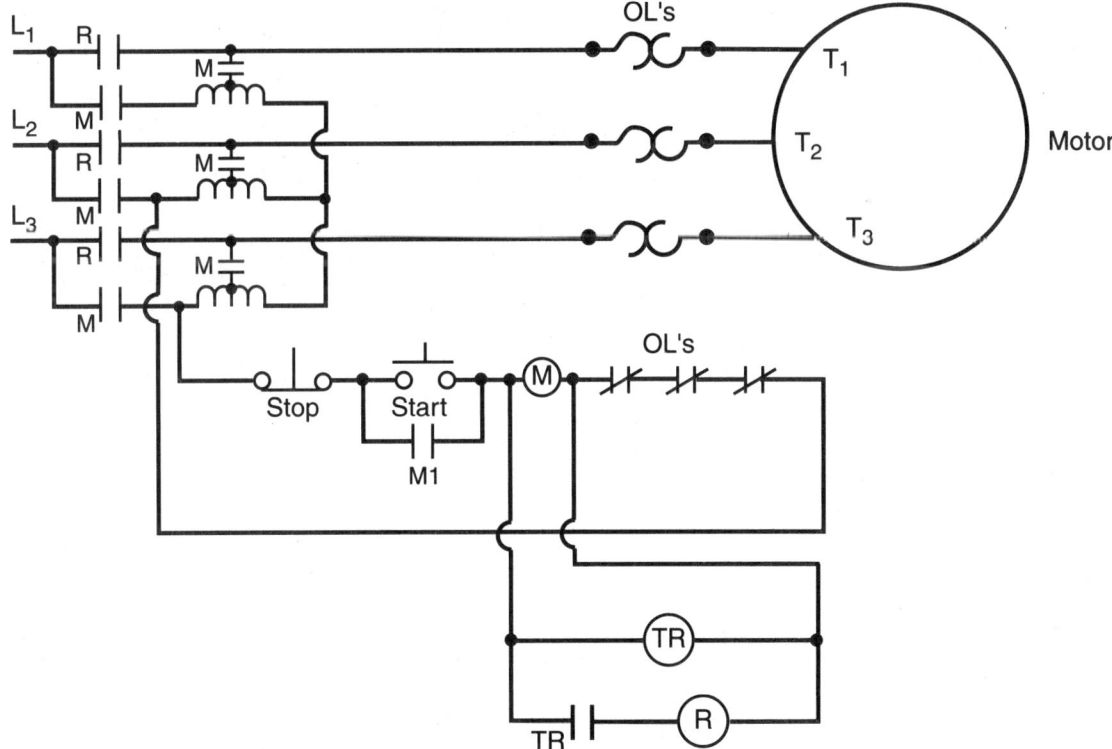

FIGURE 7–4
Autotransformer starting circuit (Courtesy of Cadick Professional Services)

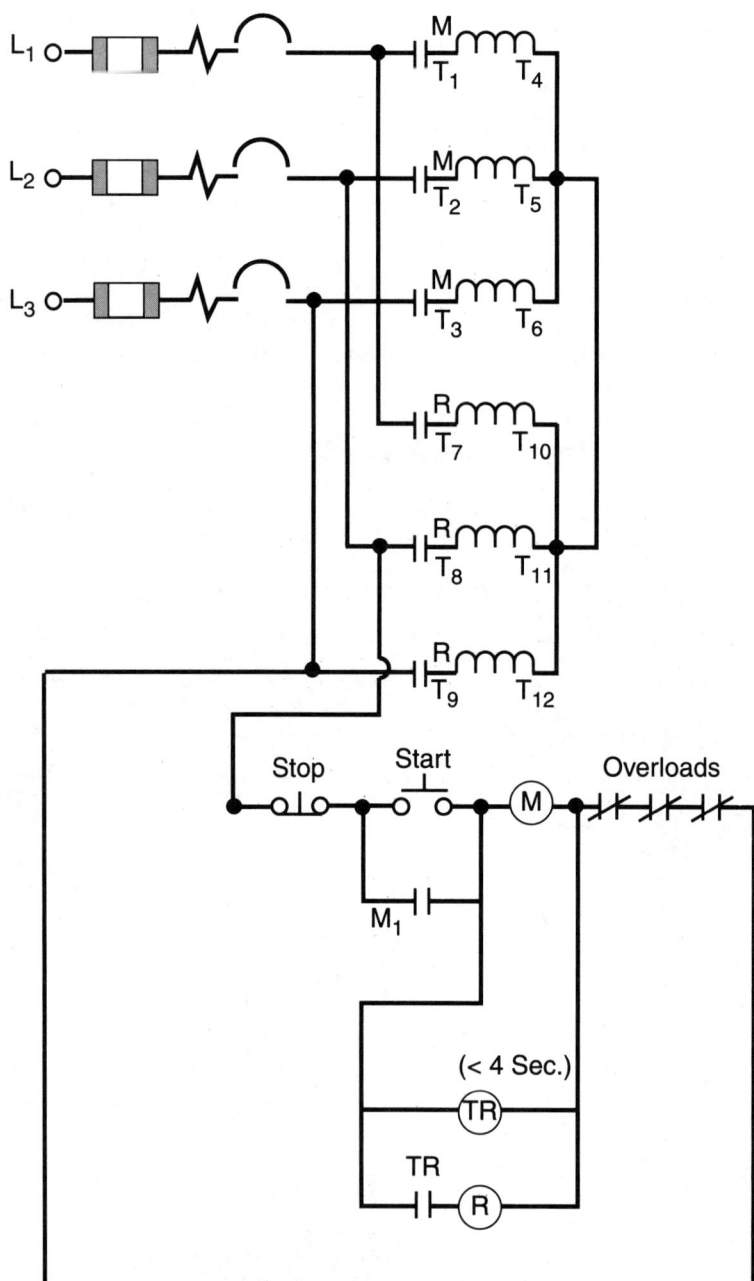

FIGURE 7–5
Two-step part winding starter for a twelve-lead motor (Courtesy of Cadick Professional Services)

AC Motor Control and Starting

A three-step starter adds one additional step. First, a set of resistors are inserted into part of the winding, then the resistors are by-passed, and finally the entire winding is closed in. Three-step starters are not used unless extremely low values of starting current are required.

Not all multi-lead motors are designed for part winding start. The user must always check with the manufacturer before employing such a system.

Wye-Delta Starters

The wye-delta starter initially applies phase-to-neutral voltage to the motor windings, then after a short time delay, modifies the connections to apply delta voltage. Figure 7–6 shows a typical wye-delta starter. When the *Start* button is pushed, the *M*, *Y*, and *TR* relays are picked up. This closes

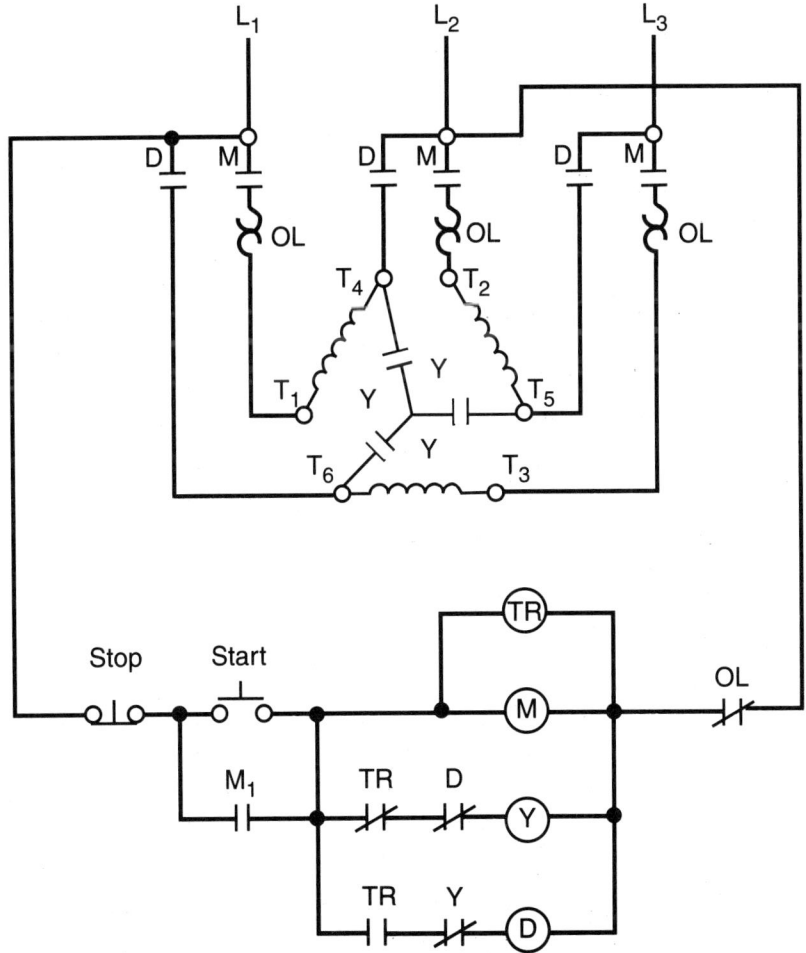

FIGURE 7–6
Wye-delta starter (Courtesy of Cadick Professional Services)

the *M* and *Y* contacts, which connects the windings in a wye configuration. Thus the phase-to-neutral voltage is applied to the motor windings for a reduced voltage start.

The *TR* relay times out and its contacts reverse position. The *Y* relay drops out, and the *D* relay picks up. The *Y* contacts open and the *D* contacts close. This connects the motor windings to the delta voltages. For example, T_1 is connected to L_1 and T_4 is connected to L_2.

Starting Methods Comparison

Table 7–1 compares the various starting methods.

MOTOR STARTERS AND MOTOR CONTROL CENTERS

General Description

A motor starter includes all necessary equipment to start, stop, control speed, protect, and otherwise operate an electric motor. This section will focus on low voltage, combination starters of standard NEMA design. Low voltage starters include power disconnect means, overcurrent protective devices, starting contactor, control relays, and pushbuttons. Figure 7–7 is a photo of a typical combination starter and Figure 7–8 is a wiring diagram for the same starter.

Motor control centers (MCC) consist of a bus and enclosure assembly with groups of motor control equipment and circuits assembled and enclosed in a steel-clad enclosure. The MCC has a three-phase bus to distribute the power to each starter.

Because of the large number of starters used in modern power systems, motor control centers are often used. Motor control centers were first introduced in the 1940s to allow the installation of many motor starters in a small, self-contained package. The MCC is designed to fit all of the various size starters. To add a new motor circuit, the user merely plugs the starter into the MCC and connects the motor leads to the starter, provided that the feeder conductor will handle the additional load. Figure 7–9 is a typical MCC.

Sizes and Ratings

Motor starters are classified into five basic categories:

1. Class A: AC air-break and oil-immersed manual or magnetic controllers for service on 600 V or less. Short circuit interrupting capability is up to ten times normal motor rating.

Table 7–1 Comparison of Five Starting Methods
(Courtesy of the Lincoln Electric Company, Cleveland, OH 44117)

Starting Method	Operation	Starting Current % of locked rotor current	Starting Torque % of locked rotor torque	Open or Closed Transition	Basic Characteristics — Advantages	Basic Characteristics — Limitations
Across-the-Line	Initially connects motor directly to power lines.	100%	100%	None	1. Lowest cost. 2. Highest starting torque. 3. Used with any standard motor. 4. Least maintenance.	1. High starting current. 2. High starting torque. 3. May shock driven machine.
Primary Resistance Reduced Voltage	Inserts resistance units in series with motor during first steps(s).	50 – 80%	25 – 64%	Closed	1. Smoothest starting. 2. Least shock to driven machine. 3. Most flexible in application. 4. Used with any standard motor.	1. High power loss because of heating resistors. 2. Heat must be dissipated. 3. Low torque per ampere input. 4. Highest cost.
Autotransformer Reduced Voltage	Uses autotransformer to reduce voltage applied to motor. Tap 50% / 65% / 80%	25% / 42% / 64%	25% / 42% / 64%	Closed	1. Best for hard to start loads. 2. Adjustable starting torque. 3. Used with any standard motor. 4. Less strain on motor.	1. May shock driven machine. 2. High cost.
Wye-Delta or Star Delta	Starts motor with windings wye connected, then re-connects them in delta connection for running.	33%	33%	Open Closed	1. Medium cost. 2. Low starting current. 3. Low starting torque. 4. Less strain on motor.	1. Low starting torque. 2. Requires Delta Wound motor.
Part Winding	Starts motor with only part of windings connected, then adds remainder for running.	70 – 80%	50 – 60% Minimum pull-up torque 35% of full load	Closed	1. Low cost. 2. Popular method for medium starting torque applications. 3. Low maintenance.	1. Not good for frequent starts. 2. May require special wound motor. 3. Low pull-up torque. 4. May not come up to speed on first step when started with load applied.

Note: The reduced starting torque (LRT) indicated for the various starting methods can prevent proper starting of high inertia loads and must be considered when sizing motors and choosing starters.

FIGURE 7–7
Typical low voltage combination motor starter (Courtesy of Cuther-Hammer)

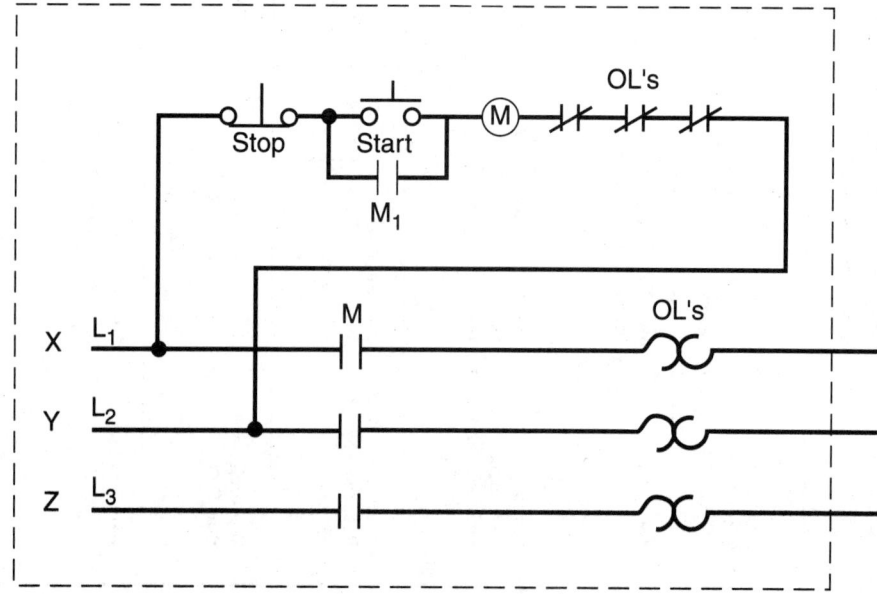

FIGURE 7–8
Wiring diagram for the starter shown in Figure 7–7

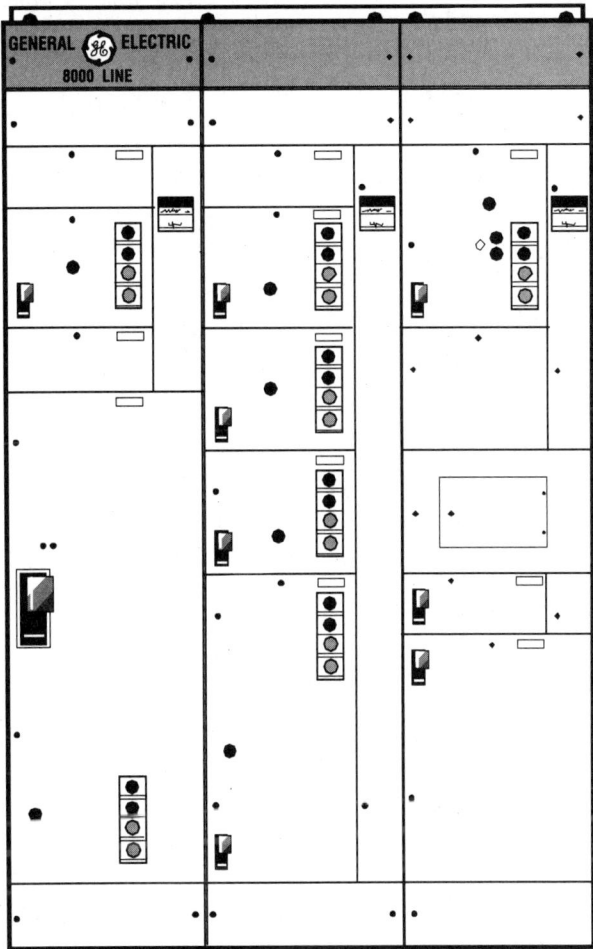

FIGURE 7-9
Typical motor control center (Courtesy of General Electric Company)

2. Class B: DC air-break manual or magnetic controllers for service on 600V or less. Their interrupting ratings are up to their overload capability.
3. Class C: AC controllers capable of interrupting faults beyond their operating overload.
4. Class D: DC controllers capable of interrupting faults beyond their operating overload.
5. Class E: AC, oil-immersed magnetic controllers for 2200 and 4600-V application. Class E1 is rated for up to 50,000 A symmetrical. Class E2 use current limiting fuses for interrupting ratings up to 200,000 A symmetrical.

Motor starters are classified by NEMA according to their voltage and horsepower capacity. Table 7-2 shows the standard classes for low voltage starters.

Table 7–2   Standard Starter Sizes (Courtesy of Cadick Professional Services)

Phase	Motor Voltage (V)	Motor Minimum Horsepower	Motor Starter Sizes (NEMA)
Single-phase contactors	115	1/3	00
		1	0
		2	1
		3	1P
		3	2
	230	1	00
		2	0
		3	1
		5	1P
		7½	2
Three-phase contactors	230	1½	00
		3	0
		7½	1
		15	2
		30	3
		50	4
		100	5
		200	6
		300	7
		450	8
		800	9
	460/575	2	00
		5	0
		10	1
		25	2
		50	3
		100	4
		200	5
		400	6
		600	7
		900	8
		1,600	9
DC contactors	230	5	1
		10	2
		25	3
		40	4
		75	5
		150	6
		225	7
		350	8
		600	9

AC Motor Control and Starting

Motor control centers are available in two classes and three types.

Class I control centers are a grouping of motor controls, feeder taps, and other components arranged in a convenient assembly. While they do include the connections from the horizontal power bus, they do not have interwiring or interlocking between units or to remotely mounted devices.

Class II control centers are similar to Class I, but include a complete control system including interwiring between units and interlocking to remote devices.

Type A units have no terminal blocks. Starters are hardwired at the factory. No external wiring is available.

Type B units have all control wiring terminating on terminal blocks. Plug-in type terminal wiring is provided.

Type C units have all the features of type B. All control and load wiring is factory installed for up to size 3, and extended from the unit terminals to master terminal blocks.

Troubleshooting Motor Starters

Table 7-3 lists the most common types of problems encountered with motors and lists common cures for these problems.

MOTOR SPEED CONTROL

Number of Poles

The speed of rotation of an AC motor is dependent on the number of poles and the frequency of the supply voltage according to the formula:

$$S = \frac{f \times 120}{P} \tag{1}$$

Where:
 S = Speed of rotation in revolutions per minute
 f = Frequency of supply voltage
 P = Number of poles in the rotating magnetic field

AC motor speed can be controlled by winding the stator in such a way that the number of poles can be changed. A multiple stage starter can then be used to switch to the pole configuration for the desired speed. Several different methods can be used to wind and control such a motor. Figures 7-10, 7-11, and 7-12 are three examples of different methods that may be used.

Note that speed controls of this type allow only discrete speed changes. If more precise controls are required, other methods must be used.

Table 7–3 Troubleshooting Techniques for Motor Starters (Courtesy of Cadick Professional Services)

Problem	Possible Cause
Main contacts close but motor does not start	• Loose or broken connections • Open overload heater coil or connection • Main contacts not making (worn or dirty) • Open resistance units (primary resistance starter) or open autotransformer (autotransformer starter) • Obstruction preventing the main contactor from moving • External problems such as open interlock circuits
Main contactor does not operate	• Open main coil or loose connections • Open or dirty stop button, start button, or overload relay contacts • Low voltage • Shorted coil • Loose or open terminal connections
Contacts open when start button is released	• M1 contacts (see Figures 7-1, 7-2, etc.) are dirty or not closing completely • Improper wiring • Shorted or grounded motor not in a starter
Fuse blows when start button is pushed	• Shorted or grounded contacts or terminations • Shorted or grounded coil
Burnt open magnet coil	• Overvoltage • Too-frequent operation • Excessive current caused by large magnetic gap, dirt, or mechanical trouble • Wrong size starter for motors (too small)
Noisy magnet	• Dirty core face • Broken shading coil or pole

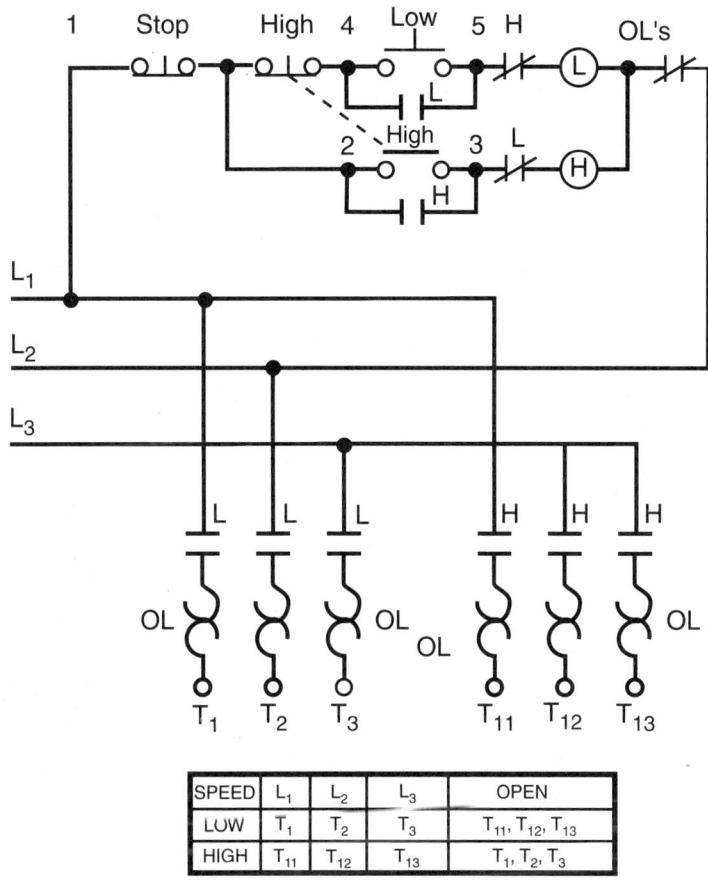

FIGURE 7-10
Elementary diagram for a two-speed, two-winding, three-phase starter (Courtesy of Delmar Publishers)

Secondary Resistance Speed Control

The wound rotor induction motor can be speed controlled by varying the resistance in the rotor windings, Figure 7-13. Speed may only be controlled from full to approximately 50%; however, such controls are inexpensive and simple. Wound rotor control is very common. Figures 7-14 and 7-15 show two types of wound rotor speed controls. Figure 7-14 represents the manual speed control which is old and obsolete.

Eddy Current Clutch

This type of control uses a clutch system attached to the shaft of a squirrel cage induction motor, Figure 7-16. The degree of clutch slip can be controlled by varying the DC voltage applied to the rotor magnet coil. This method of speed control is an older method, but is still found in many installations.

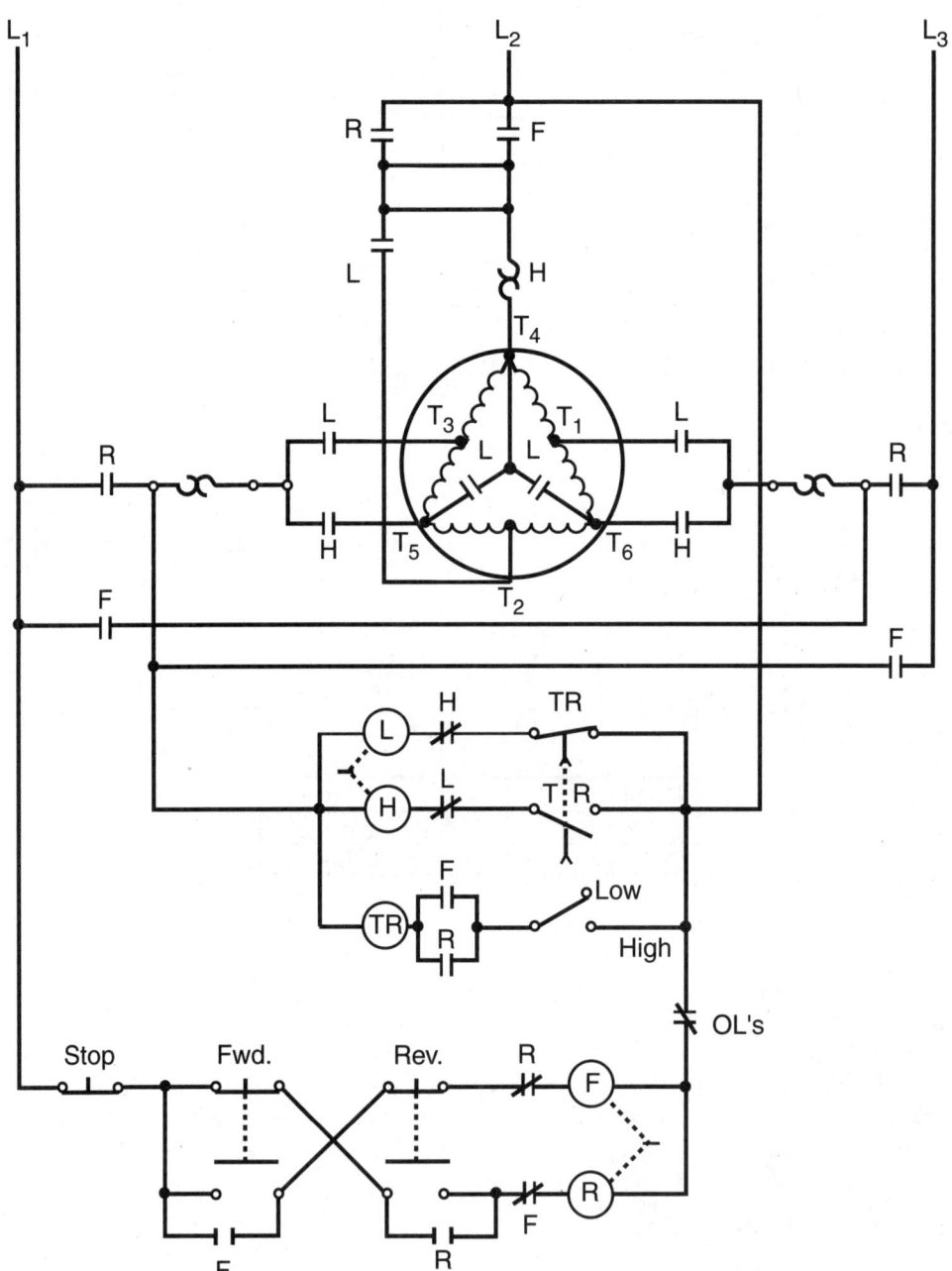

FIGURE 7–11
Elementary diagram of a two-speed reversing controller (Courtesy of Delmar Publishers)

AC Motor Control and Starting

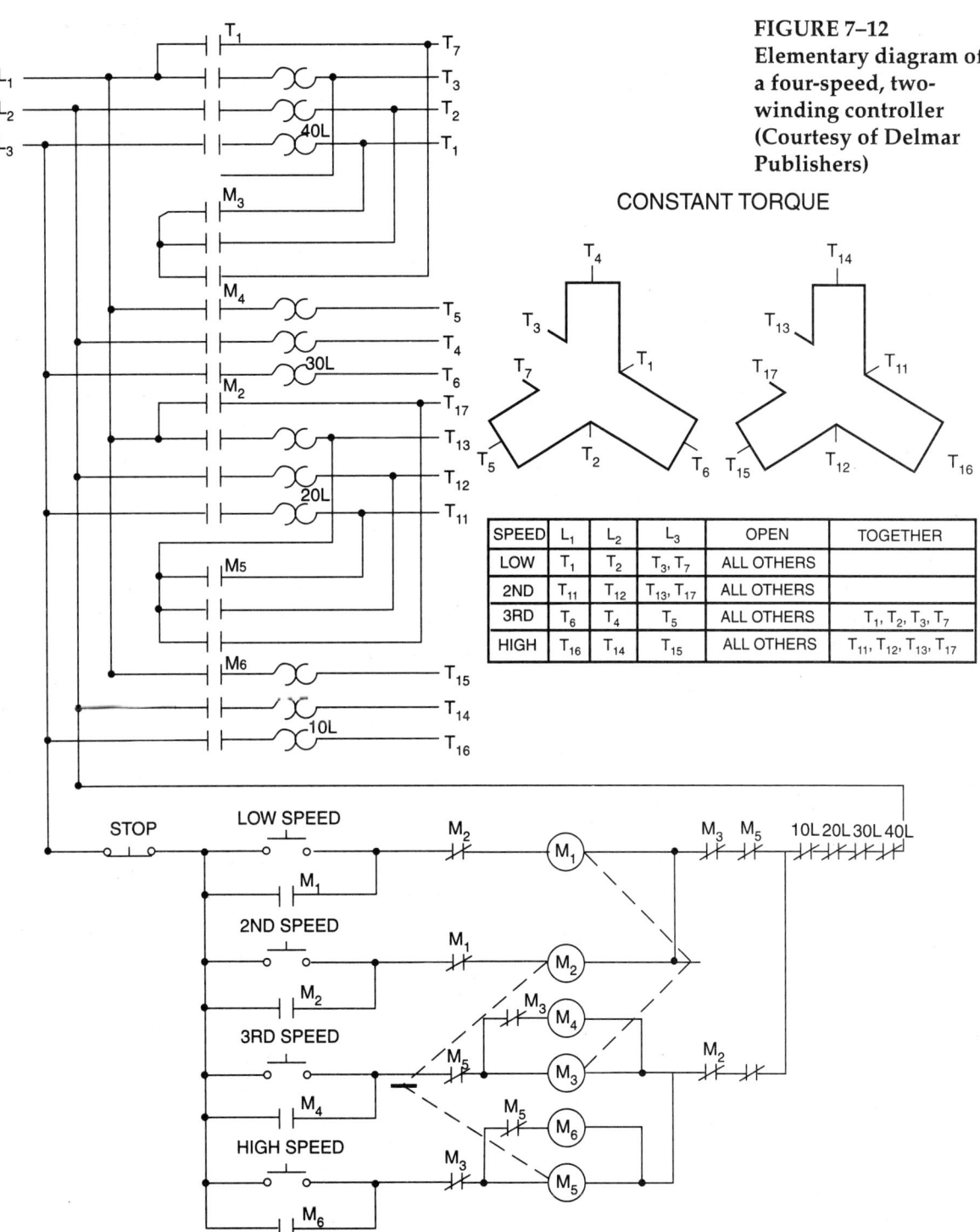

FIGURE 7–12 Elementary diagram of a four-speed, two-winding controller (Courtesy of Delmar Publishers)

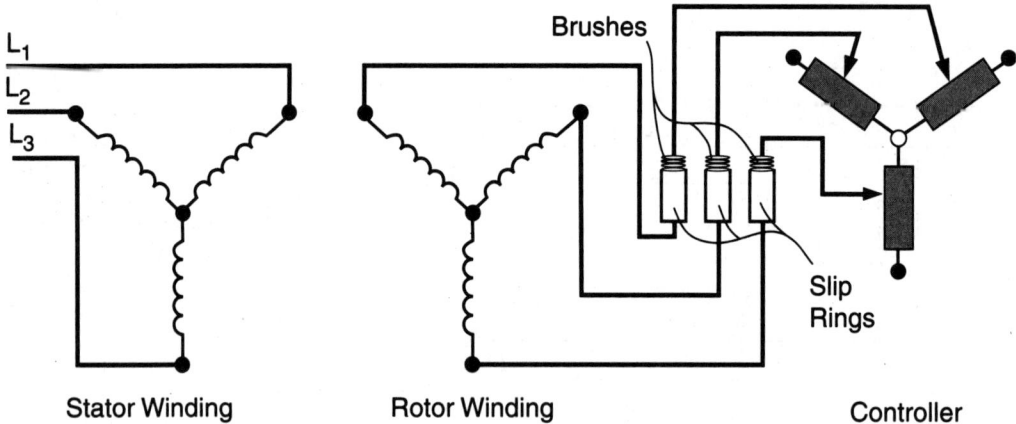

FIGURE 7–13
General diagram for controlling speed of a wound rotor induction motor
(Courtesy of Delmar Publishers)

FIGURE 7–14
Manual speed control interlocked with a magnetic starter
(Courtesy of Delmar Publishers Inc.)

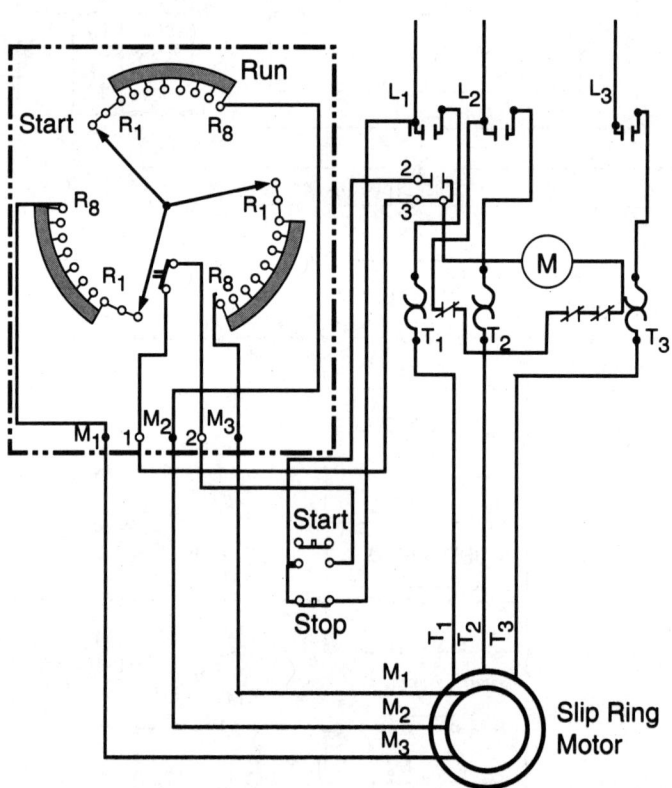

AC Motor Control and Starting

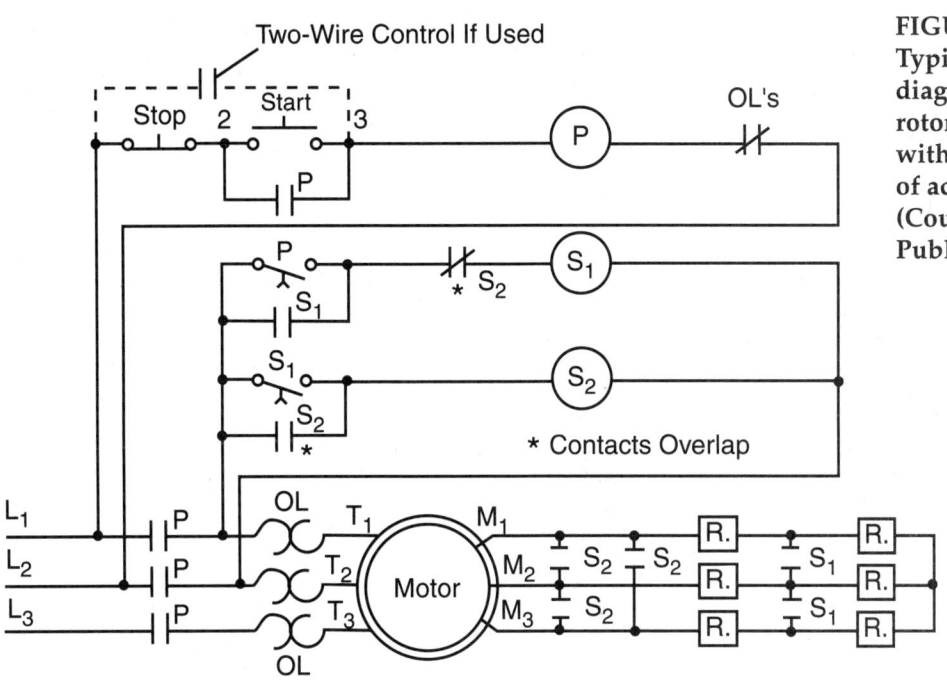

FIGURE 7–15
Typical elementary diagram of wound rotor motor starter with three points of acceleration (Courtesy of Delmar Publishers)

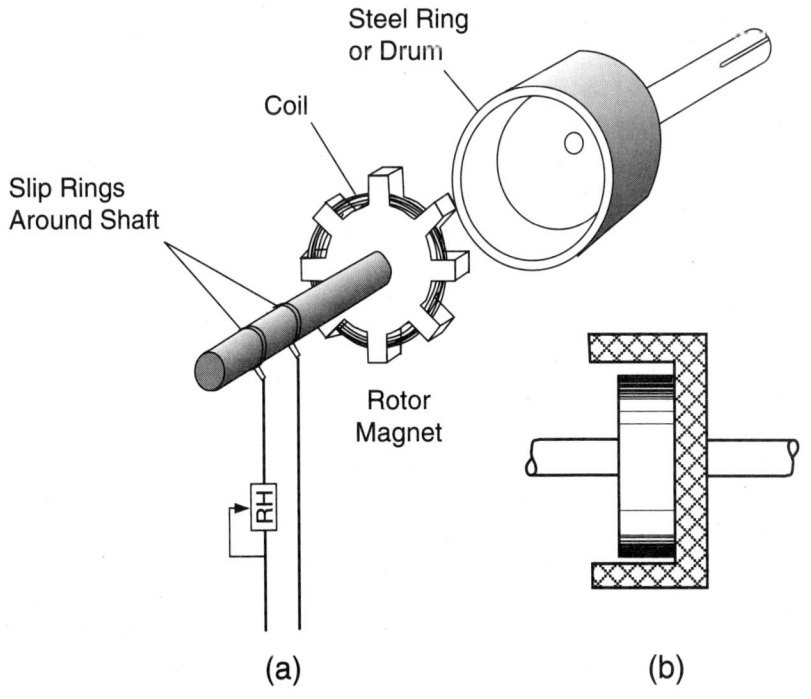

FIGURE 7–16
Diagram showing (a) open view of magnetic drive assembly; (b) Spider rotor magnet rotates within ring

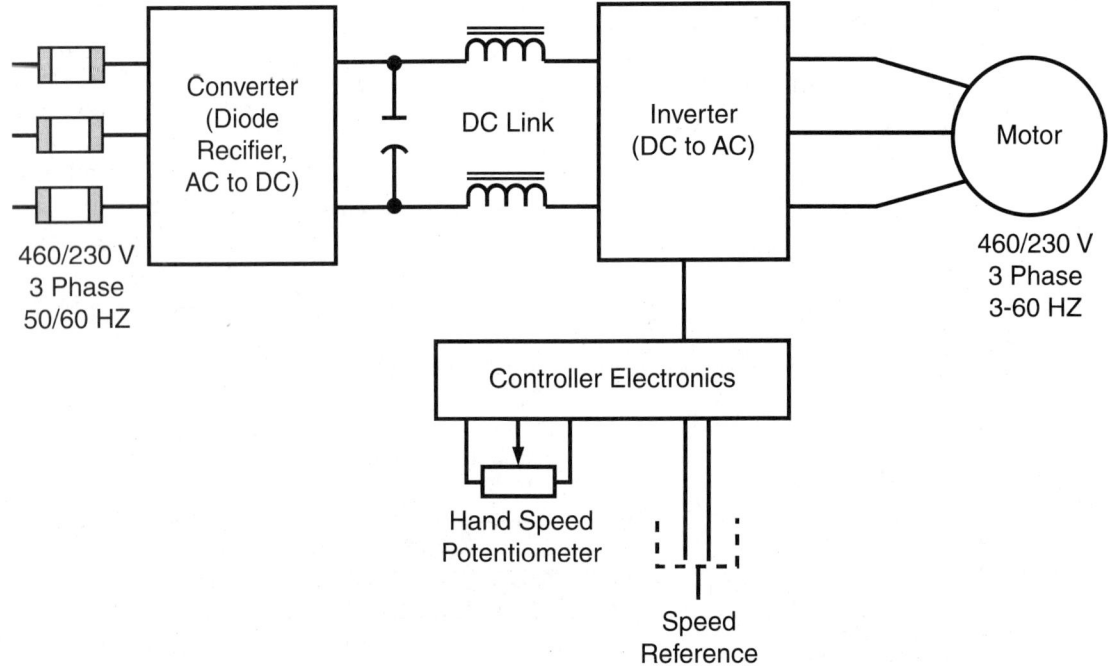

FIGURE 7–17
Block diagram of a variable frequency drive (Courtesy of Delmar Publishers)

Variable Frequency Drives

Speed can be controlled by varying the frequency of the supply voltage, formula (1). Advances in solid state technology have produced controllers that are extremely efficient and cost effective.

The basic circuit for variable frequency speed control is shown in Figure 7–17.

SUMMARY

A variety of techniques are used to start and control AC motors. The most common type of starter is the so-called "across-the-line" or full voltage starter. While it is effective, the across-the-line starter draws a significant amount of starting current, and can cause severe system voltage dips for large motors.

Many starters use reduced voltage techniques to minimize the amount of current and subsequent voltage dips. Starters such as the primary resistance, wye-delta, autotransformer, and part winding can be used for such

AC Motor Control and Starting

applications. Each of these methods offers some advantages and limitations. Table 7–1 contrasts the various reduced voltage methods.

Resistance placed in series with the rotor windings of a wound rotor motor can also be used for a "soft start" application. Because of the ability to continuously control speed, this "secondary resistance" method has been used for speed control as well as starting techniques.

All of these starting methods can be employed in standard starter packages available in low voltage sizes up to 1500 hp and more. Starter packages can be assembled into integrated packages called motor control centers. Motor control centers have all of the equipment and power required to control and operate several motor starters.

Several types of AC motor speed control are available, including multiple pole, secondary resistance, eddy current clutch, and variable frequency. Most modern speed control systems use variable frequency drives.

REVIEW QUESTIONS

1. Draw a simple schematic diagram of an across-the-line, wye-delta, autotransformer, and primary resistance starter.

2. Draw a diagram of a three-step part winding starter which combines part winding starting techniques along with a primary resistance starter.

3. You are trying to start a motor using an across-the-line starter. When you push the start button, nothing happens. List three items that could be causing the problem.

4. Draw a simple diagram of a variable frequency speed control.

Protection 8

OBJECTIVES

After studying this chapter, the student should be able to:

- *Describe the basic principles of motor overcurrent protection.*
- *Explain the operation of the various types of motor overcurrent protective devices, including fuses, melting alloy relays, bimetallic strip relays, magnetic relays, molded case circuit breakers, induction disk overcurrent relays, and electronic relays.*
- *Describe the principles employed for motor differential protection.*
- *Describe the principles employed for motor stator winding temperature protection.*

INTRODUCTION

Motors may be damaged by overload currents, short circuits, or operation in excessive ambient conditions. All three of these problems may cause the motor to be damaged due to excessive temperature. Lower values of overtemperature will cause loss of insulation life. General industry rules of thumb indicate that for each 10°C above allowed temperature that a motor is operated, its life will be cut in half. Such problems can be protected by a combination of overcurrent devices and/or temperature detectors.

When high overtemperatures are experienced, the motor may suffer immediate damage, including burnt windings. If a short circuit occurs, the magnetic and thermal damage may be so extreme that the motor is damaged beyond repair.

Because of their starting characteristics, motors are somewhat of a protection problem. This chapter will describe some of the more common types of motor protection schemes and explain their application.

OVERCURRENT PROTECTION

Basic Principles

Overview Overcurrents are classified as either overloads or short circuits. The difference between the two is the cause of the problem. Overcurrent caused by an overloaded conveyor belt, for example, is classified as a running overload. Overcurrent caused by failure of the insulation system is called a short circuit.

When an AC motor starts, it exhibits a current pattern much like the one shown in Figure 8–1. The initial inrush of current, called *locked rotor current*, lasts for a period of time called the *acceleration time*. Locked rotor current

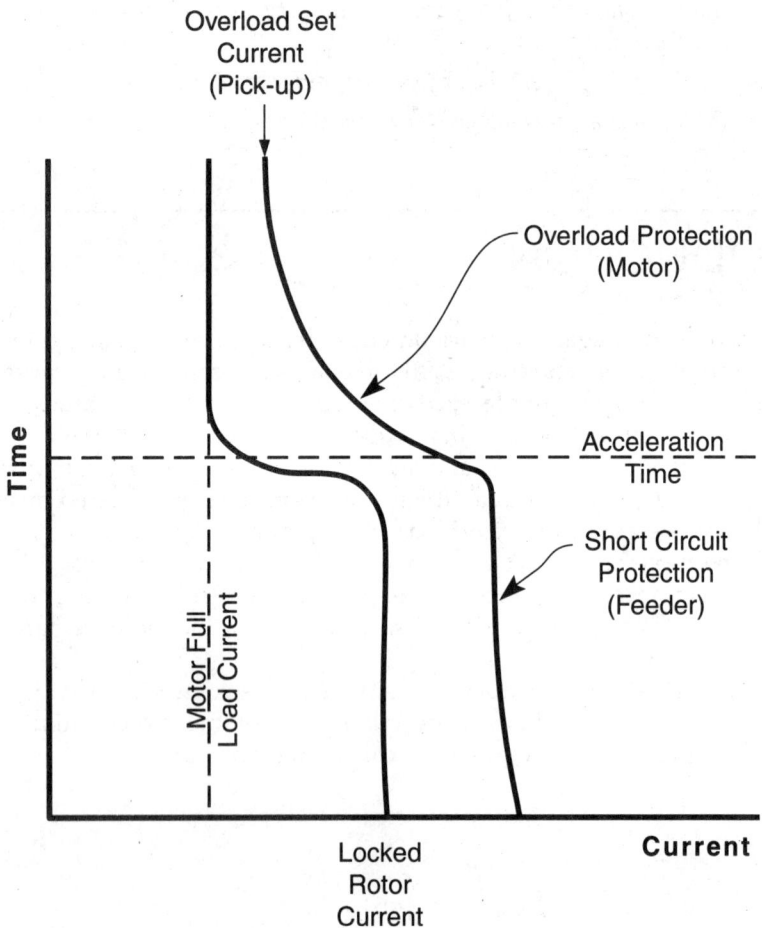

FIGURE 8–1
Motor starting curve (Courtesy of Cadick Professional Services)

may be several times the motor full load current, and acceleration time may last for a period of ten seconds or even more. After the motor reaches approximately 70%–95% of full speed, the current drops off fairly rapidly to the full load or running value.

Any overcurrent protection that is applied must allow the motor to start, and still protect the motor from damage. The overcurrent protection falls into two broad categories—overload and short circuit.

Overload The overload protection is set to start its operation at a current slightly above the motor full load current. For motors of over 1 hp, the *National Electrical Code*® allows pickup currents which do not exceed the values shown in Table 8–1.

The time curve for the overload protection must allow the motor to start; therefore, it must fall above and to the right of the motor starting curves. Low-voltage motor overload protection is usually provided by a thermal or magnetic overload relay. Medium-voltage motors normally employ an induction disk overcurrent relay or a fuse.

Older installations used only two overload devices in a three-phase circuit. This practice has been abandoned because of the possibility of excessive current on the unprotected phase when an open fuse occurs on the primary of the delta-wye transformer feeding the motor circuit. The *National Electrical Code*® requires specific numbers of overload relays for various applications. Table 8–2 outlines these requirements.

Short Circuit Protection Short circuit protection is used in the motor starter to protect the motor branch circuit from the high short circuit currents which can flow when a short circuit occurs. This protection is

Table 8–1 Motor Overload Pickup Settings
(Courtesy of Cadick Professional Services)

Motor Type	*Overload Setting (% of Full Load)*
Motors with a marked service factor not less than 1.15	125%
Motors with a marked temperature rise not over 40° C	125%
All other motors	115%

Table 8–2 Required Number of Overload Units
(National Electrical Code)

Kind of Motor	Supply System	Number and location of overload units, such as trip coils, relays, or thermal cutouts
1-phase ac or dc	2-wire, 1-phase, ac or dc ungrounded	1 in either conductor
1-phase ac or dc	2-wire, 1-phase ac or dc, one conductor grounded	1 in ungrounded conductor
1-phase ac or dc	3-wire, 1-phase ac or dc, grounded-neutral	1 in either ungrounded conductor
2-phase ac	3-wire, 2-phase ac. ungrounded	2, one in each phase
2-phase ac	3-wire, 2-phase ac, one conductor grounded	2 in ungrounded conductors
2-phase ac	4-wire, 2-phase ac, grounded or ungrounded	2, one per phase in ungrounded conductors
2-phase ac	5-wire, 2-phase ac, grounded neutral or ungrounded	2, one per phase in any ungrounded phase wire
3-phase ac	Any 3-phase	3, one in each phase*

*Exception: where protected by other approved means.

Reprinted with permission of NFPA 70-1993, the *National Electrical Code®*, Copyright © 1992, National Fire Protection Association, Quincy, MA 02269. This reprinted material is not the complete and official position of the National Fire Protection Association, on the referenced subject which is represented only by the standard in its entirety.

normally provided with a fuse or the magnetic element of a molded case circuit breaker for low-voltage motors. Fuses or instantaneous overcurrent relays are used on medium-voltage motors.

The short circuit protection must be set above the locked rotor current so the motor can start; however, some maximum protection must be provided. The *National Electrical Code®* provides guidance on the maximum values to be used, Table 8–3. There are exceptions to these settings, so be certain to check the most recent edition of the code for current protection values.

Overcurrent Protective Device Types

Fuses Fuses are the oldest electrical protective devices and are used to protect motors from short circuits and overloads. Originally, fuses were made of pieces of small diameter copper or lead alloy wire which were placed in series with the circuit to be protected. Heat generated from excess current flow melted the small diameter wire, opening the circuit. For obvious safety reasons, the use of the "open wire fuse" was discontinued.

In modern designs, fuse wire is installed in a tube-like cartridge, which is filled with a chemical to help extinguish the electric arc drawn when the fuse operates, Figure 8–2. For many years, this type of one-time fuse was known as the *NEC®* fuse because the *National Electrical Code®* set its standard dimensions. Continuous ratings for this type of fuse may be from 0 to 600 A, with interrupting ratings of up to 10,000 A at rated voltage (Interrupting rating is the amount of current a fuse will interrupt at a rated voltage). Today, the standard dimensions are set by NEMA. On modern drawings, this 10,000-A fuse is known as the type H fuse.

A variation of the type H fuse uses a fusible lead-alloy link in a tube-like cartridge without the chemical, Figure 8–3. When this type of fuse blows, the cartridge may be removed, and the fusible link replaced. This renewable link fuse has the same interrupting rating as the one-time fuse mentioned earlier; however, it is no longer permitted for use in new

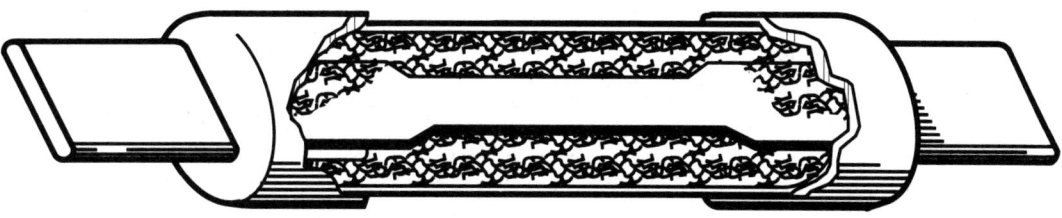

FIGURE 8–2
One-time fuse

Table 8–3 Maximum Rating or Setting of Motor Branch-Circuit, Short-Circuit, and Ground-Fault Protective Devices

(National Electrical Code)

	Percent of Full-Load Current			
	Nontime Delay Fuse	Dual Element (Time-Delay) Fuse	Instantaneous Trip Breaker	Inverse Time Breaker*
Single-phase all types				
No code letter	300	175	700	250
All ac single-phase and polyphase squirrel-cage and synchronous motors† with full-voltage, resistor or reactor starting:				
No code letter	300	175	700	250
Code letter F to V	300	175	700	250
Code letter B to E	250	175	700	200
Code letter A	150	150	700	150
All squirrel-cage and synchronous motors with autotransformer starting:				
Not more than 30 amps				
No code letter	250	175	700	200
More than 30 amps				
No code letter	200	175	700	200
Code letter F to V	250	175	700	200
Code letter B to E	200	175	700	200
Code letter A	150	150	700	150
High-reactance squirrel-cage				
Not more than 30 amps				
No code letter	250	175	700	250
More than 30 amps				
No code letter	200	175	700	200
Wound rotor				
No code letter	150	150	700	150
Direct-current (constant voltage)				
No more than 50 hp				
No code letter	150	150	250	150
More than 50 hp				
No code letter	150	150	175	150

For explanation of Code Letter Marking, see Table 430-7(b)

For certain exceptions to the values specified, see Sections 430-52 through 430-54.

* The values given in the last column also cover the ratings of nonadjustable inverse time types of circuit breakers that may be modified as in Section 430-52.

† Synchronous motors of the low-torque, low-speed type (usually 450 rpm or lower), such as are used to drive reciprocating compressors, pumps, etc., that start unloaded, do not require a fuse rating or circuit-breaker setting in excess of 200 percent of full-load current.

Reprinted with permission of NFPA 70-1993, the *National Electrical Code*®, Copyright © 1992, National Fire Protection Association, Quincy, MA 02269. This reprinted material is not the complete and official position of the National Fire Protection Association, on the referenced subject which is represented only by the standard in its entirety.

installations. Interrupting rating is the amount of current a fuse will interrupt at a rated voltage. Corrosion of the connecting device and illegal "link doubling" caused many problems.

Most modern designs employ the current limiting fuse. Such a fuse is designed to open the circuit in approximately 1/4 cycle when a high magnitude fault occurs. The design of the current limiting fuse is similar to the one-time fuse shown in Figure 8–2. The single fuse element is replaced by a single or series of links with many fine bridges, sections A and C of Figure 8–4. The links are anchored in heavy heat sinks, which keeps them from opening for normal load current. When a high magnitude fault occurs, the multiple bridges are rapidly fused, and the chemical filler around the fuses changes chemically. A fault on the order of magnitude of 16 times the rating of the fuse will be cleared in 0.004 seconds. This fast clearing time limits the amount of current flow—thus the name *current limiting*. The current limiting fuse is available in ratings of up to several thousand amperes, and has interrupting capacities of 200,000 A.

When single link fuses are properly sized to give overload protection to an electric motor and its feeder, locked rotor current may blow the fuse. Therefore, it is necessary to oversize the fuse so that the motor will start.

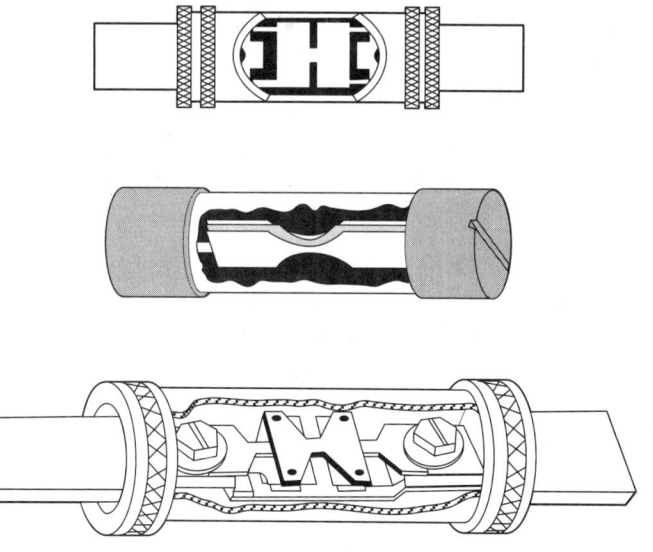

**FIGURE 8–3
Various types
of renewable
cartridge fuses**

Although allowed by the NEC®, these are being phased out by fuse companies. Fuse links are becoming so expensive, as well as being unacceptable, more than one link can be installed, making them in violation of the Code.

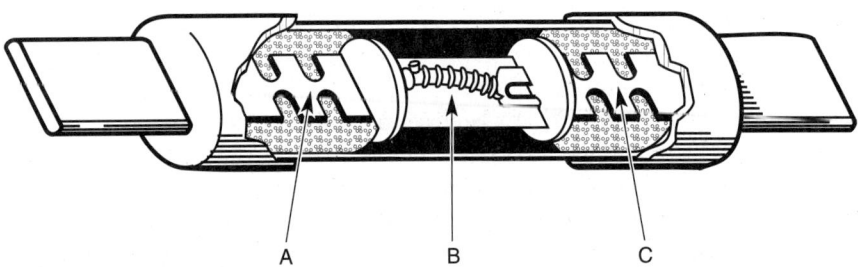

FIGURE 8-4
A typical dual element cartridge fuse with current limiting high current elements

This is done at the expense of overload protection. Such problems encouraged the development of the dual element fuse, Figure 8-4. The dual element fuse consists of two copper links, A and C, located in the two end sections of the container, and a heavy copper center strap, B.

Center strap B is held in position by a soldered connection to a spring. The soldered connection acts as a thermal element, while the copper links in the end sections provide the short circuit element. Moderate overloads will not blow the short circuit fuse element, but will open the circuit through the action of the thermal element. Center strap B is heated up to the melting point of the solder by light overloads, and the spring releases the thermal cutout portion, opening the circuit.

Because the heat coil and the center strap are quite heavy, it takes considerable time to raise the temperature of the strap sufficiently to melt the solder. This feature, together with the time required to melt the solder, gives the fuse its time-lag characteristic. On heavy overloads, the fuse link blows immediately, before the thermal cutout portion has time to function.

Melting Alloy Relays The melting alloy overload relay is a heat-sensitive, current actuated device used for overload protection of motors, Figure 8-5. Motor current passes through a small heater. A ratchet assembly is held in place by a special kind of melting alloy (solder). The solder melts rapidly when heated and hardens just as rapidly when heat is taken away. A pawl under spring tension is held by the ratchet. The pawl assembly carries half a set of contacts, which are normally closed. An overload causes the heater to melt the solder which then releases the pawl and opens the contacts. When the solder becomes hard again, usually in about two minutes, the relay can be reset.

Figure 8-6 is a typical magnetic starter showing the overload relay. Note the reset lever for the pawl and the top mounting screw for the heater element. A given relay will be used for a variety of different sized motors. Only the heater element need be changed for the different full load currents.

Protection

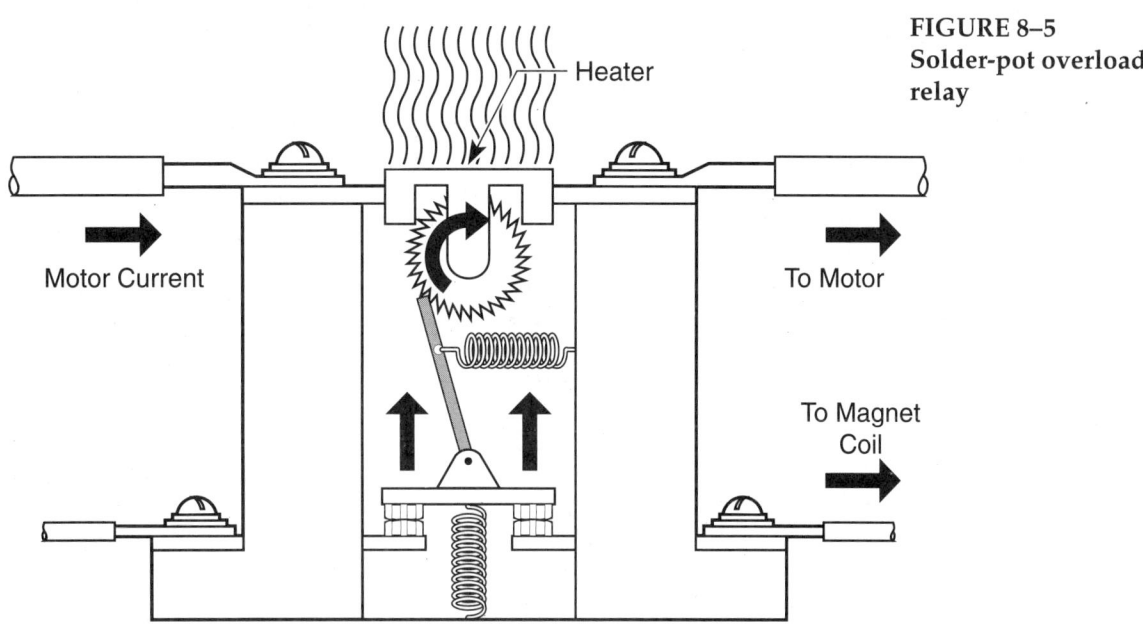

FIGURE 8–5
Solder-pot overload relay

FIGURE 8–6
AC magnetic starter with contact arcing chamber removed. Note: overload relay (lower left) and coil and magnet (lower center). (Courtesy of Square D Company)

Bimetallic Strip Relays The bimetallic strip type of relay is also a heat-sensitive, current-actuated relay, Figure 8–7. Motor current passes through the removable heater coil. Normal load currents do not generate enough heat to cause operation. When an overload occurs, the excessive current creates enough heat to cause the bimetallic strip to bend. This bending action causes the contacts to snap open, deenergizing the motor circuit.

After the relay has cooled sufficiently, the reset lever may be depressed, reclosing the contacts. Some bimetallic relays are automatic reset units. These reset table units should never be used in automatic starter circuits because of safety considerations.

Magnetic Overload Relays The magnetic overload relay, Figures 8–8 and 8–9, uses a solenoid and a dashpot assembly to provide the time delay. When the motor current rises above the solenoid set point, the piston moves through the oil. The viscosity of the oil slows the operation of the unit, resulting in a time delayed operation.

The magnetic overload relays can be adjusted for different set points by changing the spring tension on the solenoid. Different operating times are developed using different viscosity oils for the dashpot mechanism.

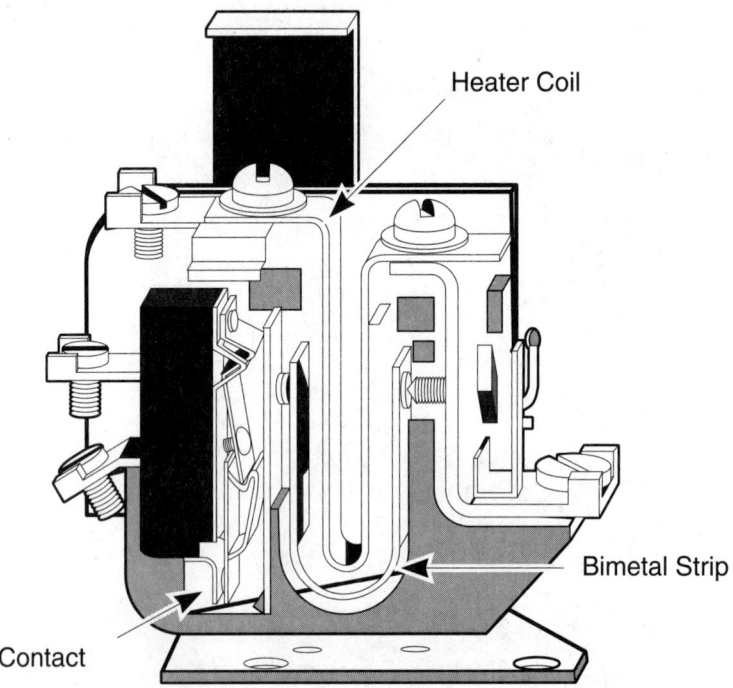

FIGURE 8–7
Bimetallic overload relay

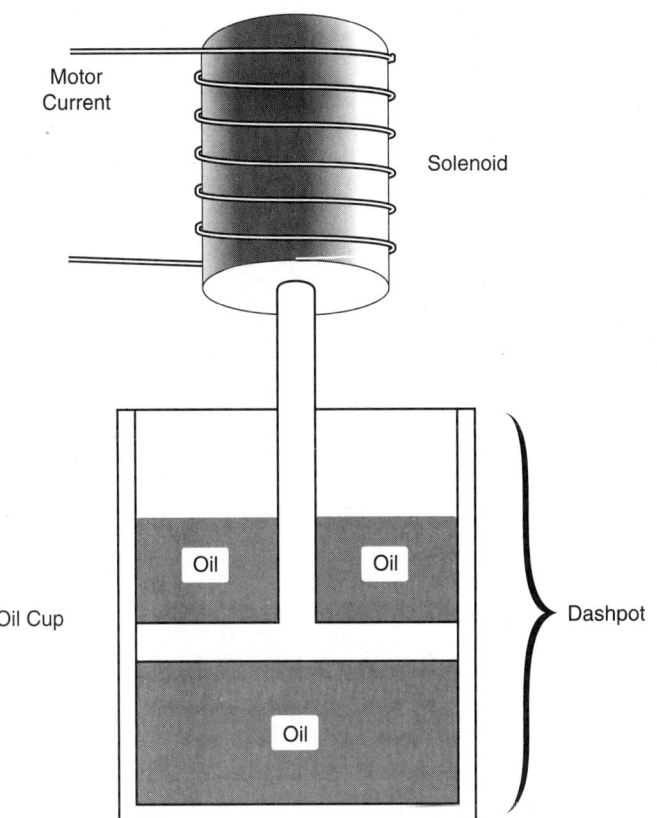

**FIGURE 8–8
Operating principle of magnetic overload relay (Courtesy of Cadick Professional Services)**

Some magnetic overloads use a bellows-type system for time delay. This is an older technique that is not used often in modern systems.

Molded-Case Circuit Breakers Molded-case circuit breakers, Figure 8–10, are low-voltage devices available in three basic types—the magnetic only, thermal-magnetic, and molded-case switch. The magnetic only, also called a motor circuit protector or MCP, has a magnetic instantaneous element used to provide short circuit protection for the motor circuit. When a short circuit occurs that is sufficient to operate the magnetic element, it strikes the trip bar and causes the breaker to open instantaneously.

The thermal-magnetic breaker has a bimetallic strip overload element as well as the magnetic element. The thermal portion of this breaker may be used to provide motor overload protection. Usually, however, an overload relay, as described previously, is employed for this purpose.

FIGURE 8–9
Magnetic overload relay (Courtesy of Square D Company)

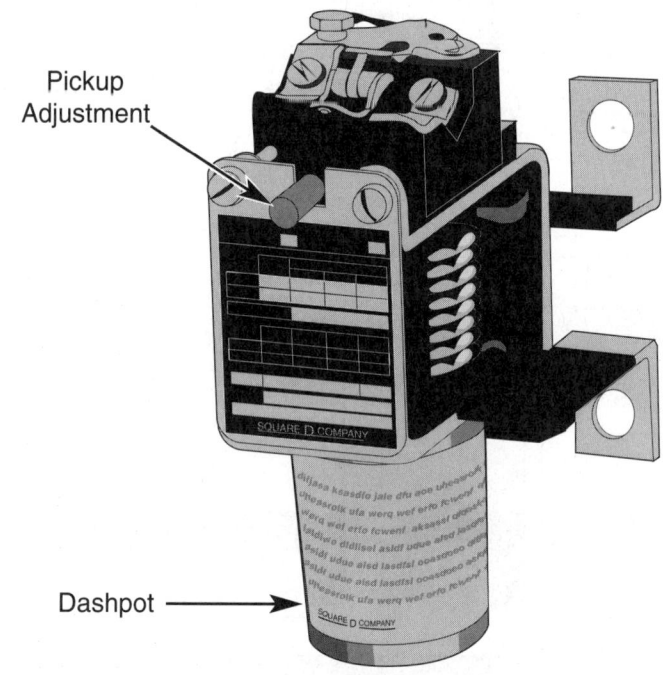

The molded case switch has no protective elements, and is used solely for disconnect purposes.

Induction Disk Overcurrent Relays All of the devices discussed previously are used in low-voltage applications. For medium-voltage applications and some critical low-voltage installations, the induction disk overcurrent relay is used to provide the overload protection.

Figure 8–11 is a drawing of a typical induction disk overcurrent relay. This relay uses a shaded pole type of movement with an aluminum disk for the rotor. It is similar in operation to a single-phase watthour meter. The two fluxes that are required to turn the rotor are developed using a shaded pole on the operating element. The spiral spring, coil taps, time dial, and damping magnet provide for a degree of adjustability exceeding that of all the devices previously discussed.

Figure 8–12 is a pictorial diagram of how an induction disk overcurrent relay is connected to protect a motor circuit. A current transformer applies the motor load current to the relay induction coil, which develops operating torque on the disk. When the current induced torque exceeds the restraint applied by the spiral spring, the disk rotates. The greater the current, the faster the rotation. When the contacts close, the circuit breaker trip circuit is completed through the battery, fuse, OCB trip coil (TC), and the

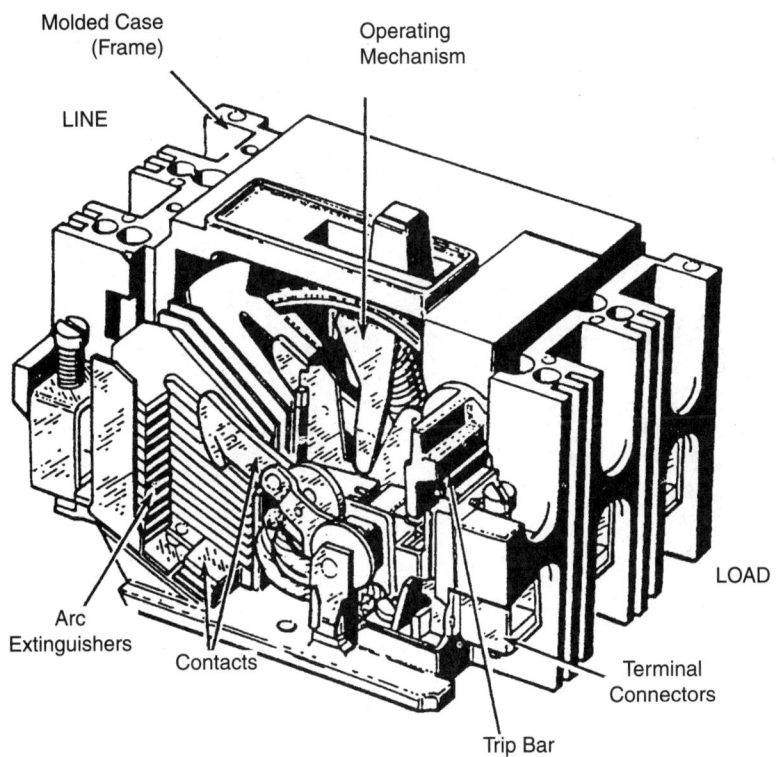

FIGURE 8–10
Principal components of a molded-case breaker

breaker *a* switch. The breaker opens, and the *a* switch opens, deenergizing the circuit. The spiral spring then returns the disk to its reset position.

The damping magnet applies a braking action to the disk which is used to control the operating time, and to limit wild swings of the disk when momentary overload currents are applied.

Electronic Protection Modern electronic relays are used to provide complete protection for motor circuits. The relay shown in Figure 8–13 provides complete overload and short circuit protection for all three phases. In addition, it can be configured to provide protection for undervoltage, abnormal power factor, phase rotation, and reverse power. It can be connected to a computer through a communications channel, so that settings and operating conditions can be monitored and changed remotely.

DIFFERENTIAL RELAY PROTECTION

Internal faults in large motors must be cleared as quickly as possible. Differential protection for generators was described in Chapter 3 and the con-

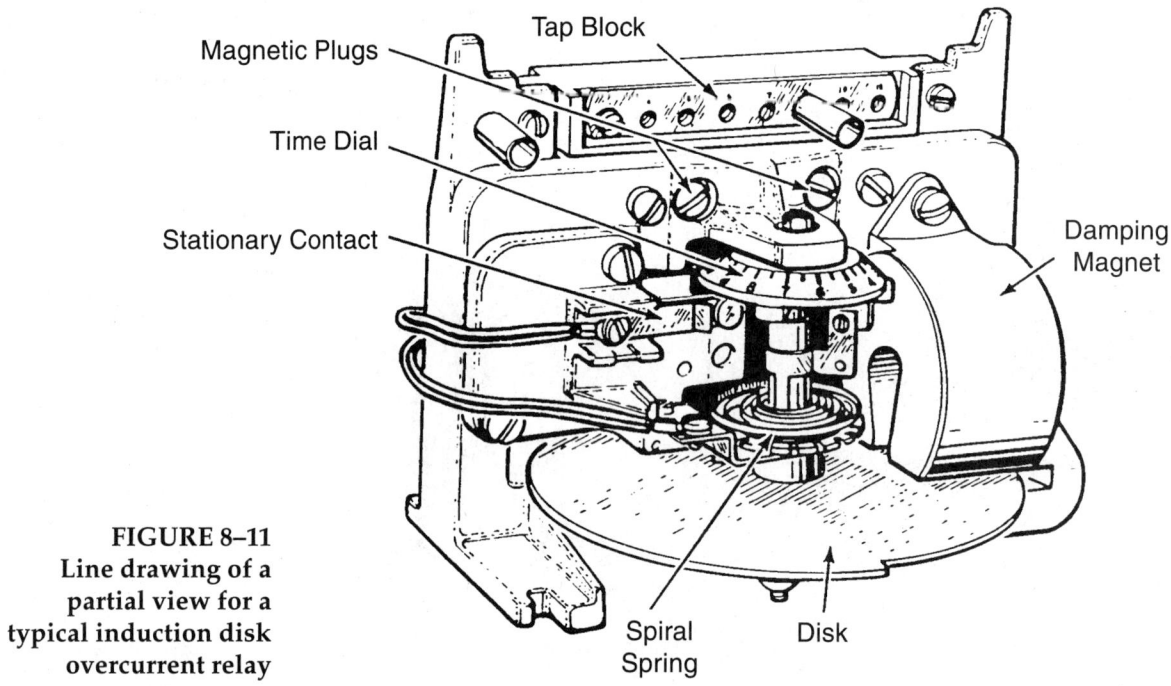

FIGURE 8–11 Line drawing of a partial view for a typical induction disk overcurrent relay

cepts and applications to AC motors are identical, Figure 8–14. As with all differential relays, an internal short circuit generates operating current, Figure 8–15, and an external short circuit does not, Figure 8–16. Notice that a motor *does* generate current into an external short circuit. When a short circuit occurs on the output of the motor, the motor "turns into" a generator for a brief time period. Because of this, the motor differential scheme operates exactly like the generator differential schemes.

The relays shown in Figures 8–14, 8–15, and 8–16 are of the percent slope variety. Such a relay will only operate when the current through the operating coil exceeds a certain percentage of the smaller of the two restraint windings. This requirement can be shown mathematically as

$$\frac{I_o}{I_{sr}} > \%\text{Slope} \tag{1}$$

The percent slope is a design characteristic of the particular relay being used. The advantage of this system is the same for motors as it was for generators. That is, small unbalances in the current transformers will not cause the differential relays to misoperate.

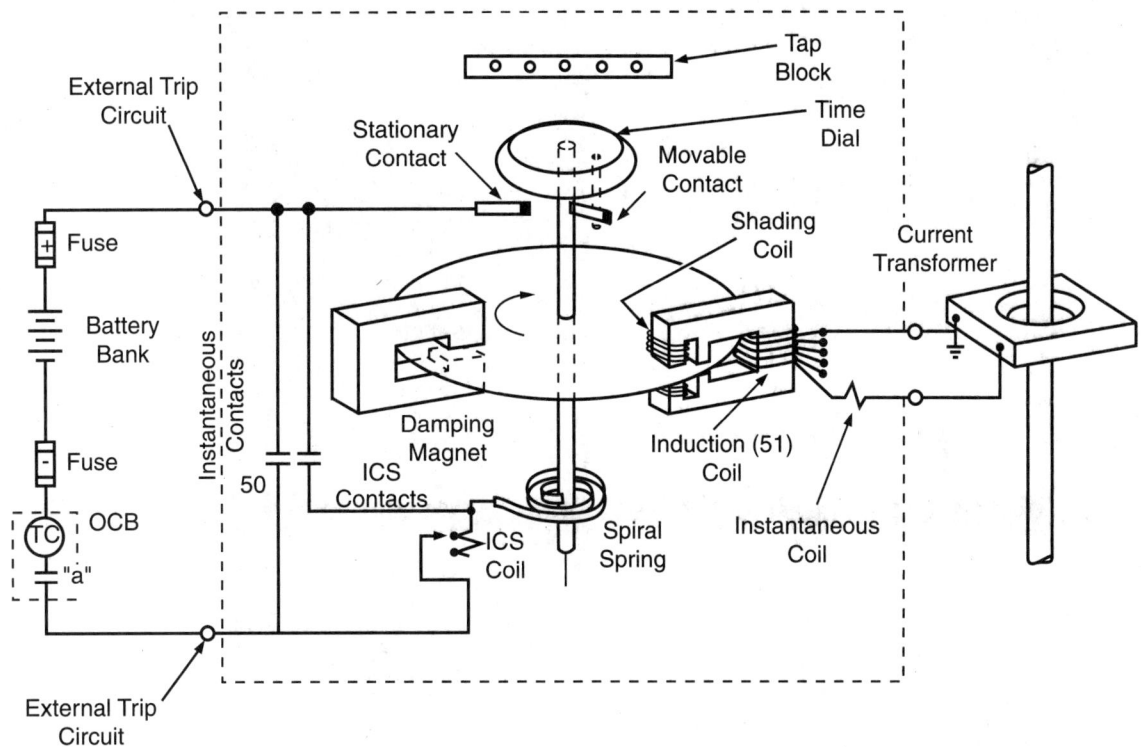

FIGURE 8–12
Typical induction disk overcurrent relay

Figure 8–17 is a diagram of a self-balancing differential scheme. Faults that occur inside the motor, or on the motor leads below the current transformers, will cause a net magnetic field to exist inside the current transformers. Faults above the current transformers will cause the two currents to go in opposite directions in the current transformers. The magnetic fields will cancel in the current transformers, and no operation will occur.

STATOR WINDING OVERTEMPERATURE

Larger motors may be protected using resistance temperature devices (RTD). Six or more RTDs are buried in the stator windings and each is connected to a bridge type relay, Figure 8–18. When the stator temperature rises above the preset value, the RTD resistance changes sufficiently to unbalance the bridge circuit in the relay. The relay operates and alarms.

FIGURE 8–13
A modern electronic motor protection relay (Courtesy of Multilin, Inc. Division of Derlan Manufacturing, Inc.)

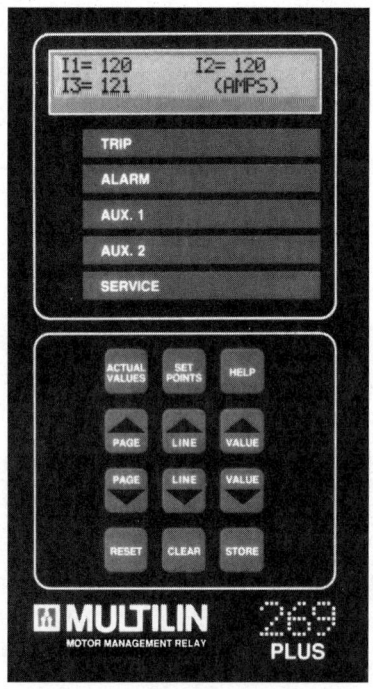

FIGURE 8–14
Motor differential relay system

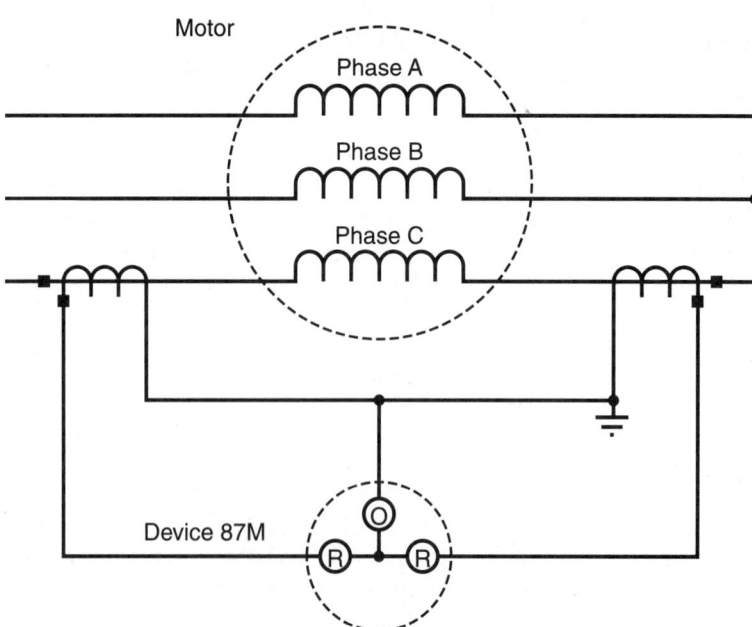

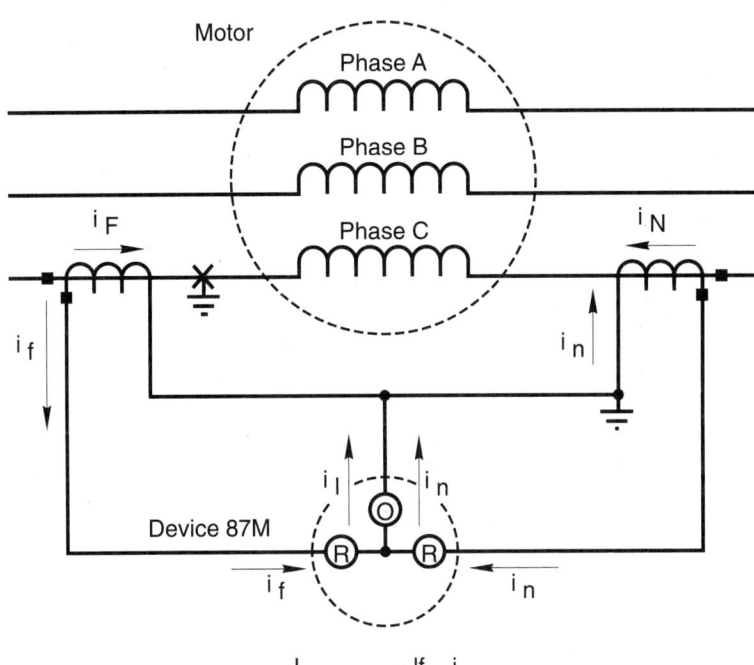

**FIGURE 8–15
Differential relay system with an internal short circuit**

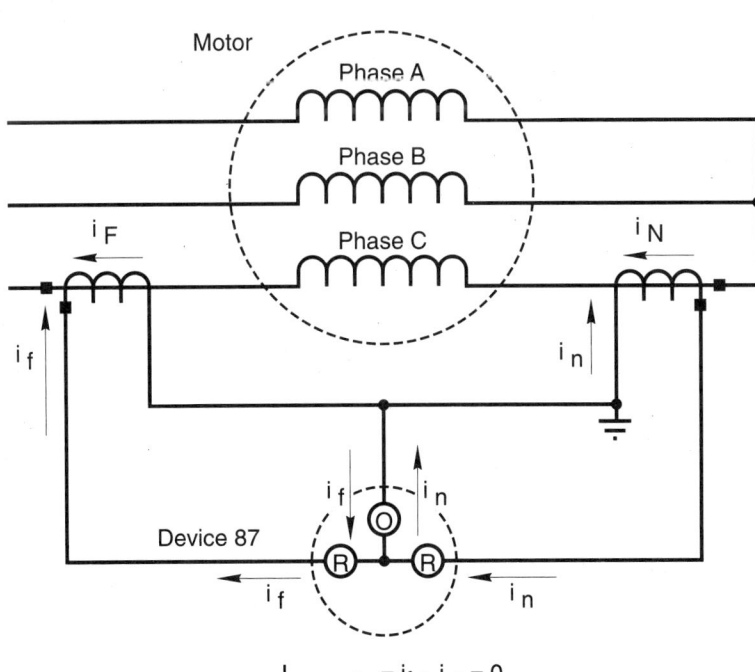

**FIGURE 8–16
Differential relay system with an external short circuit**

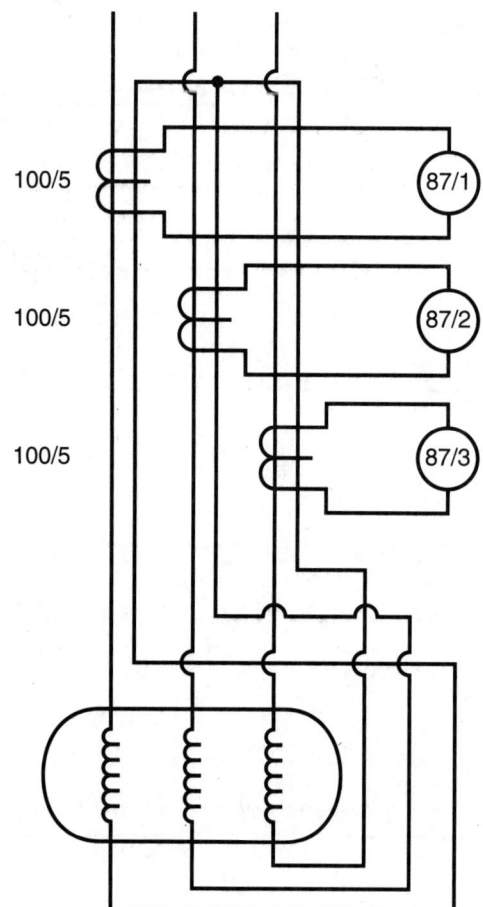

**FIGURE 8–17
Self-balancing motor
differential scheme**
(Courtesy of IEEE Std 242-1986,
IEEE Recommended Practices for
Protection and Coordination
of Industrial and Commercial
Power Systems)

OTHER TYPES OF MOTOR PROTECTION

Motors and generators are subject to many of the same types of problems, including negative sequence overheating and loss of excitation. Of course, loss of excitation applies only to synchronous motors. Motors and generators must be protected in the same way for these types of problems.

Protection

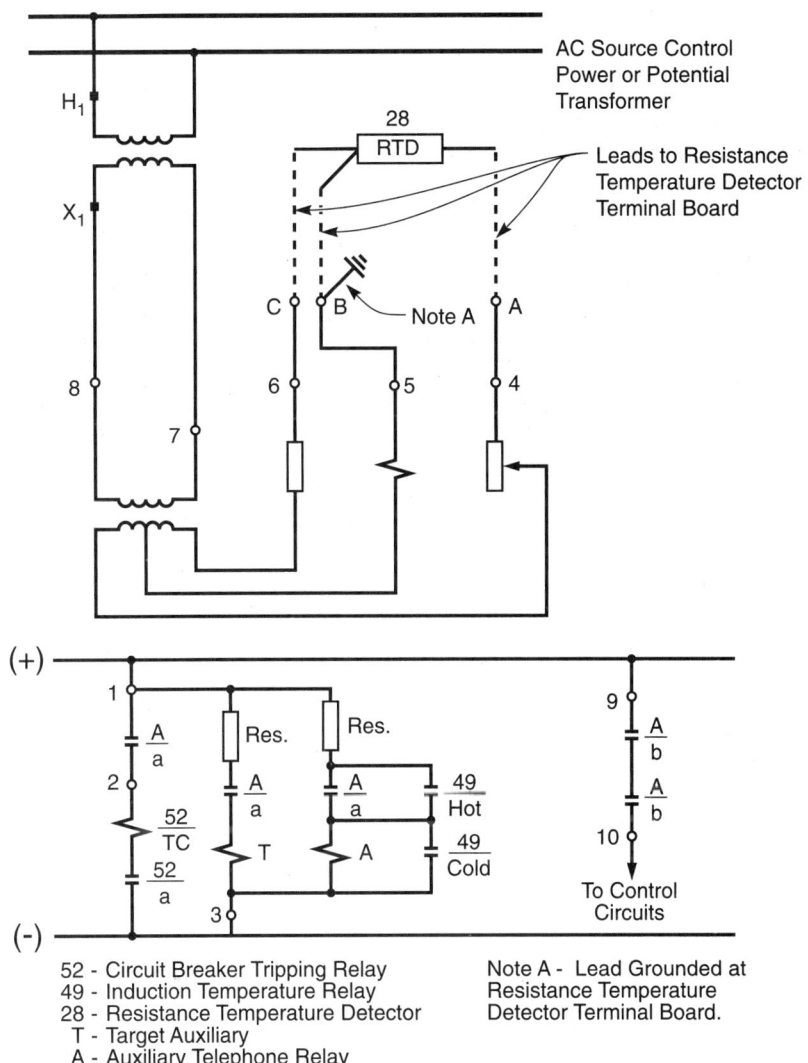

FIGURE 8–18
IRT connection diagram (Courtesy of GE Protection and Control, Malvern, PA)

SUMMARY

The two principal sources of motor damage are overloads and short circuits. Overloads are caused by drawing excessive horsepower from the motor. Short circuits are caused by insulation failure, lightning, and other such problems. Both overloads and short circuits are characterized by excessive current, overtemperature, or both.

Overcurrent devices can be used to sense both overloads and short circuits. When applied, overcurrent devices must be capable of sensing the minimum abnormal condition; however, they must not trip on normal load currents. Generally, overcurrent devices must allow locked rotor currents, and full load currents, while tripping on overcurrents or short circuits.

Many different types of overcurrent devices exist. Some, such as the melting alloy or bimetallic type, work on the basis of a thermal element. Others use solenoids and dashpots, or induction disk movements, to develop their protective characteristics. All of these devices are applied to protect motors from overcurrent conditions.

Motors are subject to many of the same protection requirements as generators, including differential, stator temperature protection, negative sequence, and loss of excitation. Motors and generators are protected in the same way for these types of problems.

REVIEW QUESTIONS

1. List three conditions that may cause motors to be damaged.
2. Overcurrents are classified as either _____ or _____.
3. Which of the following would be classified as an overload condition?
   a. 1000 lbs on a 500-lb conveyor.
   b. A wrench dropped across a 480-V bus.
   c. A liquid becomes more viscous than a pump design.
4. Draw a diagram that illustrates the proper application of motor overcurrent protection.
5. A 40-hp rotor has a service factor of 1.0 and a temperature rise of 40°C. What is the maximum pickup setting allowed for this motor's overload protection?
6. The motor in question 5 is a squirrel cage induction motor with an across-the-line starter. The motor is a code letter F. What is the maximum instantaneous trip circuit breaker that may be used for short circuit protection on this motor? (NOTE: The *NEC®* does have exceptions to the table shown as Table 8–3 in this text. Also, the breaker must be adjustable, and a part of a combination starter, to be used in this application.)
7. What is the major difference between a current limiting fuse and the one-time fuse?
8. How is a thermal overload relay "adjusted" to allow for the different motors sizes that a given starter can control?
9. In selecting protection for a 6600-V motor, would a thermal overload relay be a good choice for such a motor? Explain.
10. A differential relay has a percent slope of 10%. An external fault occurs, and the current transformers feeding the relay produce slightly different output currents. $i_l = 10$ A and $i_n = 10.5$ A (Figure 8–17). If the relay pickup is .25 A, will it operate?

AC Motor Maintenance and Troubleshooting   9

> **WARNING!!!**
> *Many of the procedures described in this chapter* **employ** *lethal voltages and/or require that the motor be energized. This material is intended for training purposes and should not be used as the basis for the development of step-by-step field procedures. For more detail, refer to manufacturer's technical literature and other industry references. Be certain to use all appropriate safety equipment and procedures.*

OBJECTIVES

After studying this chapter, the student should be able to:

- *Describe the types of visual and mechanical maintenance checks and procedures that should be performed on an AC motor.*
- *Describe the types of electrical maintenance tests that should be performed on AC electric motors.*
- *Explain the results expected from the electrical tests on AC electric motors.*
- *Describe the common types of problems that can occur on AC electric motors and possible causes of those problems.*
- *Describe the types of troubleshooting procedures that can be used for AC electric motors.*

INTRODUCTION

Motors must be tested and evaluated periodically to determine their condition and to identify minor problems before they develop into major problems. In this chapter, you will learn about the types of checks, procedures, and tests that should be performed on AC motors.

AC motors and generators are virtually identical in terms of physical construction. Because of this, all AC generator tests are identical in application and expected results when they are performed on AC motors. Please note that many of the tests described in this chapter also apply to AC generators. Some of the tests and inspections also described in this chapter apply only to larger frame motors. Items such as the air gap and magnetic center tests are not performed on small motors in which the rotor and stator are factory assembled.

VISUAL AND MECHANICAL MAINTENANCE

Physical Inspection

Damage Inspect the motor for signs of physical damage. Look for dents, burnt spots, frayed insulation, oil leaks, and other such problems. When possible, turn the rotor slowly by hand to find obvious rough spots and/or rubs.

Anchorage and Mounting Inspect the motor mounting equipment to make certain it is adequately installed. The normal operating vibration of some motors may loosen the mounting hardware over time. Check the mountings with a torque wrench to make certain they are properly installed.

Check the motor to make certain that it is level and properly aligned. Even minor variations can cause long acceleration times and excessive wear on bearings.

Grounding Verify that the motor frame is properly grounded, and the grounding conductors and hardware are undamaged and properly installed. ***This is an extremely important safety step!!*** The ground obtained through the mounting hardware and assemblies is usually not sufficient for personnel protection. Motor frames should be solidly grounded and bonded to other electrical equipment, including the starter frame and/or any metallic conduits.

Connections Carefully inspect the motor connection points. Inspections should involve two important steps:

- Infrared Scan—With the motor running and fully loaded, all connections should be inspected using an infrared scan camera. Remake any connections that are hotter than the cables to which they are connected.
- Torque Wrench—Use a torque wrench to tighten all connections, especially the bad ones that have shown up in the infrared scan.

Hot connections account for a substantial number of failures—many of them catastrophic. Be certain that surge arrester connections are never used for connections, Figure 9–1. The connection shown in Figure 9–1a uses the surge arrester connection flange to carry motor current. The flange is not rated for such use and will overheat. Figure 9–1b shows the proper way to make such a connection.

AC Motor Maintenance and Troubleshooting

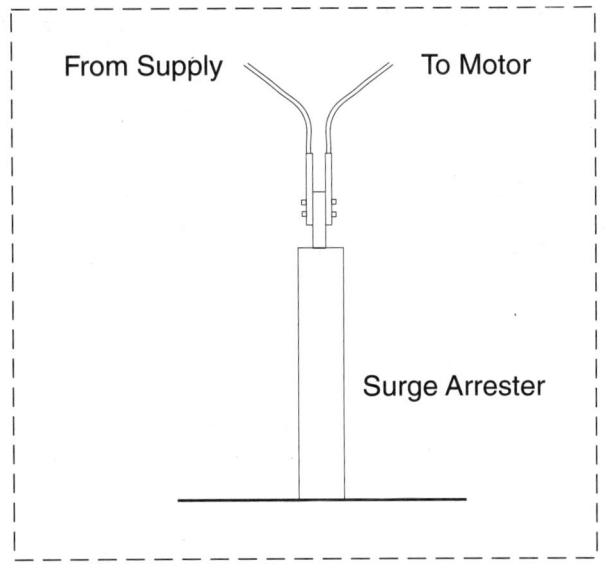

a. Wrong

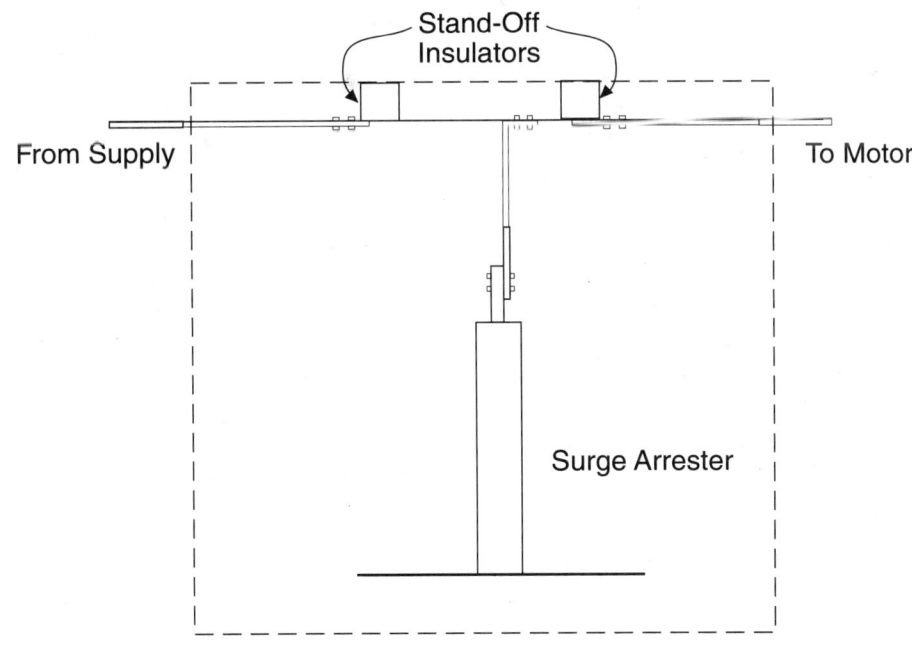

b. Correct

FIGURE 9–1
Surge arrester connections (Courtesy of Cadick Professional Services)

Lubrication Check the motor bearings to ensure that proper lubricant is present in the proper amounts. Some journal bearings are lubricated with oil rings, Figure 9–2. The oil rings are larger in diameter than the shaft, and are suspended from the shaft into a reservoir of oil. As the shaft turns, the rings also rotate, carrying oil from the reservoir to the bearings. You may need to disassemble the bearing to inspect it and ensure that the oil ring is free to move.

Oil reservoirs should be filled to the proper level with the proper oil weight and grade. The shaft must be level to ensure proper lubrication.

Ventilation Make certain all ventilation ports are clear of foreign materials that may inhibit the movement of cooling air.

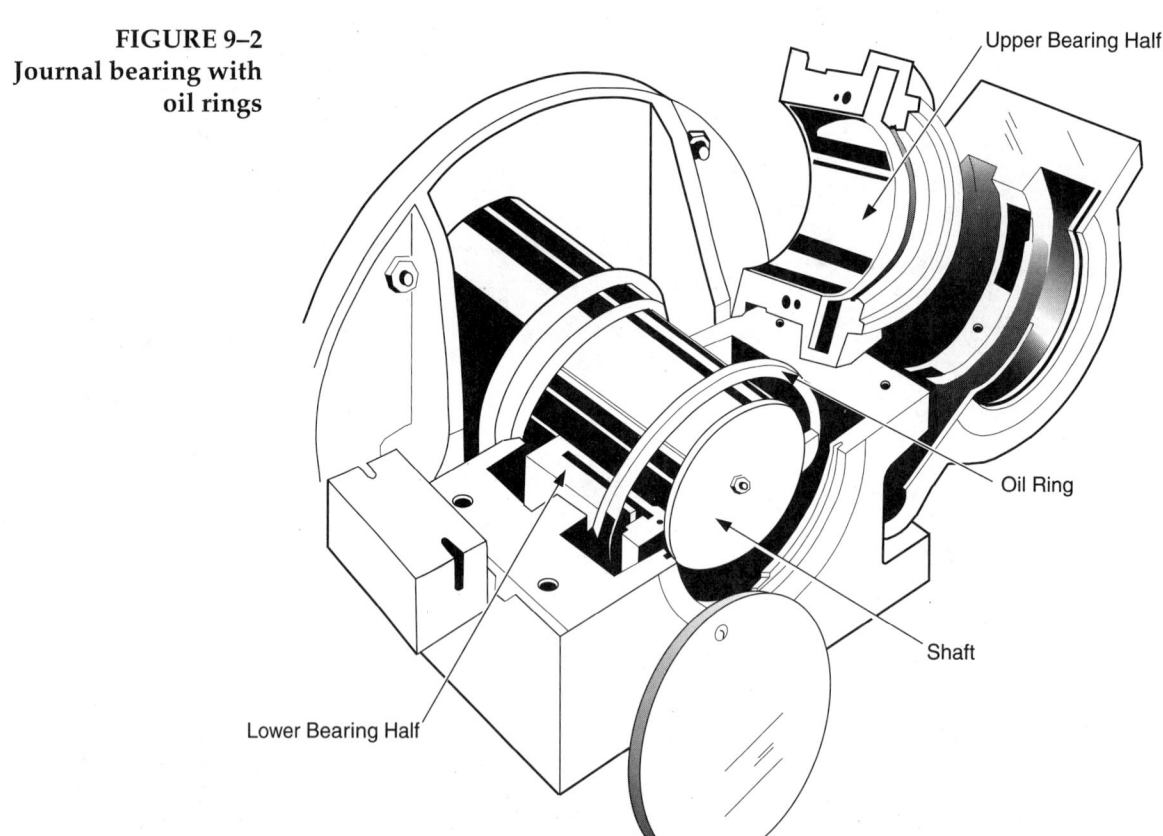

FIGURE 9–2 Journal bearing with oil rings

Rotor Air Gap

Check the rotor air gap, Figure 9–3, to make certain the rotor is properly positioned in the stator. The air gap is defined as the radial, iron-to-iron clearance between the center of a pole face and the face of a core tooth. It is measured using a tapered air gap gauge or a long feeler gauge. Check the gap at a minimum of four locations (1, 2, 3, and 4 in Figure 9–3). The readings should fall within manufacturer's specifications.

Some motors are more difficult to check than others. Refer to manufacturer's instructions for detailed procedures.

Magnetic Center

A certain amount of shaft end play is usually present when the motor is at rest. In operation, however, the rotor aligns itself magnetically within the stator. The position to which the rotor will move during operation is called the magnetic center. The motor must be installed so that no load is placed on the thrust surfaces during operation. This procedure is performed before the motor is connected to the load shaft.

On motors larger than 250 hp, manufacturers often provide markings on the rotor and stator so the magnetic center can be aligned with the stator frame. Most smaller motors do not have such markings. For motors with manufacturer's markings, the rotor and the stator are aligned so that the markings coincide. For motors without such markings, the following procedure may be used (see Figure 9–4):

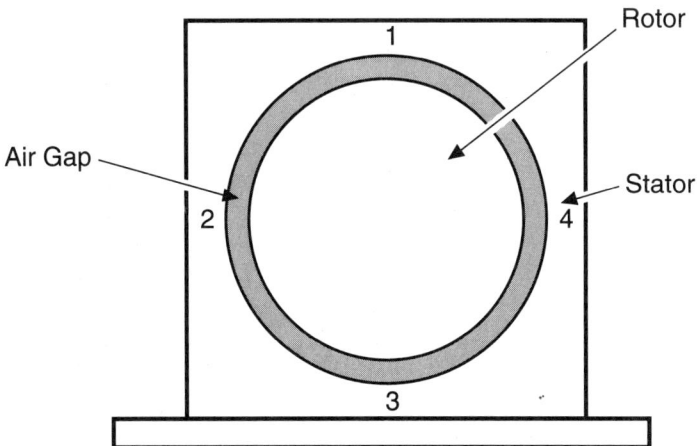

FIGURE 9–3 Measuring the rotor air gap (Courtesy of Cadick Professional Services)

FIGURE 9–4 Locating magnetic center (Courtesy of Cadick Professional Services)

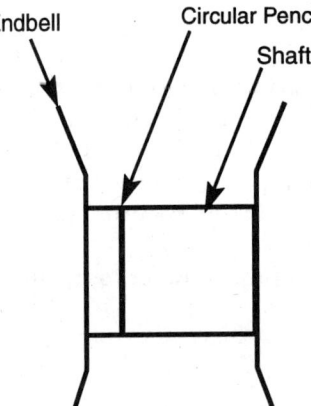

a. Motor Stationary

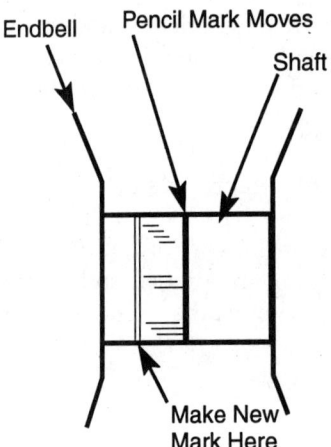

b. Motor Running

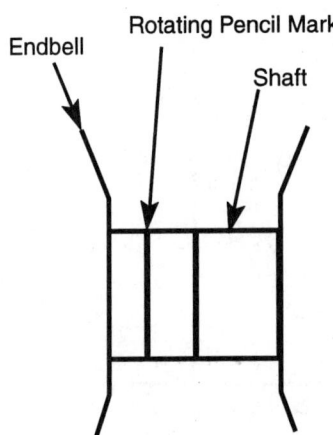

c. Motor Stationary

Rotor is repositioned so that the mark made with motor running falls on the mark with motor stationary.

AC Motor Maintenance and Troubleshooting

1. Prepare the motor for operation and place pencil marks at convenient locations on the rotor and the stator.
2. Start the motor and *carefully* note how far the rotor moved.
3. Stop the motor and reposition the rotor so the rotor magnetic center mark (made when the motor was running) aligns with the stator mark placed earlier.

ELECTRICAL TESTS ON AC INDUCTION MOTORS

Insulation Tests

All of the tests described in Chapter 4 for AC generators also apply to AC induction motors. Review Chapter 4 for that information.

Insulation Resistance of Surge Protectors

Surge protectors, also called lightning arresters, are very important for protecting a generator's insulation system from lightning and other voltage surges. Arresters should be tested periodically using a megohmmeter of suitable voltage—typically 1000 V or higher. Figure 9–5 shows connections that may be used to check a three-section surge protector.

Insulation resistance readings will vary depending on the type of the arrester. Some units may be several thousand megohms or higher, while oth-

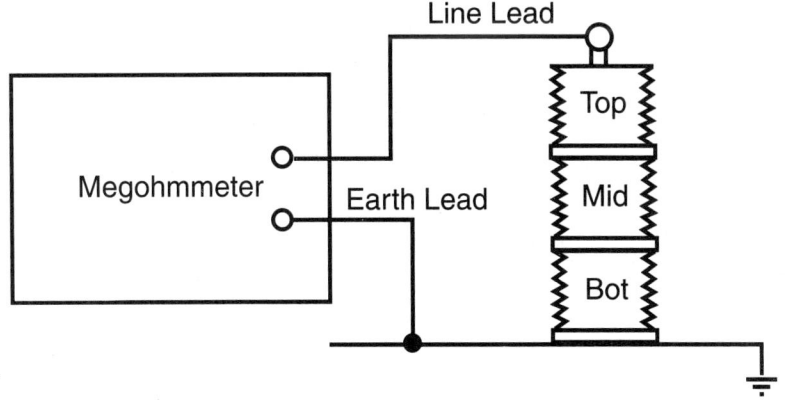

FIGURE 9–5
Insulation resistance test for a three-segment lightning arrester (Courtesy of Cadick Professional Services)

- To check top section - jumper mid and bot
- To check mid section - jumper top and bot
- To check bot section - jumper top and mid

ers may reach only 500 megohms. Evaluation of readings on any given arrester should be made based on previous readings taken on similar arresters.

Motor Starter Tests

Check the motor starter to ensure that it properly controls the motor. Start, stop, and reversing switches should properly control the motor. If the starter fails, troubleshooting techniques similar to those described in Chapter 8 may be employed.

Resistance Temperature Detectors

Check RTD circuits for continuity and proper grounding. If an accurate temperature reading of the RTD is available, measure the resistance and compare it to the manufacturer's temperature/resistance curve.

Differential Relay Checks

With the motor operating, measure the current in the operating coil of the motor differential relays. Remember, the current should be zero if the relay is properly connected. Read the current by using an AC ammeter inserted into the differential relays measuring terminals. Refer to the relay manufacturer's literature for proper measuring techniques.

> *CAUTION:*
> *This measurement is made in an active, energized current transformer circuit. Observe all appropriate safety precautions when performing it.*

Motor Space Heaters

Some motor enclosures are equipped with electric space heaters to minimize humidity and prevent condensation. These heaters may be on manual or thermostatic control. In either case, check the space heaters to ensure they operate properly. Energize the heater circuits and check the heaters for proper operation. Check thermostats for proper calibration.

Protective Devices

Test Equipment   The testing of motor protective devices requires the use of specialized equipment made for that purpose. Figures 9–6, 9–7, and 9–8 show the types of equipment used for testing protective relays.

The motor overload relay test set, Figure 9–6, provides a variable current supply and an integrated timing circuit. The motor overload heater circuit

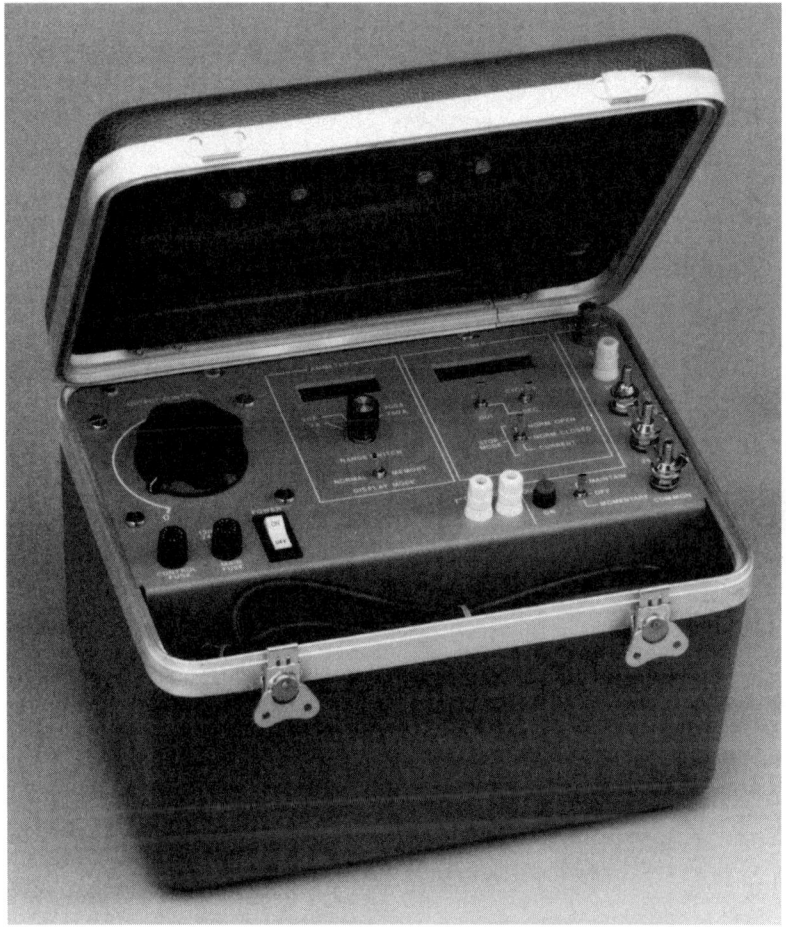

FIGURE 9–6
Motor overload relay test set (Multi-Amp® MS-2)

is connected to the current supply and is energized with a multiple of the motor full load current. The relay contacts are connected to the motor timer circuit. The timer starts when the current is initiated, and stops when the relay contacts open. The operating time is compared to published curves supplied by the manufacturer, and the relay is repaired or replaced if required.

    The universal relay test set, Figure 9–7, is used to test a wide variety of electromechanical and electronic protective relays. Overcurrent relays, differential relays, and undervoltage relays are among the types that may be evaluated. The universal relay test set has several voltage and current outputs that may be adjusted independently for any magnitude. This allows the testing of many different relay types.

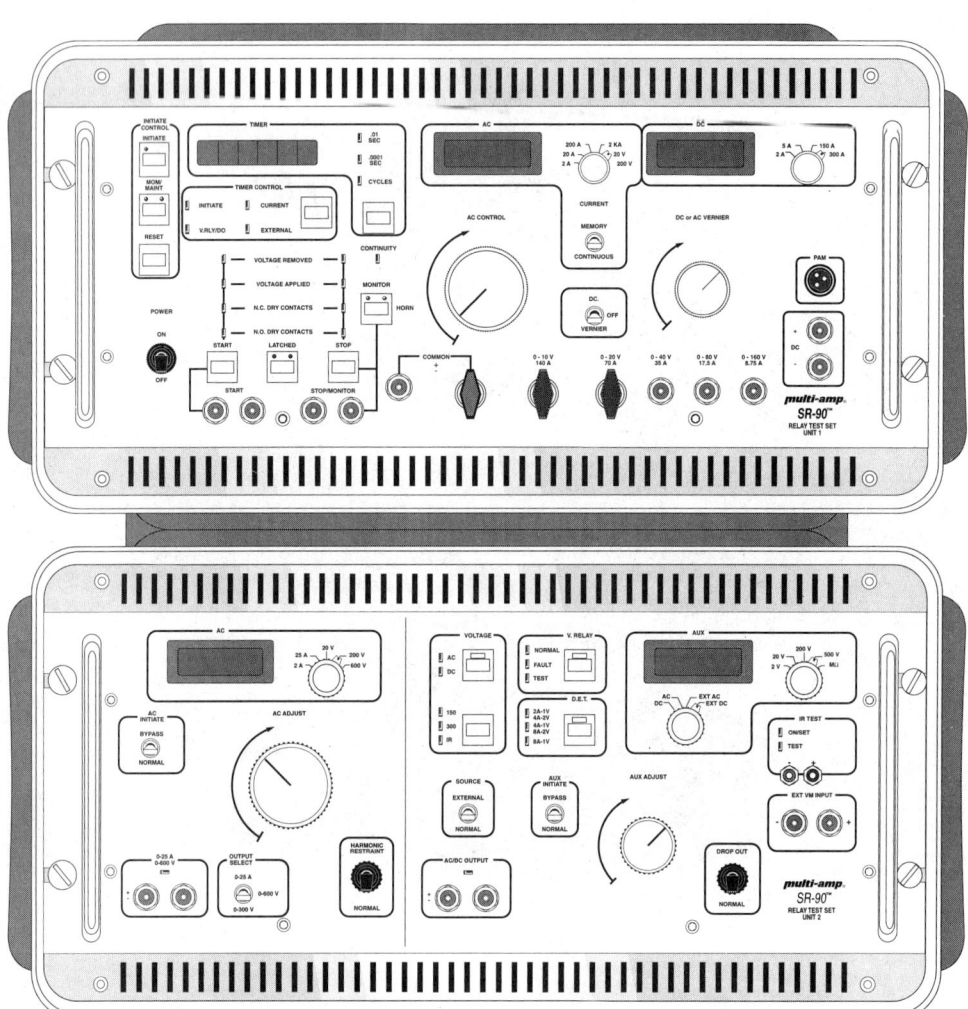

FIGURE 9–7
Universal protective relay test set (Multi-Amp® SR-90)

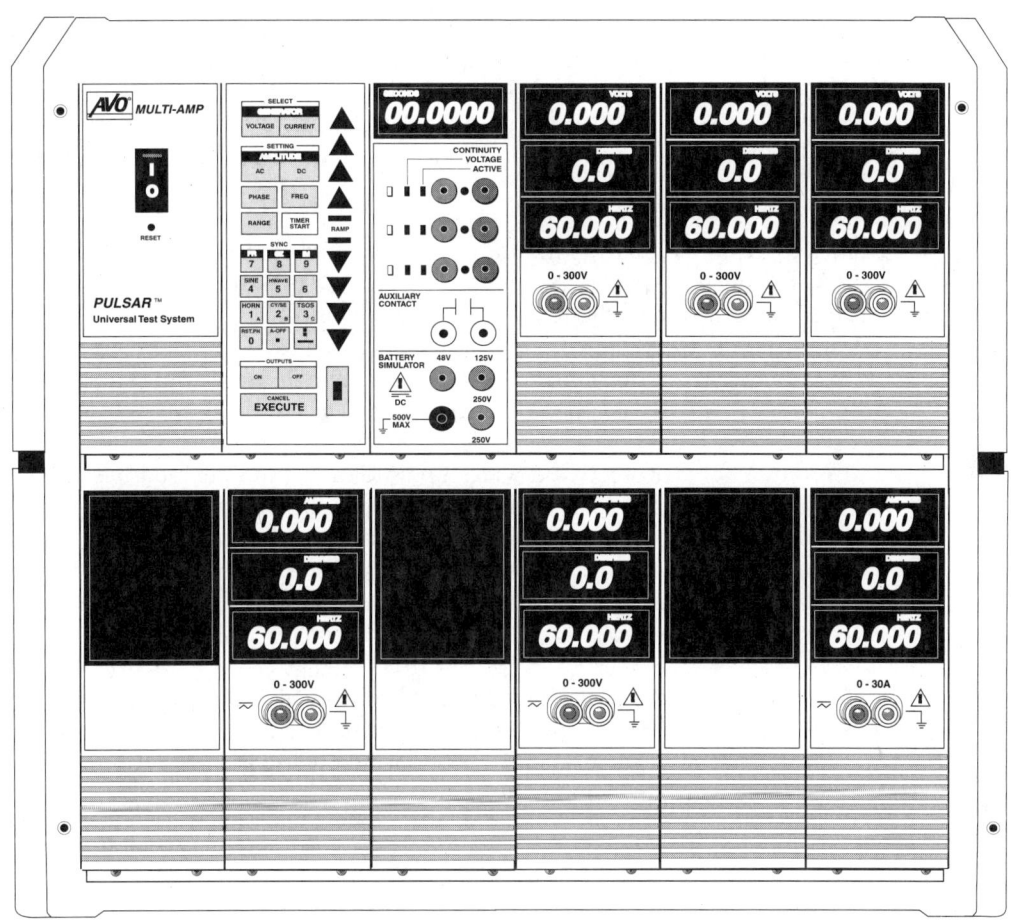

FIGURE 9–8
Modern relay test set suitable for computer-controlled testing of protective relays (Multi-Amp® Pulsar™)

Figure 9–8 shows one of the most modern of all relay test sets. This unit has the ability to provide three-phase voltage, three-phase current, and direct current in virtually any practical combination. Its outputs are digitally controlled in both magnitude and phase angle. This unit can be computer-controlled with software which allows testing virtually any protective relay.

Pickup Tests Pickup is the minimum amount of current or voltage required to operate a protective device. To test the pickup value, the test set output is connected to the relay operating coil and the continuity indicator is connected to the relay contacts circuit, Figure 9–9. Current or voltage is gradually increased until the minimum value which causes operation is determined. The device is adjusted or replaced as needed.

Timing Many protective devices operate on a time delayed basis. Figure 9–10 shows a connection for evaluating their operating time. In this test, the required test current or voltage is suddenly applied, and the timer is started simultaneously. When the device contacts close, the timer stops, and the operating time is recorded. The device is then adjusted or replaced as required.

Insulation Resistance The insulation in protective devices is subject to failure, just as any other insulation. The insulation should, therefore, be tested periodically. The tests may be performed using the insulation testing procedures described previously.

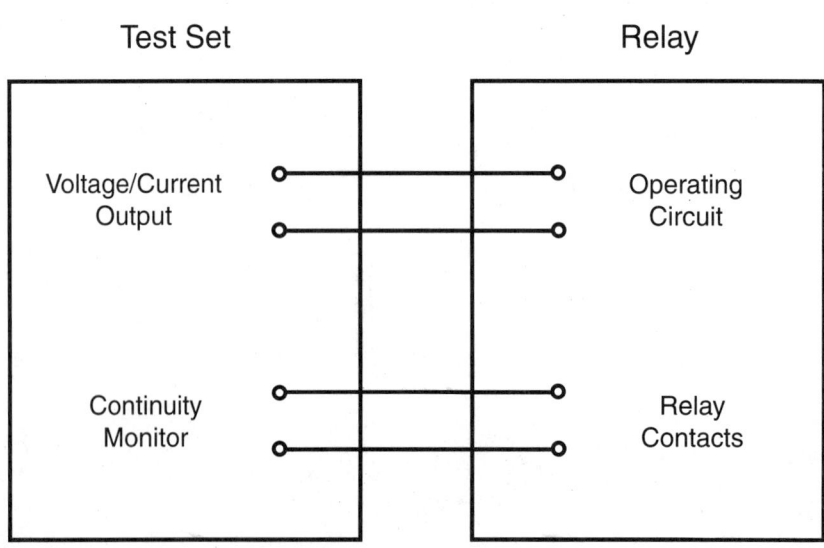

FIGURE 9–9 Connection diagram for pickup test (Courtesy of Cadick Professional Services)

AC Motor Maintenance and Troubleshooting

Special Tests Some relays have special testing requirements. For example, differential relays must be tested to make certain they operate properly when current is flowing in both circuits. Figure 9–11 shows a connection diagram used to conduct this test; it is called a slope test.

In this test, Test Current 2 is adjusted to a predetermined value—20 A, for example. Test Current 1 is then adjusted until the relay contacts just operate (somewhat like a pickup test). The ratio of the two currents is calculated and compared to the relay's nameplate percent slope characteristic. The relay is then adjusted or repaired as required.

Many other special tests may be performed on various types of relays. Always refer to the manufacturer's literature for specific information.

Rotation Test

The direction of rotation of a motor is critically important to the installation, and it is possible to install it incorrectly. Even large motors, which often have rotation direction arrows on them, can be incorrectly constructed. Because of this, the motor should be "bumped" when it is first installed. Bumping means momentarily energizing the motor so that its direction of rotation can be determined. If the rotation is incorrect, it may be reversed by one of the methods given below:

- Single-phase motors—Single-phase motors may be reversed by reversing the starting or the running windings. The connections of Figure 9–12 show the method.
- Three-phase motors—Three-phase motors may be reversed by reversing the connections to any two of the phases, Figure 9–13.

Note that when the direction of rotation is changed, cooling fans will move air in the opposite direction. It may be necessary to reverse the installation of the fan by removing it from the shaft and turning it over to reinstall it.

Running Current

The motor running current should be tested, this can be done with the motor loaded or unloaded. A clamp-on type of ammeter is employed for such readings. Single-phase motor currents should be verified to be within published tolerances. Three-phase motor magnitudes should be within tolerance and should be balanced within tolerance. If no manufacturing tolerances are given, the currents should vary by no more than ±10% from published specifications.

> *CAUTION: Taking this measurement requires that you approach energized conductors. This should not be done by anyone unfamiliar with such measurements. Always wear appropriate protective equipment and use appropriate safety procedures.*

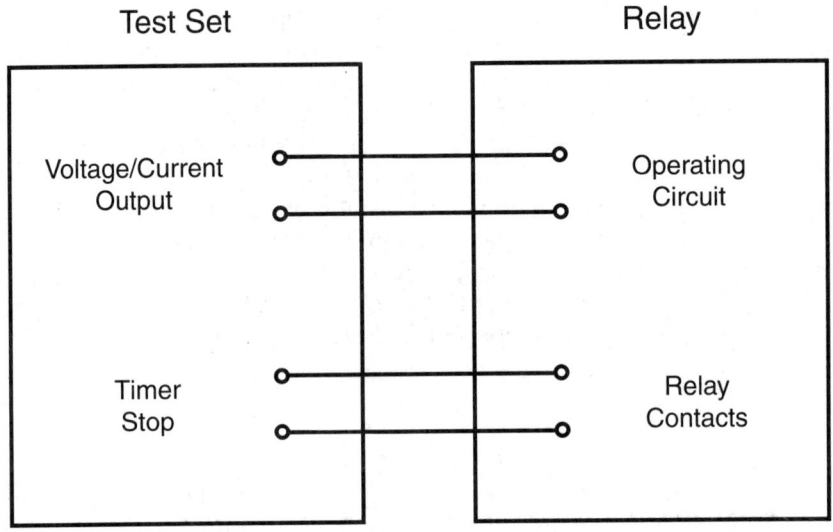

FIGURE 9–10 Connection diagram for a timing test (Courtesy of Cadick Professional Services)

Note: Timer starts automatically when test set is initiated

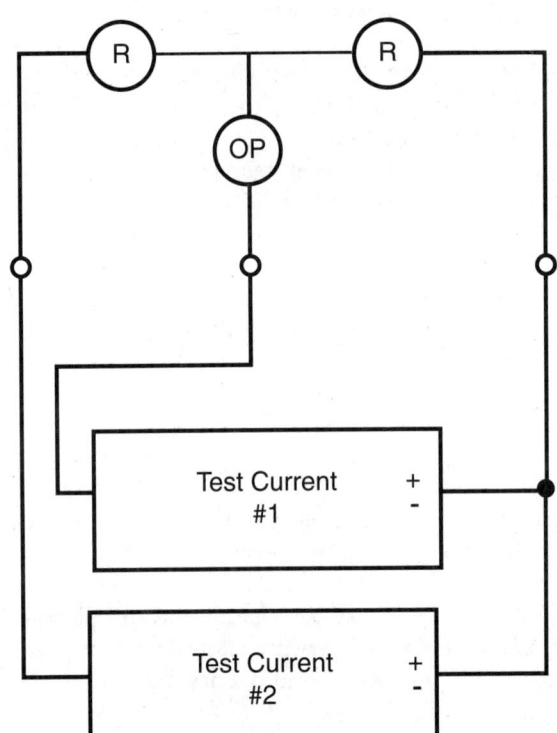

FIGURE 9–11 Test connections for differential relay slope test (Courtesy of Cadick Professional Services)

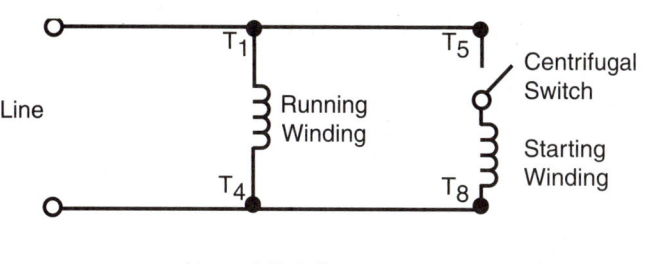

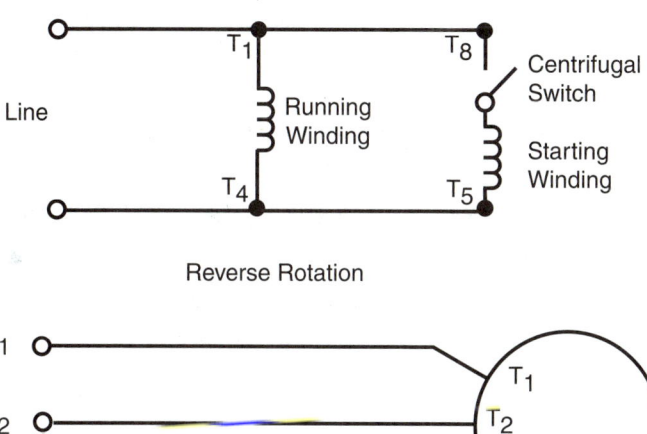

FIGURE 9–12
Reversing the rotation of a single-phase motor (Courtesy of Cadick Professional Services)

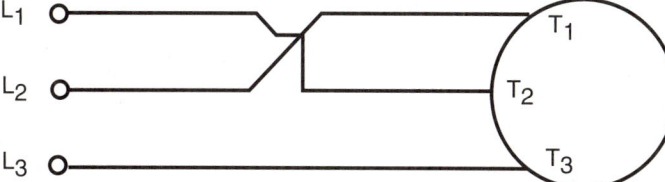

FIGURE 9–13
Reversing a three-phase motor (Courtesy of Cadick Professional Services)

Figure 9–14 shows the proper use of a clamp-on ammeter.

Vibration Tests Excessive vibration can cause a variety of problems in a motor. Vibration analysis equipment should be used to verify that vibration is within manufacturer's tolerances.

ELECTRICAL TESTS ON AC SYNCHRONOUS MOTORS

Overview

Synchronous motor stators are subject to the same battery of tests as the induction motor. There are, however, a few different tests that apply to synchronous motors. The following sections describe each of these tests.

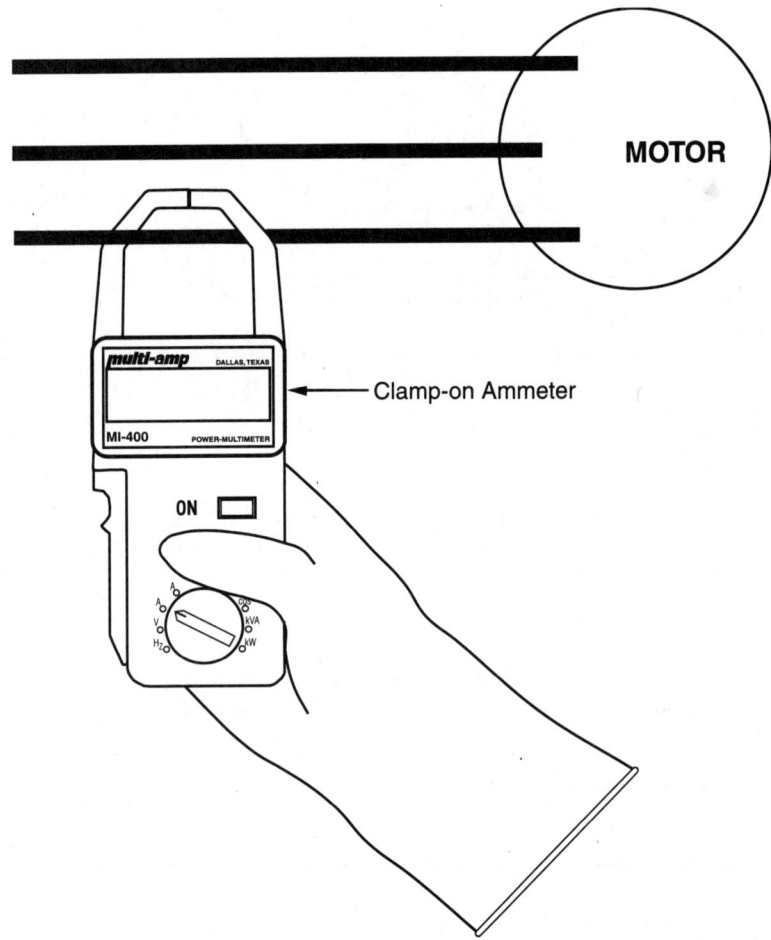

FIGURE 9–14 Using a clamp-on ammeter to measure motor running current

AC Motor Maintenance and Troubleshooting

Exciter Insulation Tests

The exciter winding insulation may be tested by using the circuit of Figure 9–15. The following steps must be taken before connecting the test set:

- Lift the brushes away from the slip rings.
- Ground the stator frame and the rotor shaft.

Field Application Relays

Synchronous motors are started as induction motors. After the rotor has reached a speed close to synchronous, the field application relays close and apply field to the motor, thus locking it into synchronous speed. The field application relays must be tested to make certain they close at the proper time and with the proper speed.

Power Factor Relays

Power factor relays are used to control the motor excitation and/or trip the motor off line in the event the power factor goes beyond the manufacturer's allowed value. Power factor relays can be tested using equipment

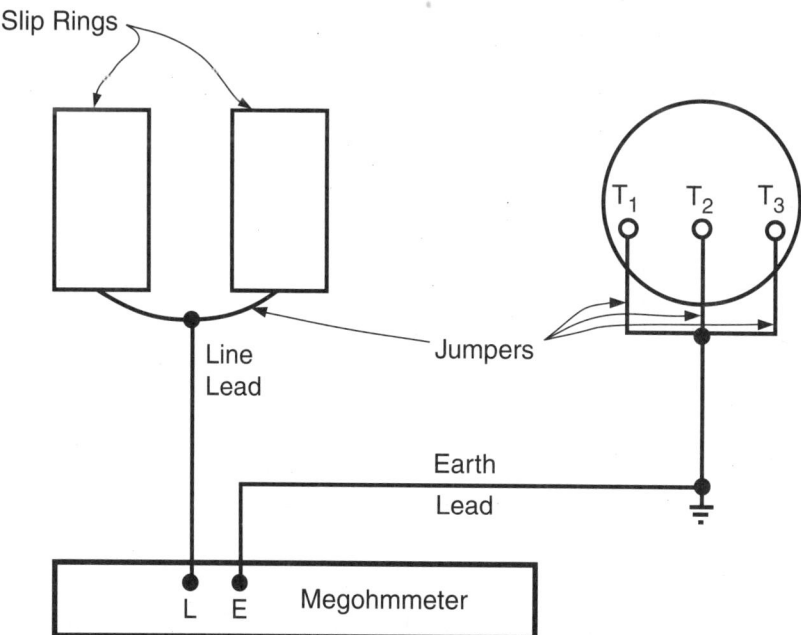

FIGURE 9–15 Insulation resistance tests on a synchronous motor exciter (Courtesy of Cadick Professional Services)

similar to that shown in Figure 9–8. Voltage and current are applied, and the phase angle is adjusted to determine the operating point of the relay. The relay is then repaired or adjusted as required.

TROUBLESHOOTING AC MOTORS

Trouble Symptoms and Causes

Several problems may cause misoperation of AC motors. The following sections identify some of these problems and their possible causes. Unless otherwise noted, the causes apply to both polyphase and single-phase motors.

Motor Fails to Start

Open supply line or fuse
Open start winding (single-phase motors)
Shorted or open starting capacitor (single-phase motors)
Open or shorting windings or coils
Grounded windings
Incorrect internal connections
Worn, damaged, or frozen bearings
Bent or misaligned shaft
Broken rotor bars (squirrel cage) or windings (wound rotor motor)
Inoperative starter

Motor Runs Slowly

Start winding stays in circuit (single-phase motors)
Run winding has a short (single-phase motors)
Reversed run winding coils (single-phase motors)
Worn or damaged bearings
Shorted winding
Reversed phase
Improper connections
Broken rotor bars (squirrel cage) or open rotor windings (wound rotor motor)

Motor Runs Hot

Short circuit between start winding and run winding (single-phase motors)
Shorted windings
Grounded windings
Loss of a single-phase (three-phase motors)
Worn bearings

> **CAUTION:** Many of the procedures in this section involve working on or around energized circuits. Be certain to employ all safety precautions, including the use of insulated clothing.

AC Motor Maintenance and Troubleshooting

Overload
Loose or broken rotor bars (open rotor windings for wound rotor motor)

Motor is Noisy

Shorted or grounded windings
Reversed windings
Worn or damaged bearings
Foreign material in windings
Broken rotor bars (squirrel cage) or open rotor windings (wound rotor)

Motor Test Circuits

Although many of the troubleshooting procedures defined in this section can be accomplished using the test equipment described earlier, some field tests can be handled quickly using the type of equipment described in the next sections.

Electrical Test Circuit The circuit shown in Figure 9–16 can be easily constructed for use in the field. Please observe the following precautions:

- Be certain to connect the line lead and the neutral as shown. Crossing them could result in a ground loop and a potentially dangerous connection.
- The resistors should be selected to provide approximately 10 to 15 additional amperes each time a switch is selected. Twelve-ohm, 1500-W resistors will provide a total current capacity of 50 A with all switches closed.

CAUTION: This circuit uses lethal voltages with lethal current capacity. Observe all appropriate safety procedures!!!

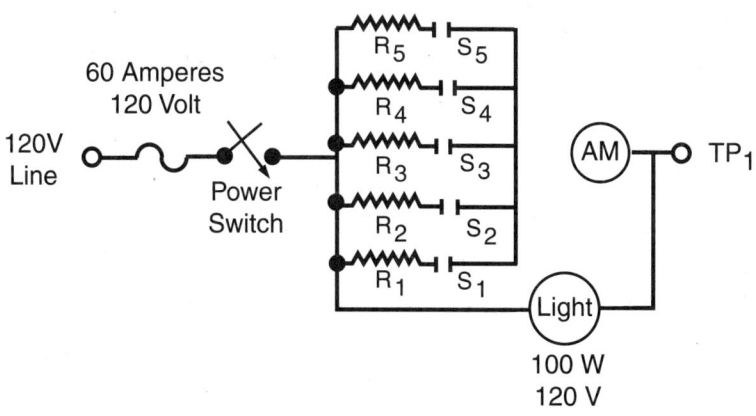

FIGURE 9–16
Motor troubleshooting circuit (Courtesy of Cadick Professional Services)

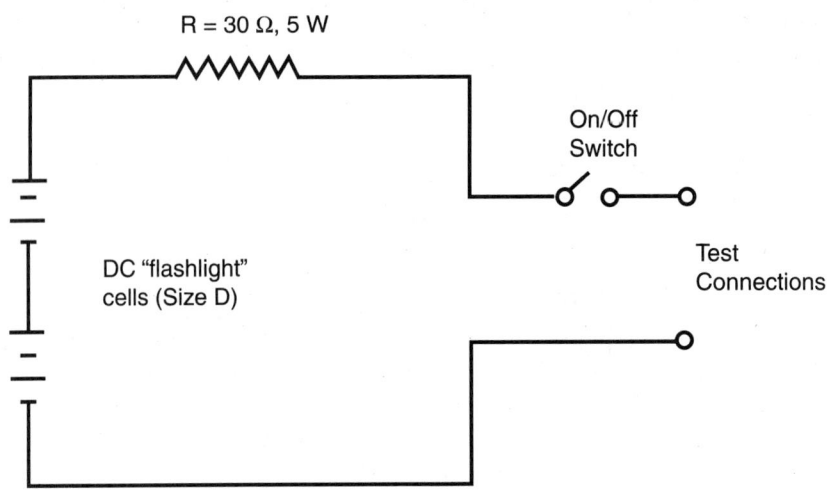

FIGURE 9–17 DC source for locating reverse coils (Courtesty of Cadick Professional Services)

- Higher currents may require a combination of a 240-V supply and/or smaller resistors. Be certain to select resistors with appropriate wattage.
- This test circuit could draw over 50 A. Be certain the power supply used has sufficient capacity.
- Calibrate the instrument by shorting the test leads, TP_1 and TP_2. Close each switch S_1 through S_5 and record the ammeter reading for each switch. This information will be used later for troubleshooting purposes.

DC Current Source Some AC motor tests require the availability of a low current DC source. The circuit of Figure 9–17 is suitable for the purpose. Two flashlight cells, size D, are placed in series with a 30-Ω, 5-W resistor, and an on/off switch.

The Growler The growler is a special piece of commercially-made test equipment Figure 9–18. As you can see, the growler is a coil of wire wrapped around a corelike-shaped set of jaws. The windings are in series with a resistor and an ammeter.

In application, the jaws of the growler behave like the primary of a transformer. When placed close to another coil, such as a rotor winding, the growler's magnetic field couples to the other winding and induces voltage into it. As you will see later, the results of this action can be used to isolate certain types of problems.

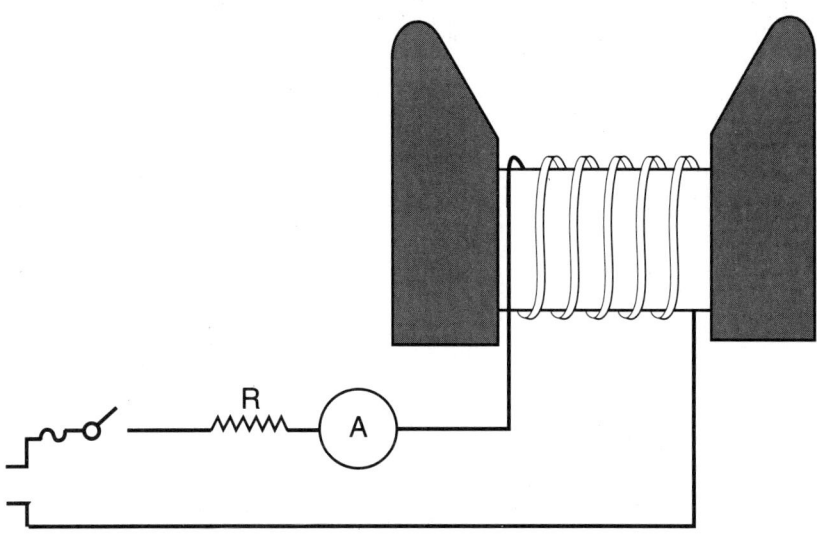

FIGURE 9–18
The growler (Courtesy of Cadick Professional Services)

Motor Troubleshooting Tests

Grounded Windings As previously discussed, most motors operate with their windings ungrounded. Occasionally, a winding may become grounded because of worn insulation, foreign materials, or other such problems. The test circuit of Figure 9–16 can be used to detect such grounds, and refer to Figure 9–19 to help locate them.

The following procedure may be used:

Disconnect and isolate all windings from each other.
Make certain all switches are open.
Connect TP_2 to the motor frame.
Connect TP_1 to the winding being investigated.
Close the power switch. If the winding is grounded, the light bulb will glow.
To locate the end of the winding closest to the ground:
Close switches 1 and 2.
Place TP_1 on T_1. Note the current reading.
Place TP_1 on T_2. Note the current reading.
The terminal with the highest current reading is the one closest to the ground.

FIGURE 9–19 Testing a motor winding for grounds (Courtesy of Cadick Professional Services)

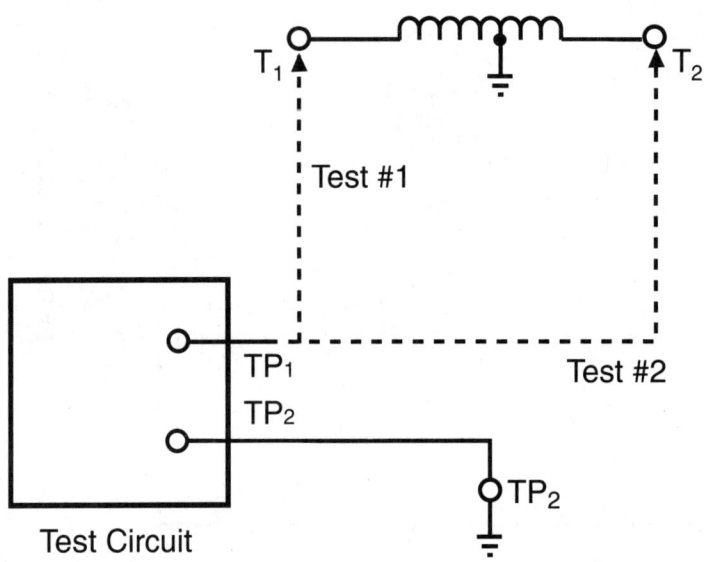

Open Windings Open windings can also be found, refer to Figure 9–20.

Disconnect and isolate all motor windings.
Make certain all switches are open.
Connect TP_1 to T_1 and TP_2 to T_2.
Close the power switch. If the winding is open, the light will not glow.

FIGURE 9–20 Locating open windings (Courtesy of Cadick Professional Services)

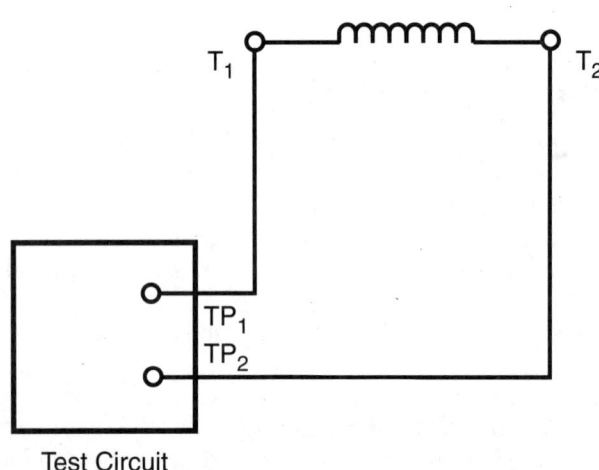

Shorted Windings Testing for shorted windings is a comparison type of test. The amount of test current through the winding is compared to other windings and/or current that flows when the test leads are clipped together, Figure 9–21. The following procedure may be used:

Connect TP_1 and TP_2 to the winding being tested. Make sure that the winding is isolated from other windings.
Close one or more of the resistor switches (R_1–R_5).
Close the power switch. If the current that flows is not significantly different from the current that flows when the test leads are shorted together, the winding is shorted.

If you wish to isolate the particular winding or coil group that is short-circuited, use the following procedure:

Disassemble the motor, disconnect the windings, and disconnect the coil groups.
Connect TP_1 and TP_2 to each end of the first coil group.
Close one or more of the resistor switches (R_1–R_5).
Close the power switch and record the current.
Open the power switch and re-connect TP_1 and TP_2 to a different coil group.
Close the power switch and record the current again.
Continue this process until all coil groups have been measured. The coil group with the highest current flow is the shorted group.

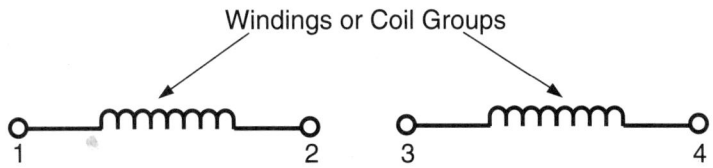

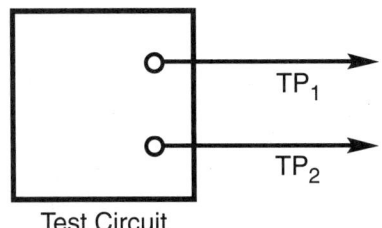

FIGURE 9–21
Testing for shorted windings (Courtesy of Cadick Professional Services)

Reversed Coils The easiest way to find a reversed coil is to connect the motor to a low voltage, low current source of direct voltage, as previously described and shown in Figure 9–17.

To perform this test, do the following:
Remove the rotor from the motor.
Connect the motor stator winding to the DC supply, as shown in Figure 9–22. Note that to use this method for a three-phase motor, you must have access to both ends of the coil.
Close the power switch on the DC supply.
Move a compass around the inside of the stator slowly. The compass needle should reverse as it moves into the influence of each successive coil. That is, it will read a north pole followed by a south pole and so on.
If a three-phase motor is being tested, re-connect the DC supply to each of the other two phases and repeat the test.

Open Rotor Bars Open rotor bars can be diagnosed using the test setup shown in Figure 9–23. The growler couples an AC field to the rotor, causing

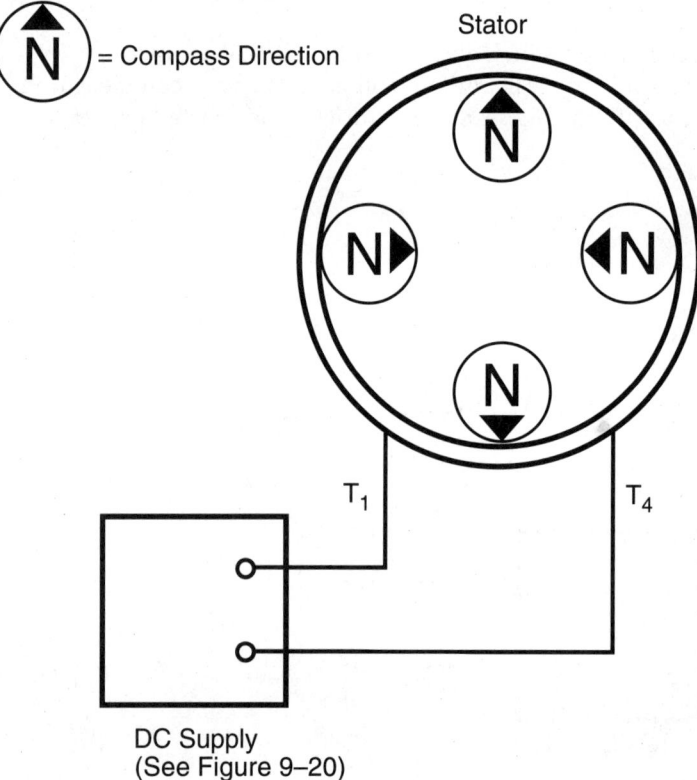

FIGURE 9–22 Testing for reversed stator coils (four-pole motor shown) (Courtesy of Cadick Professional Services)

current to flow in all of the good rotor bars. The iron filings will be attracted to all of the rotor bars with current in them. Open rotor bars, however, will have no current and, therefore, the filings will not be attracted. To perform the test, the rotor is turned in the jaws of the growler until a position is found where the filings are not attracted. This indicates a broken rotor bar.

An alternative is to use the ammeter as an indicator. As the broken rotor bar moves close to the growler, the ammeter reading will drop dramatically.

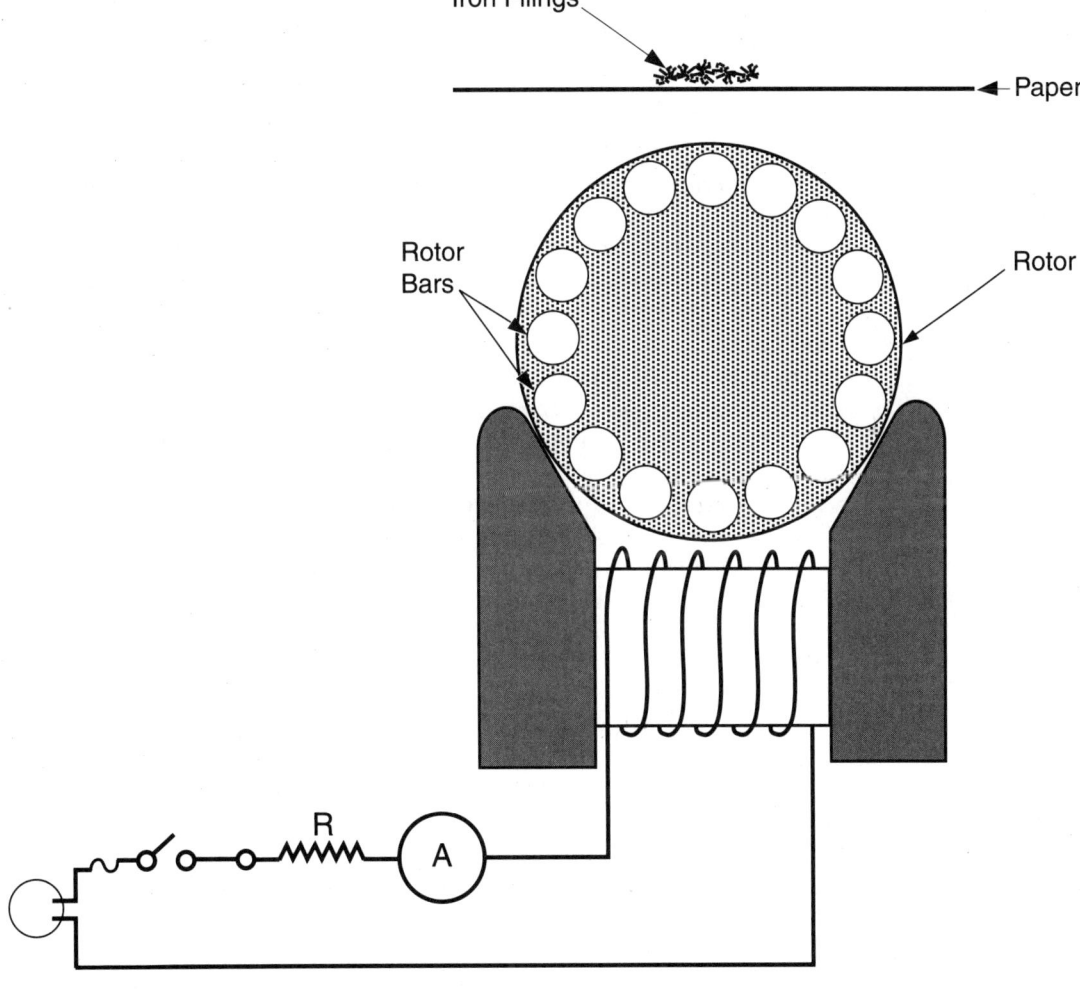

FIGURE 9–23
Using the growler to find open rotor bars (Courtesy of Cadick Professional Services)

For very large rotors, the growler may be moved while the rotor is held stationary.

Defective Capacitors Capacitors are used for either starting or power factor correction. In either case, they can be checked for shorts or opens using the test circuit of Figure 9–16. The following steps can be used to check a capacitor:

Close one or two resistor switches (Figure 9–16).
Connect the capacitor terminals to TP_1 and TP_2.
Close the power switch.
If no current flows, the capacitor is open.
If the same current flows as when TP_1 and TP_2 are shorted together, the capacitor is shorted.

SUMMARY

To minimize unexpected failures and problems with electric motors, they should be tested periodically. These preventive maintenance procedures can identify impending problems, and isolate them for correction before the motor fails.

AC motors and AC generators are virtually identical in terms of general construction principles. Because of this, many of the same tests and procedures that are performed on generators are also performed on motors. Preventive maintenance procedures fall into two general categories: visual mechanical maintenance and electrical tests.

Visual mechanical tests and inspections are performed to determine the general mechanical condition of the motor and its associated equipment, such as physical damage, loose or improper mounting, poor or non-existent grounds, loose connections, improper lubrication, and malfunctioning ventilation hardware.

Electrical tests are performed to evaluate insulation condition, motor starter operation, differential relays, and other protective devices. Such tests are performed by applying voltage and/or current to the equipment being tested, and then evaluating the results compared to equipment standards or previous results.

Even the best maintained motor will occasionally fail; consequently, the motor mechanic must have an arsenal of troubleshooting tools at his or her disposal. Commercially available equipment, such as that used for maintenance testing, is not always readily available. Simple, "home-made" equipment may be used to perform troubleshooting tests for such problems as shorted, open, or grounded windings, bad capacitors, open switches, and other such failures.

REVIEW QUESTIONS

1. List at least three items that should be searched for during a visual mechanical inspection.

2. What can happen if an oil ring is jammed and will not turn?

3. The position at which a rotor aligns itself is called the _____.

4. Insulation resistance readings on a surge arrester are performed using a megohmmeter of 1000-V rating or more. The readings vary from one arrester to another and may be as high as _____ or as low as _____.

5. A differential relay is being tested for its percent slope characteristic (see Figure 9–11). The restraint current is set to 30 A. The operating current is then slowly increased and the relay operates at 3 A. What is the percent slope of this relay?

6. How is the rotation direction of a single-phase motor reversed?

7. If no other tolerances are given, a motor operating current should not deviate from its nameplate value by more than the plus or minus _____ percent.

8. List five (5) things that may prevent a motor from starting.

9. The growler is like the _____ winding of a transformer.

10. If DC current is passed through a motor winding, and a compass is then moved around inside the stator, the compass needle should reverse direction as it approaches each coil. What does it mean if the needle does not reverse direction?

Part IV

DC Machines

DC Generator Theory and Construction   10

OBJECTIVES

After studying this chapter, the student should be able to:

- *Describe the operating principles of DC generators.*
- *Define the terms* commutator, neutral point, ripple, voltage droop, *and* voltage regulation.
- *Describe the physical construction of "real" DC generators, including the frame, field windings, armature, commutator, brushes, and brush holders.*
- *Describe the ratings of DC generators as defined in NEMA Standard MG 1.*
- *Describe the operation and characteristics of separately-excited, series-connected, shunt-connected, and compound-connected generators.*

INTRODUCTION

Although direct-voltage systems are becoming increasingly less common, DC still offers some advantages for power distribution. Power factor and phase angle are not an issue with DC; DC motors still offer some of the highest torque-to-horsepower ratios of all motors. DC circuit calculations are substantially simpler than AC, because reactance is not an issue.

Direct voltage can be generated either by rectifying alternating voltage or by generating it with a DC generator. (As you will see, even a DC generator creates DC by rectification.) In this chapter, you will learn "the basics" of DC generators: how they are constructed, how they work, and how they are connected internally and externally.

OPERATING THEORY

Commutation

The fundamental construction of the DC generator is shown in Figure 10–1. Notice that this machine is very similar to the basic AC generator shown in Figure 2–1. The DC generator, however, has a special type of assembly called a *commutator*.

The commutator is a high speed, rotary switch. As the generator turns, the commutator connects each side of the armature winding to each brush. Consequently, the positive half of the armature loop is connected to the same brush all the time.

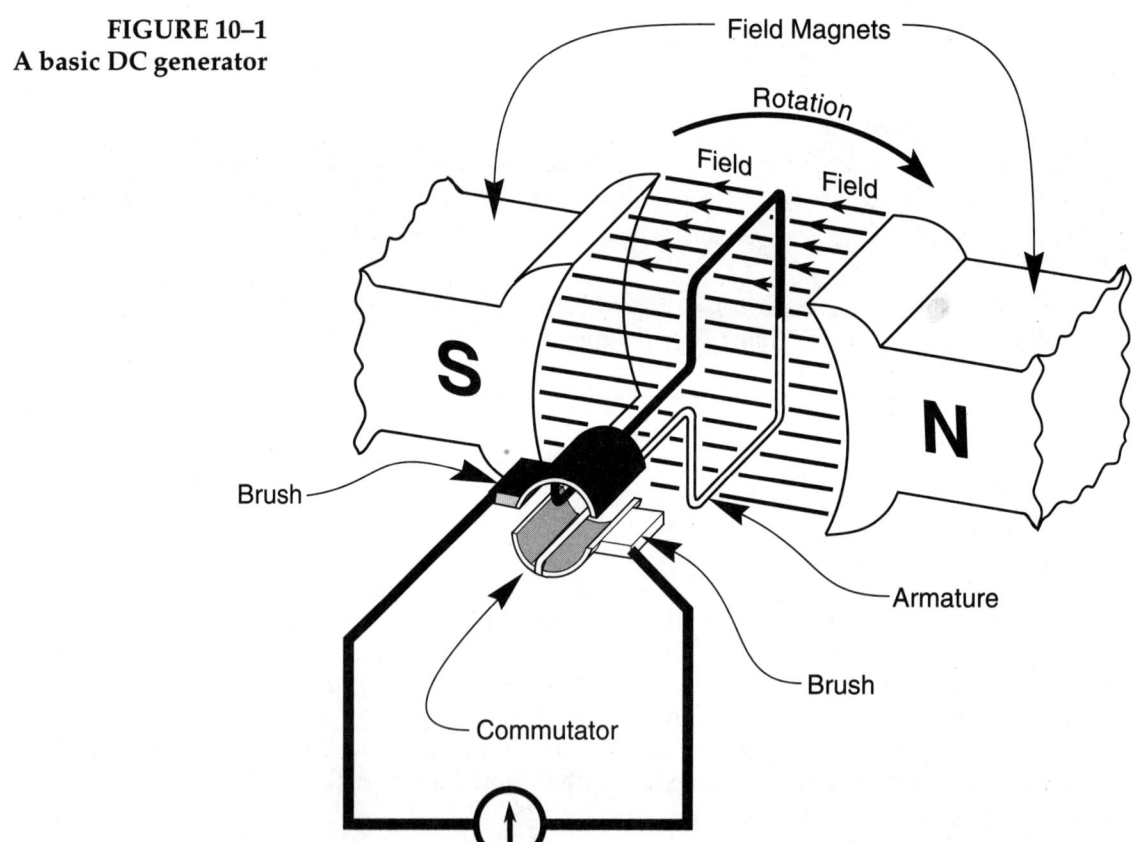

FIGURE 10–1
A basic DC generator

The output of such an assembly for one complete revolution is shown in Figure 10–2. As the armature turns through its first half revolution, the output voltage, as indicated by the meter, is positive. As the armature enters the second half revolution, the commutator switches the armature loops so they are connected to the opposite brushes. Thus, as the loop completes its first full revolution, the output is again positive on the meter.

The overall result of this is a pulsing, DC output voltage, as shown in Figure 10–2. Clearly, the DC generator is simply an AC generator whose output is rectified, or commutated.

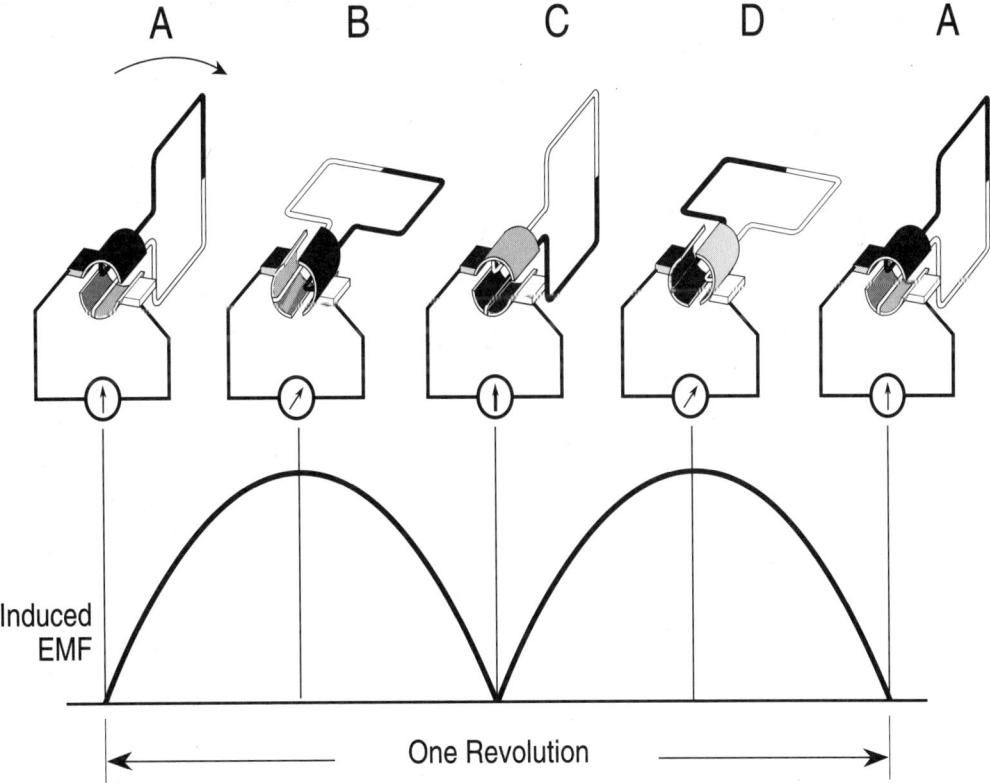

FIGURE 10–2
Output voltage from the generator of Figure 10–1

The Neutral Point

The relative position of the brushes and the commutator segments is critical to the proper operation of the DC generator. Since the brushes are slightly wider than the slots between the commutator segments, they will be contacting both segments for a brief instant as they pass from one to the other. To avoid arcing at the brushes, the voltage on the two commutator segments must be equal when the brushes are contacting them both. This point is called the *neutral point*.

For a two-segment commutator, as shown in Figure 10–1, the neutral point is when the voltage between the two segments is equal to zero. If the brushes are rotated so they commutate at some value other than zero, they will be shorting the segments when there is a voltage across them. The resulting short circuit will cause severe arcing at the brushes with the resultant damage.

DC generators with more than one armature loop, as in Figure 10–3, will have a corresponding increase in the number of commutator segments. The neutral point of Figure 10–3 is when the voltages on the two loops are equal. This occurs at 90° intervals for each revolution, or in this case at 45°, 135°, 225°, and 315°. The brushes will be in neutral when they commutate

FIGURE 10–3
Voltage output from a DC generator with two armature windings

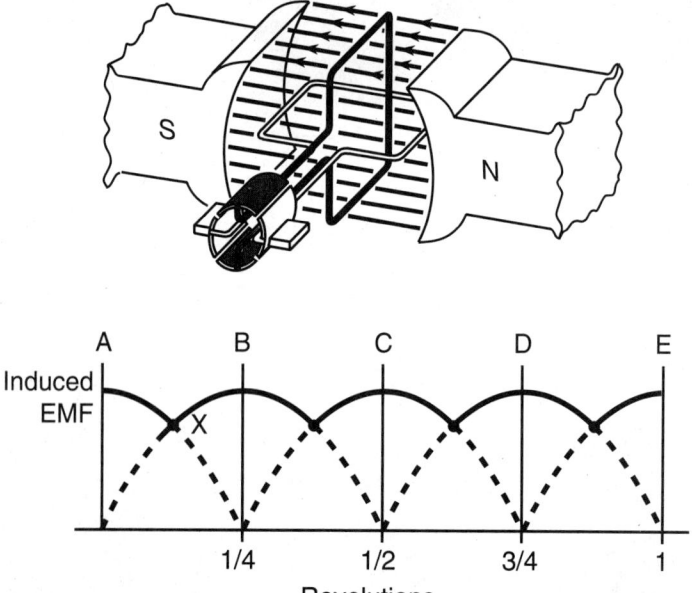

DC Generator Theory and Construction

for these four positions of the armature. Zero degrees in this case is when the black loop is vertical as shown.

Note that the voltage on the loops at neutral is 70.7% of the peak voltage for each loop and that the total output of the generator is shown by the solid lines in Figure 10–3.

Generator Ripple

The output of the DC generator shown in Figure 10–1 is very bumpy. Rather than a steady direct voltage, the output is composed of a series of current pulses starting at zero volts, rising to maximum and returning to zero. Such an output is said to be direct voltage with an AC ripple. The voltage is direct because it has an average value greater than zero. It has an AC ripple because it pulsates. The frequency of the single loop DC generator, Figure 10–1, will be twice the rotational frequency of the armature. Thus, if the armature is making 60 revolutions per second, the ripple frequency will be 120 Hz.

Notice that the pulses of the output of Figure 10–3 are substantially smaller than those of Figure 10–1. Each individual loop voltage varies the same in both cases; however, since the commutator of Figure 10–3 commutates before the pulse has a chance to drop to zero, the resulting total output is considerably smoother.

The greater the number of loops and commutator segments, the smoother the output and, therefore, the smaller the ripple value. Also, the ripple frequency will be greater. The ripple frequency of a generator, such as that shown in Figure 10–3, is four times the frequency of rotation instead of two times, as shown in Figure 10–1. This higher ripple frequency is easier to filter when the use of the generator calls for very pure DC outputs.

Commercial DC generators will have many loops with the necessary number of commutator segments. The outputs of such machines have very low A magnitude and high frequency ripple.

Voltage Droop and Voltage Regulation

For most generators, the output voltage drops as the load current increases. The amount of voltage drop that naturally occurs on a generator is referred to as *voltage droop*. Note that series-connected DC generators actually increase their output voltage as the load current increases. Series generators will be discussed later in this chapter.

The use of external circuits to compensate for droop is referred to as *voltage regulation*. Regulation and droop for various types of DC generators will be illustrated later in this chapter.

CONSTRUCTION

Frame

Figure 10–4 shows the main construction elements of DC generators. The frame serves two purposes. First, it provides mechanical support for the magnetic main poles and the interpoles. Second, it provides a path for the completion of the field flux lines.

Frames are usually made of electrical grade steel with the alloy content and heat treatment optimized for maximum permeability. The physical construction of the frame varies from one manufacturer to another. Figure 10–5 shows one method used to construct the frame and the magnetic cores for the field poles.

Field Windings

Main Poles The simple generators shown in Figures 10–1 and 10–3 use permanent magnets for generating the magnetic field flux. Although some small, commercial generators use permanent magnets, most generators develop their magnetic fields electromagnetically. A typical field pole is shown in Figure 10–6.

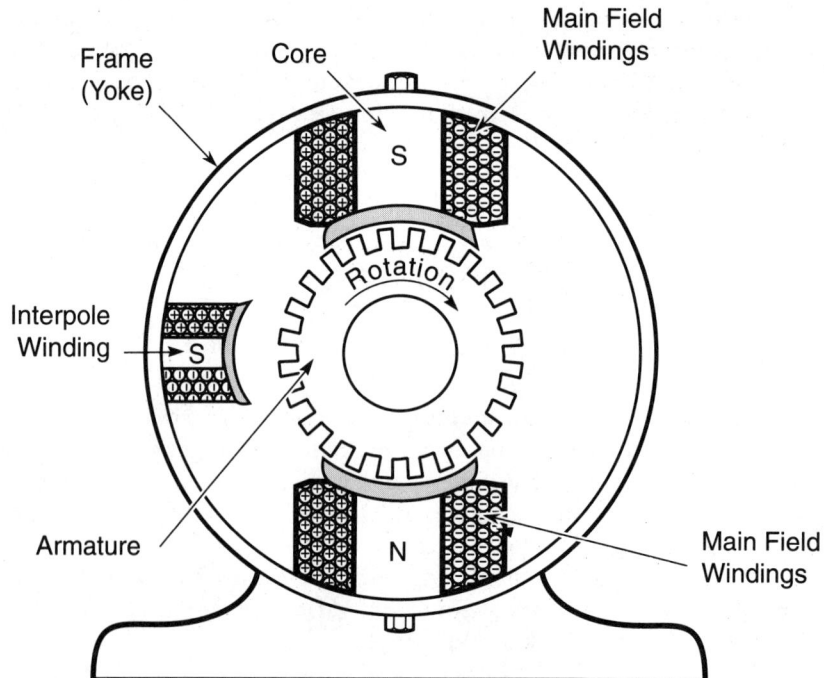

FIGURE 10–4
Major components of a two-pole DC generator with interpole

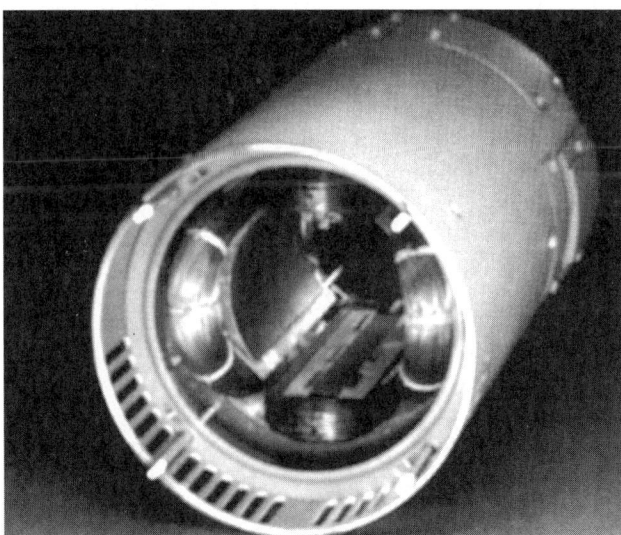

FIGURE 10–5
DC generator frame (Courtesy of General Electric Company)

Field windings are wound with electrically conductive wire wrapped around the pole core. Copper wire is normally used, although aluminum is occasionally employed. The pole cores are usually a laminated construction to eliminate eddy currents set up by the pulsating flux created by the armature. Remember that, although the output is DC, the current in the actual armature windings is alternating.

The field windings are energized with DC, and are wound so they produce alternating north and south poles as shown. The wire is insulated with varnish, lacquer, or spiral wrappings of cloth or paper. The entire assembly is impregnated with an insulating material, such as varnish or lacquer.

The DC supply for the main poles may be supplied from an external source or from the armature itself. If supplied from an external source, the DC generator is referred to as a *separately-excited* generator. If connected to the armature, the generator is referred to as a *self-excited* generator. The main poles create flux in the yoke, pole pieces, air gap, and the core.

When connected for self-excitation, the main poles may be connected either in series or in shunt with the armature. Main poles that are connected in shunt are wound with many turns of relatively small wire. Series field windings are wound with a smaller number of turns of relatively large wire.

The generators shown in Figures 10–1 and 10–3 are two-pole generators; that is, they have one north main pole and one south main pole. Four-pole generators are also commonly used. They have four main poles creating alternate north and south poles.

Interpoles Interpoles are connected in series with the armature. They are wound and connected in such a way that they produce the same polarity flux as the main pole just ahead of them in the direction of rotation, Figure 10–4. The interpole has the effect of reducing sparking at the commutator by magnetically suppressing the sudden change in the flux caused when the commutator opens an armature coil.

FIGURE 10–6
Field pole

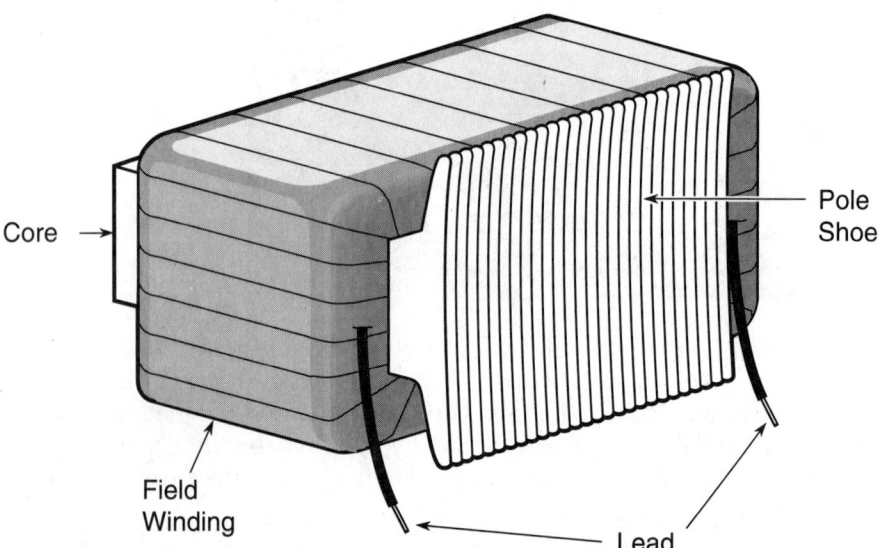

Interpoles are not as powerful as the main poles. A DC generator will usually have the same number of interpoles as main poles; however, half the number can be used with very little reduction in efficiency.

Armature

The armature consists of conductive coils which are rotated through the field set up by the field poles. Circular in cross section, the armature is made of many laminated sheets of soft iron. The laminated sheets are slotted to provide a place for mounting the coil windings. The laminations are pressed onto the shaft, as shown in Figure 10–7b.

The windings are then wound into the slots, and held in place with wooden or fiber wedges, Figure 10–7a. Each end of each winding termi-

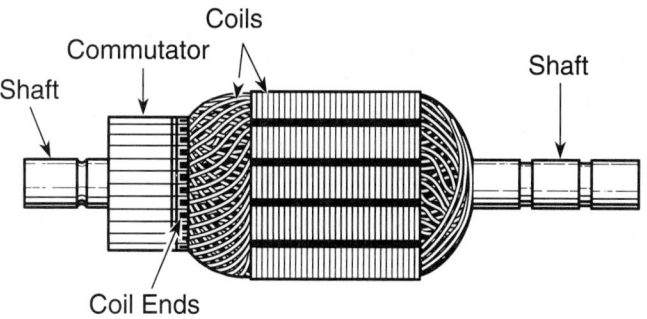

(a)
Complete Armature

FIGURE 10–7
Armature construction

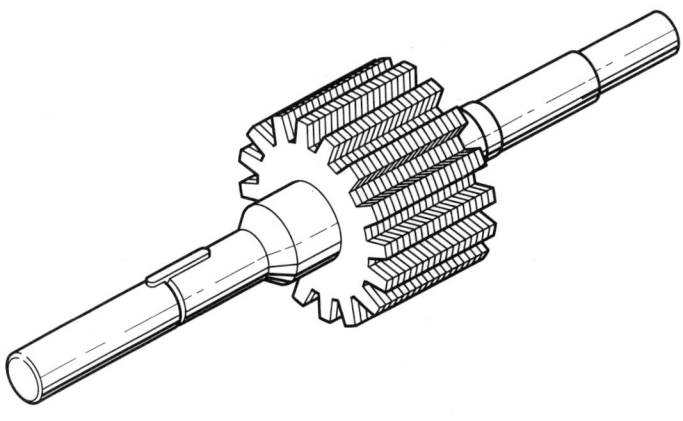

(b)
Unwound Armature Core on a Shaft

nates at the commutator. Each winding may have multiple loops of wire. Most DC generators are constructed with *form wound* armatures. In these machines, the windings are made to proper shape, insulated, and then installed on the rotor. *Random windings* are formed directly on the armature core by winding insulated wire into the slots. Random windings are more commonly used on very large DC generators.

Commutator

As previously discussed, the commutator is a high speed, rotary switch. It is used to switch from one armature winding to the next, to keep the output at the brushes at the same DC polarity. Figure 10–8 is a typical multi-segment commutator. The commutator is mounted on the armature shaft, and each segment is connected to one end of an armature winding. Thus, the commutator will have a pair of segments called *bars,* for each winding. The segments are insulated from each other by thin strips of mica.

Commutator bars are held in place by steel v-rings or clamping flanges. The commutator sleeve is keyed to the armature shaft and a mica collar insulates the commutator bars from the shaft.

The commutator bars usually have risers or flanges to which the leads from the associated armature windings are soldered. These risers serve as a shield for the soldered connections when the commutator becomes worn. When risers are not provided, the armature winding leads must be soldered to short slits in the ends of the commutator bars.

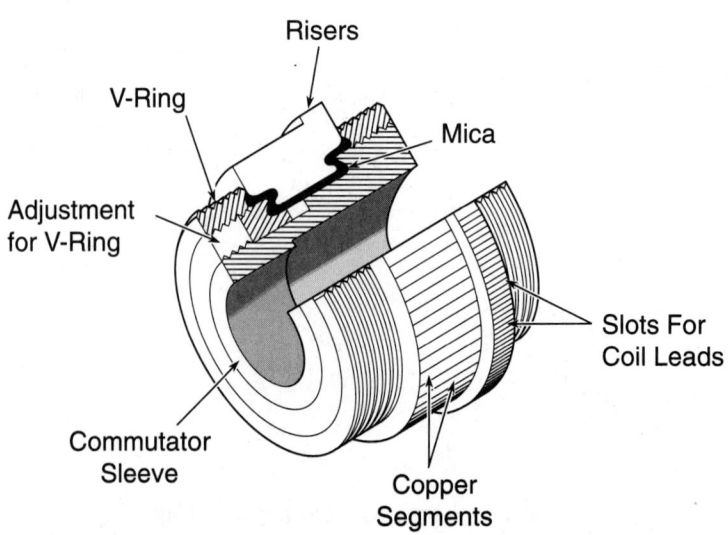

FIGURE 10–8 Commutator construction

Brushes and Brush Holders

Brushes are the points of contact for the motor to the load. They are usually made of a mixture of carbon in a graphite form with some metallic power added for strength and electrical conductivity. The brushes are held in place by mounting them in brush holders. An entire brush and brush holder assembly is shown in Figure 10–9. The proper brush pressure is maintained by mounting springs, and the brush is electrically connected through flexible, braided copper pigtails.

The brush holders are attached to the frame of the generator with an assembly called a brush *rigging*. The rigging is insulated from the generator frame by an insulating pad. The brush holders are electrically connected to the external DC circuit.

Because most DC generators have multiple pairs of north and south poles and multiple armature coils to provide a more constant level of current, there is generally more than one pair of brushes. A pair of brushes is usually required for each pole pair. Each pair of brushes consists of a positive and negative brush. The polarity of the brushes alternates about the commutator. The positive brushes are connected to the positive output terminal and the negative brushes to the negative output terminal.

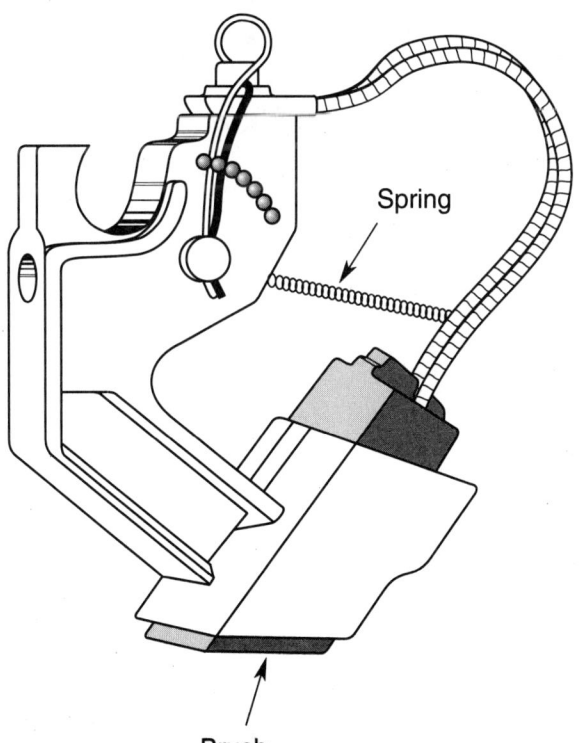

FIGURE 10–9
Brush and brush holder

240   CHAPTER 10

Figure 10–10 shows an exploded view of a typical DC generator. Note that this manufacturer refers to the interpoles as commutating poles. Figure 10–11 shows three coils and different assembled DC generators.

Ratings

Small and Medium-Sized Generators   NEMA defines small and medium-sized generators as those that are smaller than 1.0 kw per rpm. NEMA defines the ratings for such generators in a number of areas including kilowatts, speed, voltage, time rating, ambient temperature, insulation class, and nameplate markings.

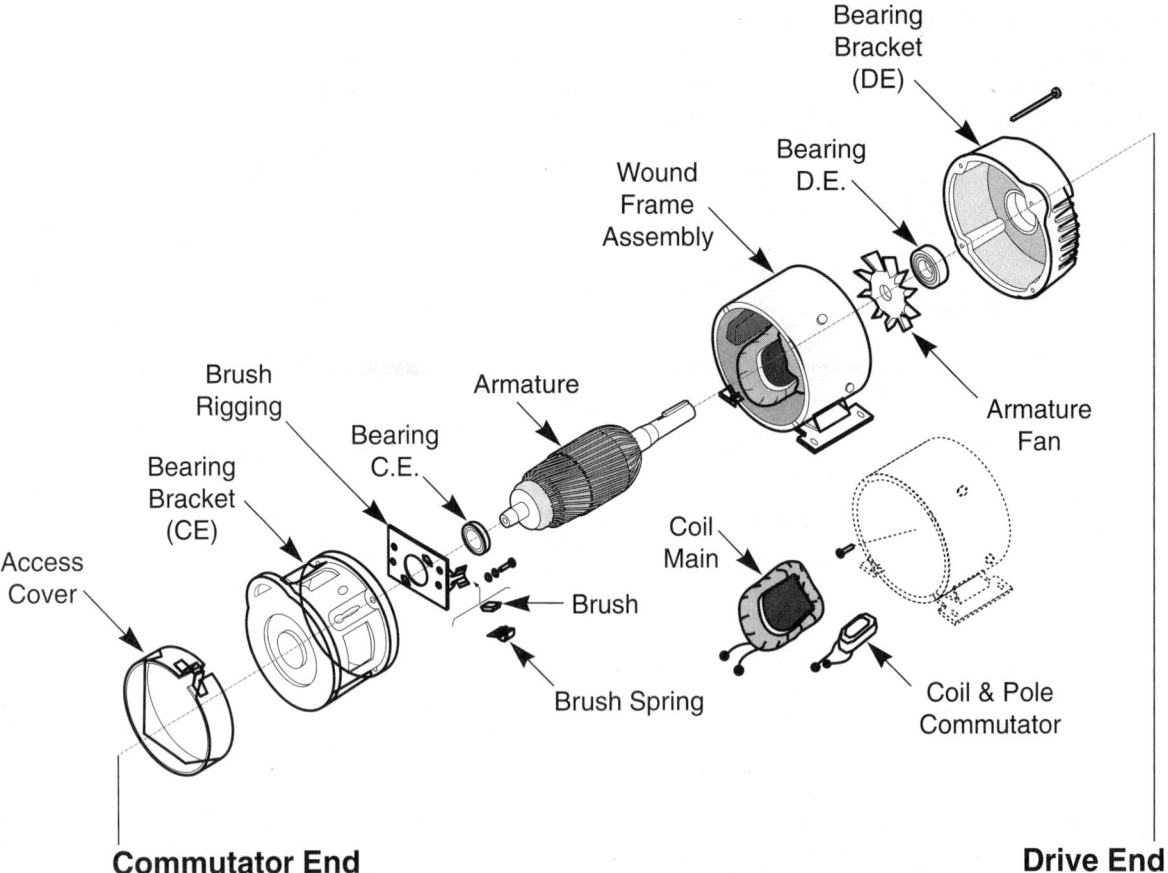

FIGURE 10–10
Exploded view of a typical DC generator (Courtesy of General Electric Company)

DC Generator Theory and Construction

- Kilowatt, speed, and voltage ratings, Table 10–1;
    Table 10–1 defines the NEMA standard kilowatt, speed, and voltage ratings for small and medium-sized generators.
- Time Ratings
    All DC generators are to be rated on a continuous basis.
- The maximum ambient temperature of a generator is 40°C unless otherwise noted. Table 10–2 gives standard temperature rise information for the various insulation classes.
- Nameplates
    NEMA requires a minimum of ten pieces of information on the nameplate of a DC generator as follows:

Manufacturer's type designation and frame number
Kilowatt output

FIGURE 10–11
Typical DC generators (Courtesy of General Electric Company)

Table 10–1 Kilowatt, Speed, and Voltage Ratings for Small and Medium-Sized Generators
(Courtesy of NEMA)

Rating kW			Speed, Rpm				Rating Volts
¾	3450	1750	1450	1150	850	...	125 and 250
1	3450	1750	1450	1150	850	...	125 and 250
1½	3450	1750	1450	1150	850	...	125 and 250
2	3450	1750	1450	1150	850	...	125 and 250
3	3450	1750	1450	1150	850	...	125 and 250
4½	3450	1750	1450	1150	850	...	125 and 250
6½	3450	1750	1450	1150	850	...	125 and 250
9	3450	1750	1450	1150	850	...	125 and 250
13	3450	1750	1450	1150	850	...	125 and 250
17	3450	1750	1450	1150	850	...	125 and 250
21	3450	1750	1450	1150	850	...	125 and 250
25	3450	1750	1450	1150	850	...	125 and 250
33	3450	1750	1450	1150	850	...	125 and 250
40	3450	1750	1450	1150	850	...	125 and 250
50	3450	1750	1450	1150	850	...	125 and 250
65	...	1750	1450	1150	850	...	250
85	...	1750	1450	1150	850	...	250
100	...	1750	1450	1150	850	...	250
125	...	1750	1450	1150	850	...	250
170	...	1750	1450	1150	850	...	250
200	...	1750	1450	1150	850	720	250 and 500
240	...	1750	1450	1150	850	720	250 and 500
320	...	...	1450	1150	850	720	250 and 500
400	...	...	...	1150	850	720	250 and 500
480	...	...	...	...	...	720	500
560	...	...	...	...	850	720	500
640	...	...	...	...	850	720	500
720	...	...	...	...	850	720	500
800	...	...	...	1150	850	...	500

Table 10-2 DC Generator Temperature Rise
(Courtesy of NEMA)

	Totally-Enclosed and Non-Ventilated and Totally-Enclosed Fan-Cooled Generators, Including Variations Thereof				Generators with All Other Enclosures			
Class of Insulation System .... (see MG) 1-1.65)	A	B	F	H	A	B	F	H
Time Rating								
Temperature Rise, Degrees C								
1. Armature windings and all windings other than those given in items 2 and 3 – resistance............................	70	100	130	155	70	100	130	155
2. Multi-layer field windings – resistance............................	70	100	130	155	70	100	130	155
3. Single-layer field windings with exposed uninsulated surfaces and bare copper windings – resistance.........	70	100	130	155	70	100	130	155
4. The temperatures attained by cores, commutators, and miscellaneous parts (such as brushholders, brushes, pole tips, etc.) shall not injure the insulation or the machine in any respect.								

Suggested Standard for Future Design 11-17-1955, revised 6-24-1957, NEMA Standard 11-12-1959, revised 1-21-1964; 11-12-1964; 11-11-1965; 1-25-1972; 11-8-1973; 9-17-1981.

Note I — Abnormal deterioration of insulation may be expected if ambient temperature of 40°C is exceeded in regular operation.

Note II — The foregoing values of temperature rise are based upon operation at altitudes of 3300 feet (1000 meters) or less. For temperature rises for generators intended for operation at altitudes above 3300 feet (1000 meters), see MG 1-14.04.

Note III — When the rated maximum ambient temperature is other than 40°C, the rated value should be 50°C or 65°C. The maximum temperature rise values for each machine part should be the value selected from the appropriate table with adjustment as follows:

Rated Maximum Ambient Temperature Degrees C	Adjustment to Temperature Rise Values, Degrees C
50	Subtract 10
65	Subtract 25

Time rating (continuous in all cases)
Maximum ambient temperature
Insulation system designation
Rated speed in rpm
Rated load voltage
Rated field voltage if different than the armature
Rated current in amperes
Winding: series, shunt, or compound

Large Generators Many of the same ratings are provided for large generators. For example, Table 10–3 shows the kilowatt, speed, and voltage rating for large generators.

- Time ratings and temperature rises are determined by the manufacturer. The maximum ambient temperature is 40°C. The temperature rises based on insulation class are shown in Table 10–4.
- Nameplate information for large generators is very similar to the nameplate information for small generators:

Manufacturer's name and serial number
Kilowatt output
Time rating
Temperature rise for rated continuous load
Overload
Time rating of overload
Temperature rise for overload
Rated speed in rpm
Voltage rating
Rated current in amperes
Winding: series, shunt, or compound

SEPARATELY-EXCITED AND PERMANENT MAGNET DC GENERATORS

Some very small generators derive their magnetic fields from a permanent magnet or from a separate supply of DC. Permanent magnet generators are usually used as exciter generators for the excited units separately.

Separately-excited generators, Figure 10–12, derive their field current from another generator, batteries, or some other source.

The droop curve shows how the generator output voltage varies as a function of armature current. This curve is a droop curve, so it assumes that no external regulation is being employed; that is, the field excitation voltage is held constant.

Table 10–3   Kilowatt, Speed, and Voltage Ratings for Large DC Generators
(Courtesy of NEMA)

KW Ratings	\multicolumn{13}{c}{Speed, RPM}
	900	720	600	514	450	400	360	327	300	277	257	240	225	200
125	For smaller ratings, see MG 1-15.10.				...	...	...	...	...	...	...	...	...
170				...	...	...	...	...	...	...	...	...	...
200	...	...	...	...	...	...	...	...	...	...	...	...	...
240	...	...	...	...	...	...	...	...	...	...	...	A	A
320	...	...	...	...	...	...	...	...	A	A	A	A	A	A
400	...	...	...	...	...	...	A	A	A	A	A	A	A	A
480	...	...	...	...	A	A	A	A	A	A	A	A	A	A
560	...	...	...	A	A	A	A	A	A	A	A	A	A	A
640	...	...	B	B	B	B	B	B	B	B	B	B	B	B
720	...	...	B	B	B	B	B	B	B	B	B	B	B	B
800	...	B	B	B	B	B	B	B	B	B	B	B	B	B
1000	C	B	B	B	B	B	B	B	B	B	B	B	B	B
1200	C	B	B	B	B	B	B	B	B	B	B	B	B	B
1400	C	C	B	B	B	B	B	B	B	B	B	B	B	B
1600	C	C	C	B	B	B	B	B	B	B	B	B	B	B
1800	...	C	C	C	B	B	B	B	B	B	B	B	B	B
2000	...	C	C	C	C	C	B	B	B	B	B	B	B	B
2400	...	...	C	C	C	C	C	C	C	C	C	C	C	C
2800	...	...	...	D	D	D	D	D	D	D	D	D	D	D
3200	...	...	...	D	D	D	D	D	D	D	D	D	D	D
3600	...	...	...	...	D	D	D	D	D	D	D	D	D	D
4000	...	...	...	...	...	D	D	D	D	D	D	D	D	D
4800	...	...	...	...	...	...	D	D	D	D	D	D	D	D
5600	...	...	...	...	...	...	...	...	D	D	D	D	D
6400	...	...	...	...	...	...	...	...	D	D	D	D	D

"A" indicates voltage rating at either 250 or 500 volts.
"B" indicates voltage rating at either 250, 500, or 700 volts.
"C" indicates voltage rating at either 500 or 700 volts.
"D" indicates voltage rating at 700 volts only.

Table 10-4 Allowable Temperature Rises for Large Generators
(Courtesy of NEMA)

Observable Temperature Rises, Degrees C*

Item	Machine Part	Method of Temperature Determination†	General Industrial Service — Semi-enclosed Continuous Rated 100% Load — Insulation Class				General Industrial Service — Totally-enclosed Continuous Rated 100% Load — Insulation Class					Metal Rolling Mill Service — Metal Rolling Mills, Excluding Reversing Hot Mills — Forced-ventilated or Totally-enclosed Water-air-cooled — Continuous Rated 100% Load — Insulation Class				Metal Rolling Mill Service — 2 Hours‡ 125% Load — Insulation Class				Metal Rolling Mill Service — Reversing Hot Mills — Forced-ventilated or Totally-enclosed Water-air-cooled — Continuous Rated 100% Load — Insulation Class			
			A	B	F	H	A	B	F	H		A	B	F	H	B	F	H	B	F	H
1	Armature windings and all other windings other than those given in items 2 and 3	Thermometer Resistance	50 70	70 100	90 130	110 155	55 70	75 100	95 130	115 155		40 60	60 90	75 110		55 80	75 110	95 135	50 70	70 100	90 130
2	Multilayer field windings	Resistance	70	100	130	155	70	100	130	155		60	90	110		80	110	135	70	100	130
3	Single-layer field windings with exposed uninsulated surfaces and bare copper windings	Thermometer Resistance	60 70	80 100	105 130	130 155	65 70	85 100	110 130	135 155		50 60	70 90	90 110		65 80	85 110	110 135	60 70	80 100	105 130
4	Commutator and collector rings	Thermometer	65	85	105	125	65	85	105	125		55	75	90		65	85	105	65	85	105

5 The temperatures attained by cores, commutators, and miscellaneous parts (such as brushholders, brushes, pole tips, etc.) shall not injure the insulation or the machine in any respect.

*For generators which operate under prevailing barometric pressure and which are designed not to exceed the specified temperature rise at altitudes from 3300 feet (1000 meters) to 13000 feet (4000 meters), the temperature rises, as checked by tests at low altitudes, shall be less than those listed in the foregoing table by 1 percent of the specified temperature rise for each 330 feet (100 meters) of altitude in excess of 3300 feet (1000 meters).
†Where two methods of temperature measurement are listed, a temperature rise within the values listed in the table, measured by either method, demonstrates conformity with the standard.
‡Temperature limits apply at end of 2-hour operation at 125-percent load following operation at rated load long enough to reach a stable temperature.

NOTE I—See 1.66 for description of classes of insulation.
NOTE II—Abnormal deterioration of insulation may be expected if the ambient temperature of 40°C is exceeded in regular operation.

DC Generator Theory and Construction

As the armature current increases, the output voltage is reduced by two factors:

- The voltage drop across the armature windings and the brushes increases and subtracts from the load voltage.
- The armature reaction voltage drop is caused by the change and distortion of the air gap magnetic field caused by the armature load current.

Both of these effects cause a decrease in the generator output voltage as the current increases. To compensate for these drops, the field current can be increased. The increased field current will cause an increased armature voltage and will offset the normal droop characteristic. Figure 10–12 shows one crude but effective way of controlling the field current by using a rheostat in series.

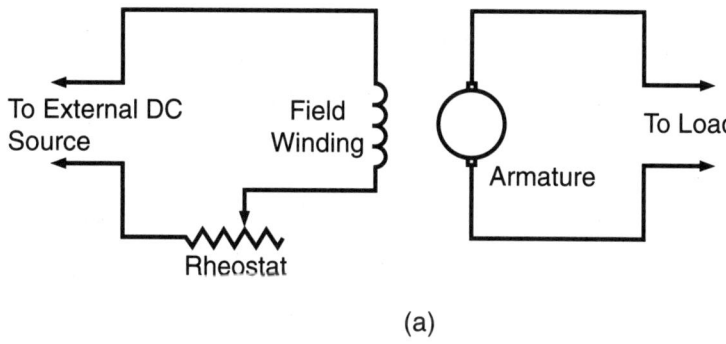

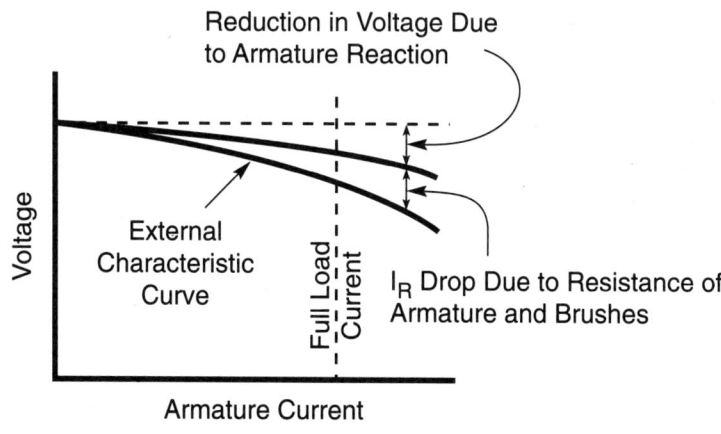

FIGURE 10–12 Separately-excited DC generator and its associated voltage droop curve

SELF-EXCITED DC GENERATORS

Overview

The simplest and most cost effective way for a DC generator to obtain field current is from its own output. If the field current is derived from the generator itself, no external supply is needed. Of course, some current is required to start the generator in the beginning. This is usually supplied by residual magnetism in the generator iron; however, occasionally the generator must be started by *flashing* the field. This means putting a pulse of voltage onto the field long enough to get the generator output started.

There are only a limited number of ways in which a generator may be connected. The following paragraphs describe each connection, and explain a little about how the generator is used.

Series-Connected Generators

The series-connected generator, Figure 10–13, is named because the field winding is connected in series with the armature. Thus, the load, the field, and the armature currents are the same. The field windings must be low resistance windings; therefore, it is wound with relatively few turns of large wire. The terminal designations for the armature are A_1 and A_2. The series field connections are S_1 and S_2.

Series-connected generators, which are seldom used in modern installations, operate best as constant speed machines that supply series lighting systems.

The droop curve is unique for the series type of generator. Since the output voltage increases as the field increases, and the field increases as the load current increases, the output voltage actually increases as the load current increases. Since there is some armature reaction and some brush resistance, the droop curve does have some voltage loss.

Because the terminal voltage of a series generator varies under changing load conditions, the generator normally is connected in a circuit that demands constant current. Used this way, it is sometimes referred to as a constant current generator, even though it does not tend to maintain a constant current on its own. Constant current can be achieved by connecting a variable resistance in parallel with the field coil.

Shunt-Connected Generators

When the field windings are connected in parallel with the armature, the generator is shunt-connected, Figure 10–14. The shunt-connected generator has a very slight droop characteristic, and is therefore essentially a constant voltage generator. Field coils in this type of generator are connected across

DC Generator Theory and Construction

the armature, and may have a series rheostat which can be used for voltage control. Since the shunt field is a voltage coil, it is composed of many turns of relatively small wire.

Since the field voltage is connected across the armature, the reduction of the voltage with increased current is amplified by the reduction in the field voltage and therefore the field current.

Shunt-generators are sometimes used with automatic voltage controls as exciters for large AC generators, and in applications where the prime mover speed is not constant, such as automobile and aircraft engines.

The shunt field terminals are designated F_1 and F_2 respectively.

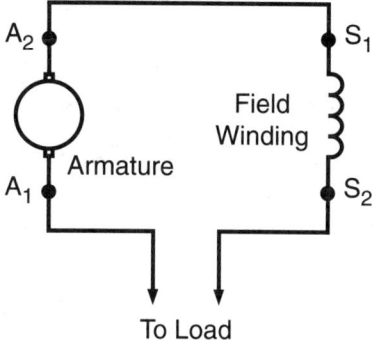

FIGURE 10–13
Series-connected DC generator and its associated droop characteristic

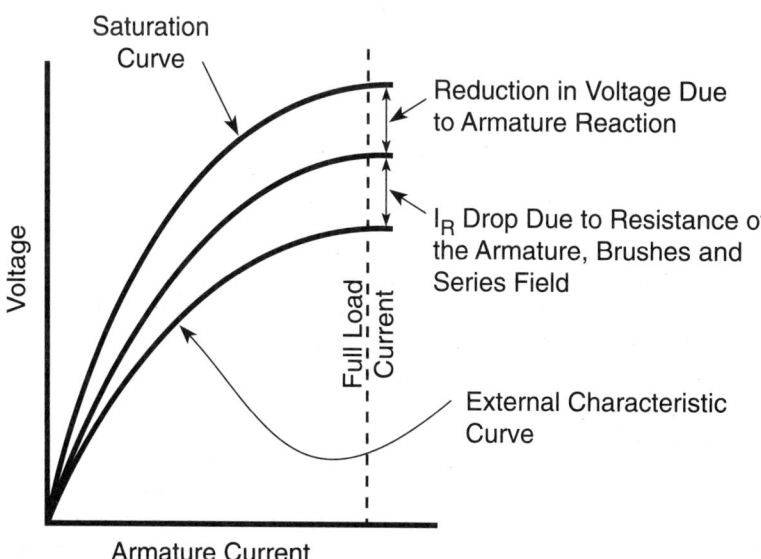

FIGURE 10–14 Shunt-connected DC generator and its associated droop curve

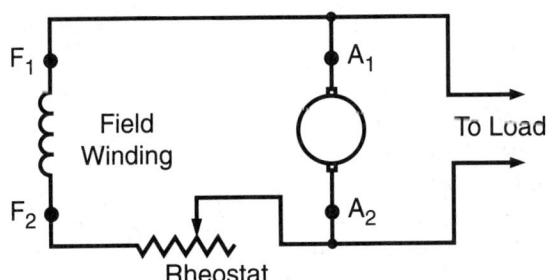

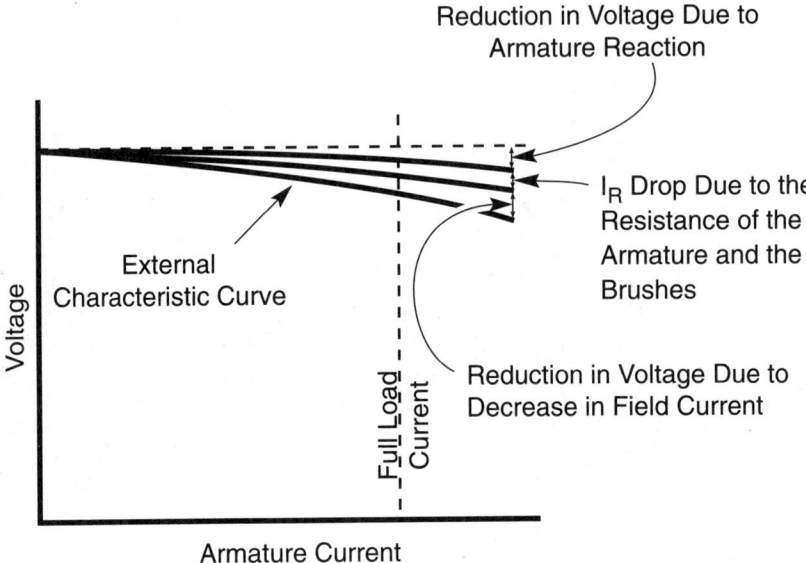

Compound-Connected Generators

Compound-connected generators Figure 10–15, employ both series and a shunt-connected field. It has droop characteristics between the series-connected and the shunt-connected fields. The generator shown in Figure 10–15 is called a short shunt connection, because the shunt field is paralleled only with the armature. If the shunt field is connected in parallel with the armature and the series field, the machine is called a long shunt connection.

The compound generator is the most widely used DC generator. By changing the number of turns in the series field, it is possible to obtain

DC Generator Theory and Construction

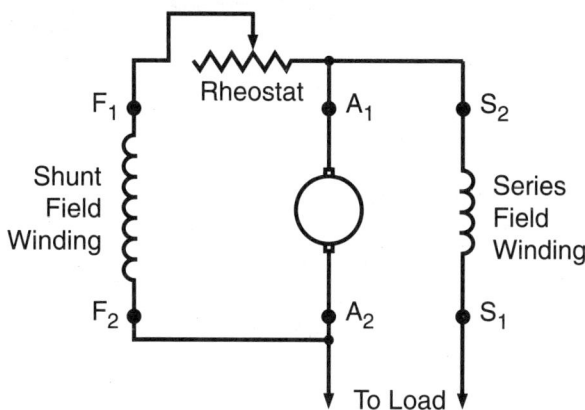

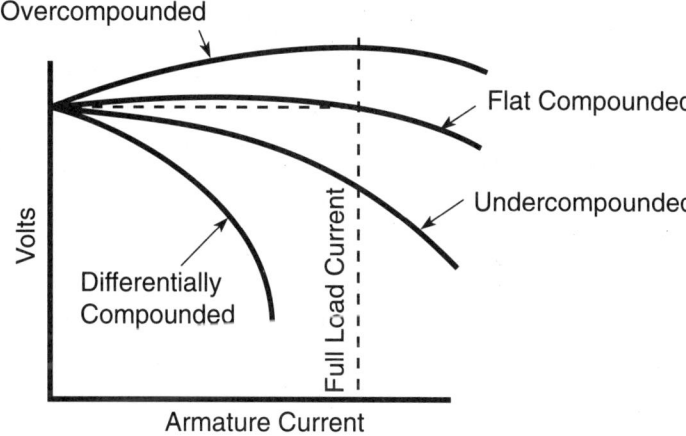

FIGURE 10–15
Compound-connected generator and its associated droop curve

three distinct types of compound generators: over-compounded, flat-compounded, and under-compounded.

An over-compounded generator is one in which there are more turns in the series field. Because of this, the compound generator behaves more like a series generator. This type of characteristic is desirable when power is to be transmitted over a long distance. The higher generated voltage compensates for the voltage loss of the transmission line.

A flat-compounded generator keeps its output voltage essentially constant over its entire range of usable loads. This type of generator is as close to self-regulated as any generator.

An under-compounded generator is one in which the series field does not have as many turns. Such a generator behaves more like a shunt-connected generator. When the shunt and series fields are connected to oppose each other, the output voltage drops off very rapidly as the load current increases. Such a generator is referred to as a differentially-compounded generator, and has been used in applications prone to short circuits, such as arc welders.

SUMMARY

The construction of a basic DC generator is very similar to an AC generator. Instead of bringing the armature voltage out through slip rings, however, the DC generator connects each end of the armature to a commutator. The commutator is sort of a split slip ring. As the armature rotates, the brushes are connected, alternately, to each side of the armature loop. This has the effect of keeping the brushes at the same DC potential, and produces a direct voltage output.

The commutator segments are positioned in such a way that the brushes change from one segment to the next when the voltage on the segments are equal. This prevents arcing and excessive brush and commutator wear. The point at which the segment voltages are equal is called the neutral point. To create a smoother DC output, more armature loops and commutator segments are used. This produces a higher percentage of direct voltage and a lower percentage of AC ripple output.

"Real" DC generators have multiple windings on the armature, and a field created by an electromagnet. There are two types of field poles: the main poles and the interpoles. The main poles are used to create the output. The interpoles are used to help in the commutation process, and minimize arcing at the brushes. The commutator is composed of copper segments insulated from each other with mica. The brushes are a carbon-based composition. There is a set of brushes for each main pole. Most DC generators have either two or four poles.

NEMA defines the standard ratings for DC generators including items like kilowatt ratings, speed, and voltage. The standards provide a level playing field so that purchasers can compare features and characteristics.

Generators may be of the permanent magnet, separately-excited, series-connected, shunt-connected, or compound-connected type. Each type has its own specific features and capabilities. The compound-connected generator is the most commonly used.

Generators have voltage/current characteristics called droop. For all but the series-connected, the output voltage drops as the load current increases. The compound-connected generator can be wound in such a way that its droop characteristic is essentially flat across the range of usable load currents. Other generators may be regulated by controlling the field voltage. As the load current increases, a voltage regulator increases the field voltage. This action increases the output voltage and regulates it to the desired value.

DC Generator Theory and Construction

REVIEW QUESTIONS

1. If the brush rigging is not properly positioned at the neutral point, _____ will occur when the brushes momentarily connect one commutator segment to the adjacent segment.

2. AC ripple can be decreased by _____ the number of armature windings and commutator segments.

3. The amount of voltage drop that occurs naturally as load current increases is called _____.

4. The frame of a generator provides mechanical _____ for the pole structures and a _____ for the magnetic flux lines.

5. The generator main poles may be connected in _____ or in _____ with the armature.

6. Interpoles are used to suppress arcing at the brushes. To do this, they are connected in series with the armature and wound in such a way that they produce the _____ polarity magnetic pole as the main pole that follows them in the direction of rotation.

7. The commutator is a _____ _____, _____ switch that rectifies the output of the armature windings.

8. The brushes are held against the armature with _____ .

9. What is the lowest standard voltage generator that will produce 480 kw at 450 rpm?

10. A compound-connected DC whose output drops off very rapidly with an increase in load current is called a _____ compounded generator.

DC Motor Theory and Construction 11

OBJECTIVES

After studying this chapter, the student should be able to:

- *Describe the operating principles of DC motors.*
- *Describe the physical construction of "real" DC motors, including the frame, field windings, armature, commutator, brushes, and brush holders.*
- *Describe the ratings of DC motors as defined in NEMA Standard MG-1.*
- *Describe the operation and characteristics of separately-excited, series-connected, shunt-connected, and compound-connected motors.*

INTRODUCTION

DC motors are very similar to DC generators. In fact, DC generators can be operated as DC motors by simply connecting a supply of rated DC voltage to the output terminals. In a similar fashion, a DC motor can be operated as a DC generator by simply connecting a prime mover to the shaft and a load to the DC terminals.

DC motors offer some of the highest torque to horsepower ratios of all motors. In this chapter, you will learn "the basics" of DC motors—how they are constructed, how they work, and how they are connected internally and externally.

OPERATING THEORY

Rotation

The fundamental construction and operation of the DC motor is shown in Figure 11–1. Notice that this machine is very similar to the basic DC generator shown in Figure 10–1.

The DC motor, Figure 11–1, works as follows: (Note that all directions are referenced looking into the commutator end of the armature.)

The DC supply produces a current that flows into the left-hand commutator segment and out of the right-hand commutator segment. This produces a magnetic field with counterclockwise polarity around the left-hand side of the armature loop, and a clockwise field around the right-hand side of the armature loop.

The magnetic field poles create a field flux, with a north to south polarity from left to right.

The magnetic field around the armature loop bends the field flux as shown in Figure 11–1.

The bent flux creates a net "up" force on the left-hand side of the armature loop and a net "down" force on the right-hand side of the armature loop. This causes the armature to turn in a clockwise direction as shown.

FIGURE 11–1 Basic DC motor operation and torque

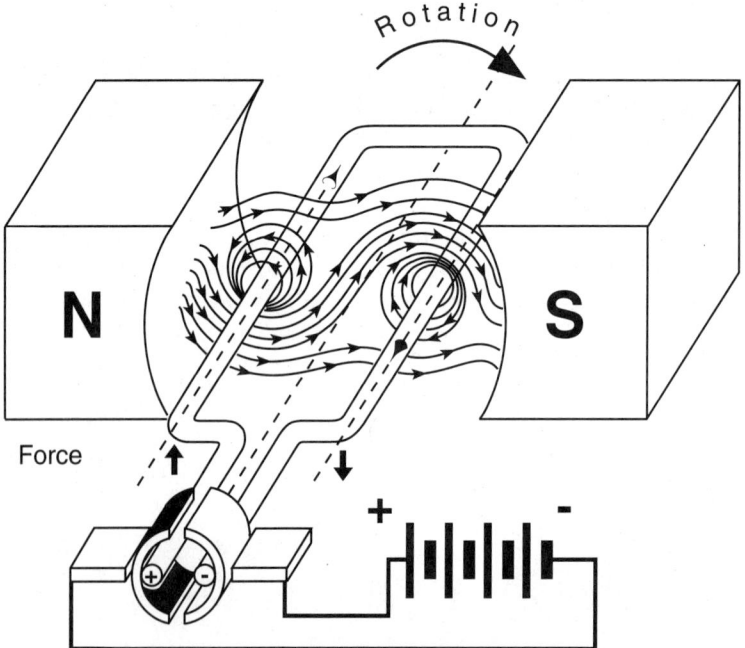

When the loop has made a 90° turn, the commutator switches the brushes so that the current in the armature reverses. This reversal allows the rotation to continue in a clockwise direction for another 90°.

When the armature has reached a point that is 270° from the starting position, the commutator again reverses the current and the rotation continues.

The operation of the motor can also be explained using Figure 11–2. The DC current flow through the armature creates a net magnetic field, as shown in the figure. To prove this fact, consider the armature to be a single loop coil. Looking down on the top of the loop shows the current flowing in a clockwise direction. Apply the left-hand rule for coils, and you will see that the north magnetic pole will be up.

The operation of the motor, using Figure 11–2, is explained as follows:

The north magnetic pole of the field opposes the north magnetic pole created by the armature current. In a similar fashion, the south magnetic pole of the armature opposes the south magnetic pole of the field. This opposition pushes the armature into a clockwise rotation as shown.

After the armature has rotated 90°, the commutator reverses the current flow; consequently, the armature magnetic poles reverse position. This reversal puts the armature south pole adjacent to the field south pole. The field pushes the armature on around, and continues the rotation.

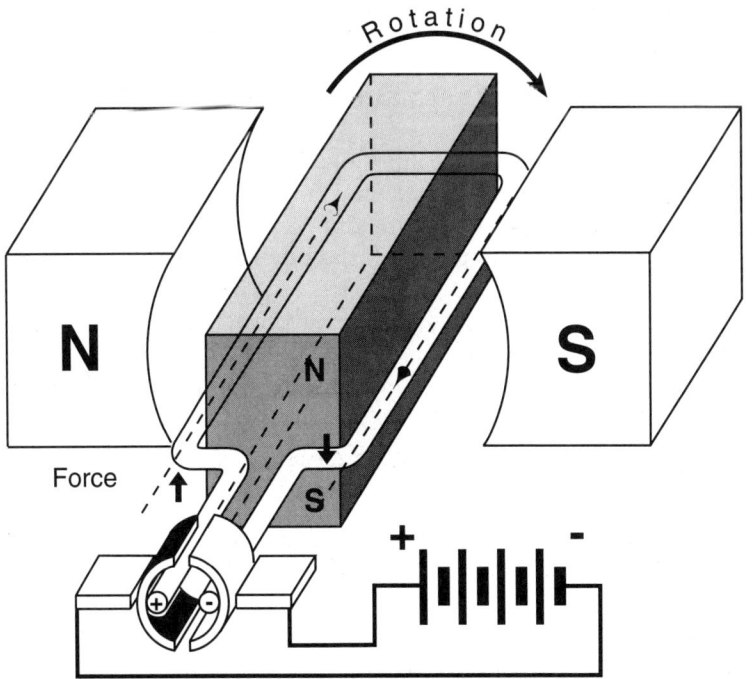

FIGURE 11–2
Alternative method of understanding DC motor operation

At the 270° point, the commutator again reverses the magnetic poles, and the motor rotation continues.

The Neutral Point

The relative position of the brushes and the commutator segments is as critical to the proper operation of the DC motor as it was to the DC generator. If the brushes are not located at the neutral point, the motor will run very roughly, because the magnetic field is being switched before it should. This creates a magnetic "bump" in the motor rotation and a consequent vibration.

Improper brush position will also create sparking at a DC motor commutator. If the armature field is reversed improperly, the armature magnetic field will cut through the field magnetic flux and cause a voltage arc at the commutator segments.

For a two-segment commutator, as shown in Figure 11–1, the neutral point is when the armature loop is exactly vertical. In other words, the neutral point of a DC motor occurs at exactly the same place as the neutral point of a DC generator.

DC motors with more than one armature loop have a corresponding increase in the number of armature segments. In fact, the neutral point for any DC motor behaves in exactly the same way as the neutral point for a DC generator.

DC Motor Torque, Power, and Speed

Each current-carrying conductor in the armature is in the form of a loop. Two sides of the loop lie in the magnetic field set up by the field poles, Figure 11–1. The force acting in one direction, on one side of the loop, and the force acting in the other direction, on the other side, combine to cause the coil to turn on its axis. The loop, thus, acts as if it were a lever with a turning force at each end. Due to this lever arrangement, the force at each end is magnified by its distance from the center. The overall turning effect on the loop is called the motor *torque*. Torque is the combined force on the two sides of the loop, multiplied by the distance of the conductors from the axis. Stated in a simple equation, the torque created on a single conductor is given as:

Torque = Force on conductor x radius of the loop
or
$$T = F \times r \tag{1}$$

The torques on all the individual armature conductors add up to the total motor torque. Torque is measured in foot-pounds. The greater the motor torque, the more turning force the motor has to drive a mechanical load. If it does not have enough torque to pull the load, it stalls.

The total torque of a DC motor can be determined from the equation:

$$T = K \times I_a \times \phi \qquad (2)$$

Where:
   T = the motor torque, K = a constant, I_a = armature current, and ϕ = field flux
   Note that the constant, K, is a function of the motor design.

Counter EMF

When any conductor cuts through magnetic lines of force, a voltage (emf) is induced into it. If the circuit is closed, such as an armature loop, a current will be induced in the loop. Finally, from Lenz's law, the current produced will be in such a direction that it opposes the motion that produced it.

This concept is illustrated in Figure 11–3. The line voltage, E, is applied to the armature, and causes current flow around the coil, as shown by the solid arrows. This current creates a magnetic field that causes the armature to turn as described earlier. Immediately after rotation starts, an induced emf is produced, because the armature is cutting through the flux created by the field poles. By Lenz's law, this induced emf produces a current that flows in the opposite direction in the armature. Thus, the rotation induced emf in the motor armature opposes the applied voltage. The induced emf in a motor is therefore called a *counter emf*, abbreviated *cemf*. The magnitude of this voltage is given by the equation:

$$E_c = K \times \phi \times S \qquad (3)$$

Where:
   K = a constant whose value depends on the design details of machine (length and number of armature conductor, number of field poles, and so on).

   ϕ = The flux from each pole
   S = The speed of rotation of the armature.

NOTE: The value of *K* in this equation is different from that given in equation (2).

This formula indicates that the cemf will be proportional to the speed of rotation and to the field strength. Thus, when the load on the motor is light, the motor rotates at high speed and a large cemf is developed. This limits the armature current to a low value. When the motor is heavily loaded, the

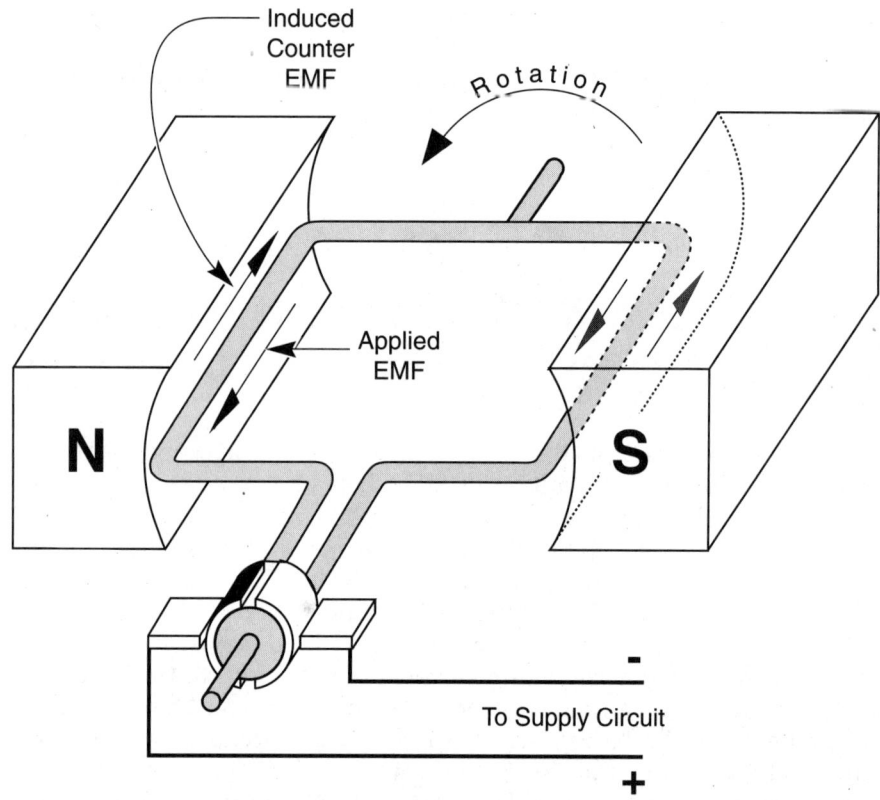

FIGURE 11–3 Counter EMF in a motor

speed decreases and the cemf decreases. This permits more current flow in the armature, thus producing greater torque.

Note also that when the speed is low, the cemf will also be low. This is the reason why the locked rotor current of all types of motors is very high.

CONSTRUCTION

Frame

Figure 11–4 shows the main construction elements of DC motors. Notice that Figure 11–4 is identical to Figure 10–4 with the exception of the interpole location. This likeness is because DC generators and DC motors are identical. The frame serves two purposes: first, it provides mechanical support for the magnetic main poles and the interpoles, and second, it provides a path for the completion of the field flux lines.

Frames are usually made of electrical grade steel with the alloy content and heat treatment optimized for maximum permeability. The physical

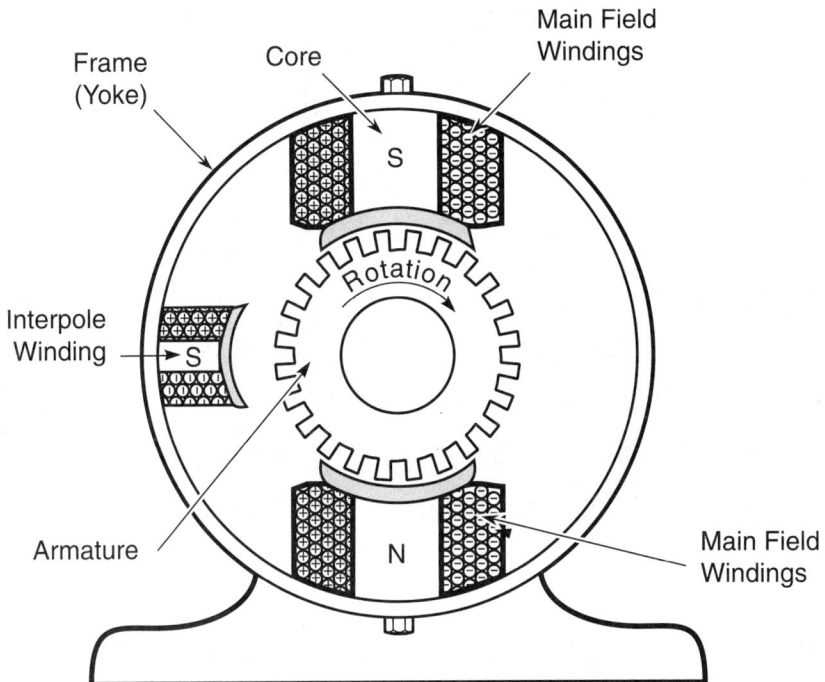

FIGURE 11–4
Major components of a two-pole DC motor with interpole

construction of the frame varies from one manufacturer to another. Figure 11–5 shows one method used to construct the frame and the magnetic cores for the field poles.

Field Windings

Main Poles The simple motor shown in Figure 11–1 uses permanent magnets to generate the magnetic field flux. Although some small, commercial motors use permanent magnets, most motors develop their magnetic fields electromagnetically. A typical field pole is shown in Figure 11–6.

Field windings are wound with electrically-conductive wire wrapped around the pole core. Copper wire is normally used, although aluminum is occasionally employed. The pole cores are usually a laminated construction to eliminate eddy currents set up by the pulsating flux created by the armature. Remember that although the output is DC, the current in the actual armature windings is alternating.

The field windings are energized with DC and are wound so they produce alternating north and south poles as shown. The wire is insulated with varnish, lacquer, or spiral wrappings of cloth or paper. The entire assembly is impregnated with an insulating material, such as varnish or lacquer.

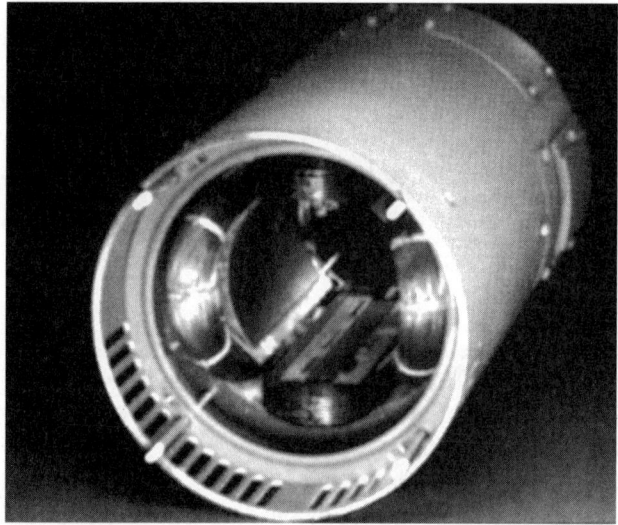

FIGURE 11–5
DC motor frame (Courtesy of General Electric Company)

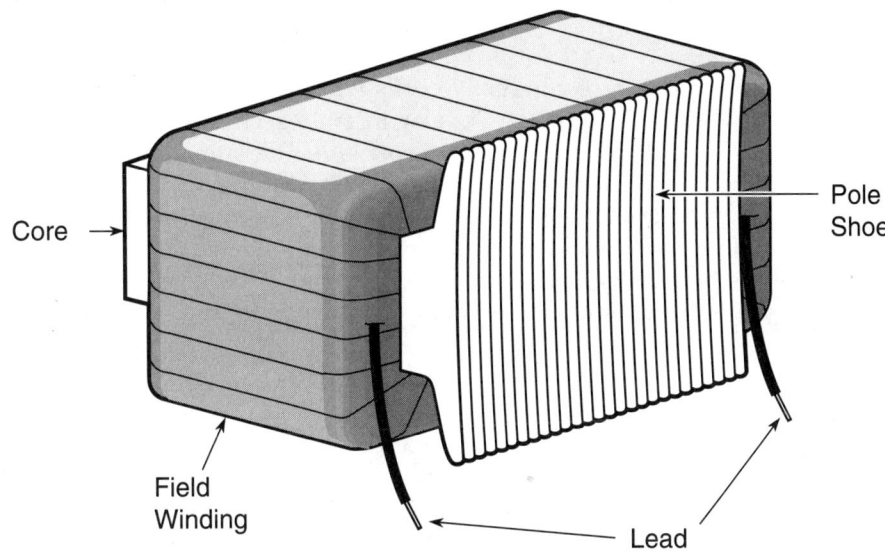

FIGURE 11-6
Field pole

The DC supply for the main poles may be supplied from an external source or from the armature itself. If supplied from an external source, the DC motor is referred to as a *separately-excited* motor. If connected to the armature, the motor is referred to as a *self-excited* motor. The main poles create flux in the yoke, the pole pieces, air gap, and core. Virtually all DC motors are of self-excited design.

When connected for self-excitation, the main poles may be connected either in series or in shunt with the armature. Main poles that are connected in shunt are wound with many turns of relatively small wire. Series field windings are wound with a smaller number of turns of relatively large wire.

The motor shown in Figure 11-1 is a two-pole motor: that is, it has one north main pole and one south main pole. Four-pole motors are also commonly used. They have four main poles creating alternate north and south poles.

Interpoles Interpoles are connected in series with the armature. Motor interpoles are connected with a polarity that is opposite a DC generator. That is, they are wound and connected in such a way that they produce the same polarity flux as the main pole just **behind** them in the direction of rotation, Figure 11-4. The interpole has the effect of reducing sparking at the commutator by magnetically suppressing the sudden change in the flux caused when the commutator opens an armature coil.

Interpoles are not as powerful as the main poles. A DC motor will usually have the same number of interpoles as main poles; however, half the number can be used with very little reduction in efficiency.

Armature

The armature consists of conductive coils which are rotated through the field set up by the field poles. Circular in cross section, the armature is made up of many laminated sheets of soft iron. The laminated sheets are slotted to provide a place for mounting the coil windings. The laminations are pressed onto the shaft Figure 11–7a.

The windings are then wound into the slots, and held in place with wooden or fiber wedges Figure 11–7b. Each end of each winding terminates at the commutator. Each winding may have multiple loops of wire.

FIGURE 11–7
Armature construction

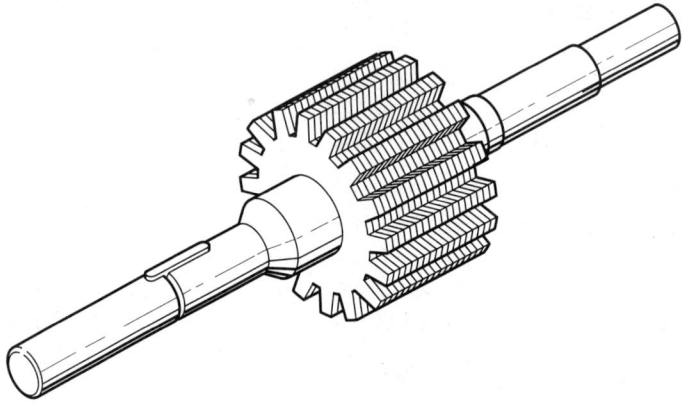

(a)
Unwound Armature Core on a Shaft

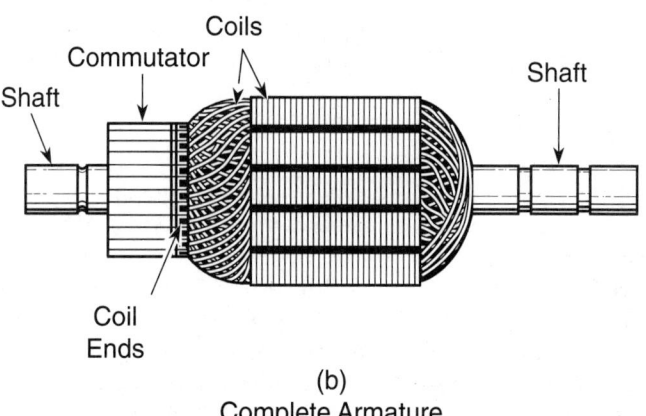

(b)
Complete Armature

Most DC motors are constructed with *form wound* armatures. In these machines, the windings are made to proper shape, insulated, and then installed on the rotor. *Random windings* are formed directly on the armature core by winding insulated wire into the slots. Random windings are more commonly used on very large DC motors.

Commutator

As previously discussed, the commutator is a high speed, rotary switch. It is used to switch from one armature winding to the next, to keep the output at the brushes at the same DC polarity. Figure 11–8 is a typical multi-segment commutator. The commutator is mounted on the armature shaft, and each segment is connected to one end of an armature winding. Thus the commutator will have a pair of segments, called *bars*, for each winding. The segments are insulated from each other by thin strips of mica.

Commutator bars are held in place by steel v-rings or clamping flanges. The commutator sleeve is keyed to the armature shaft and a mica collar insulates the commutator bars from the shaft.

The commutator bars usually have risers or flanges to which the leads from the associated armature windings are soldered. These risers serve as a shield for the soldered connections when the commutator becomes worn. When risers are not provided, the armature winding leads must be soldered to short slits in the ends of the commutator bars.

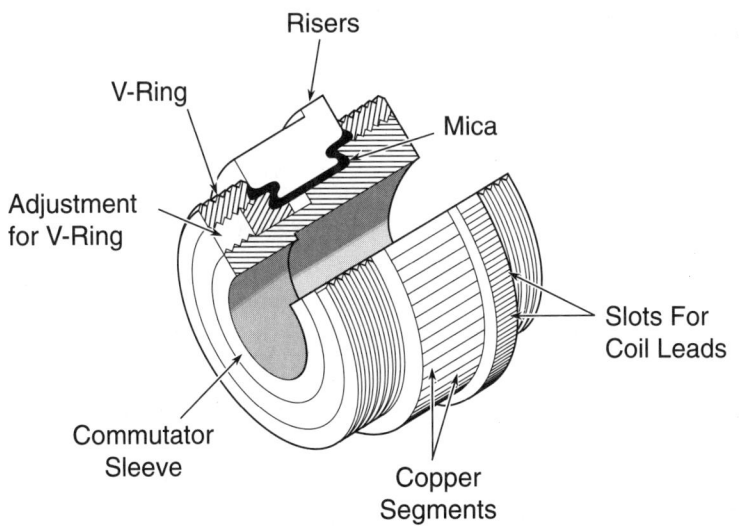

FIGURE 11–8 Commutator construction

Brushes and Brush Holders

Brushes are the points of contact for the motor to the load. They are usually made of a mixture of carbon in a graphite form with some metallic power added for strength and electrical conductivity. The brushes are held in place by mounting them in brush holders. An entire brush and brush holder assembly is shown in Figure 11–9. The proper brush pressure is maintained by mounting springs, and the brush is electrically-connected through flexible, braided copper pigtails.

The brush holders are attached to the frame of the motor with an assembly called a brush *rigging*. The rigging is insulated from the motor frame by an insulating pad. The brush holders are electrically-connected to the external DC circuit.

Because most DC motors have multiple pairs of north and south poles and multiple armature coils, there is generally more than one pair of brushes. A pair of brushes is usually required for each pole pair. Each pair of brushes consists of a positive and negative brush. The polarity of the brushes alternates about the commutator. The positive brushes are connected to the positive supply terminal, and the negative brushes to the neg-

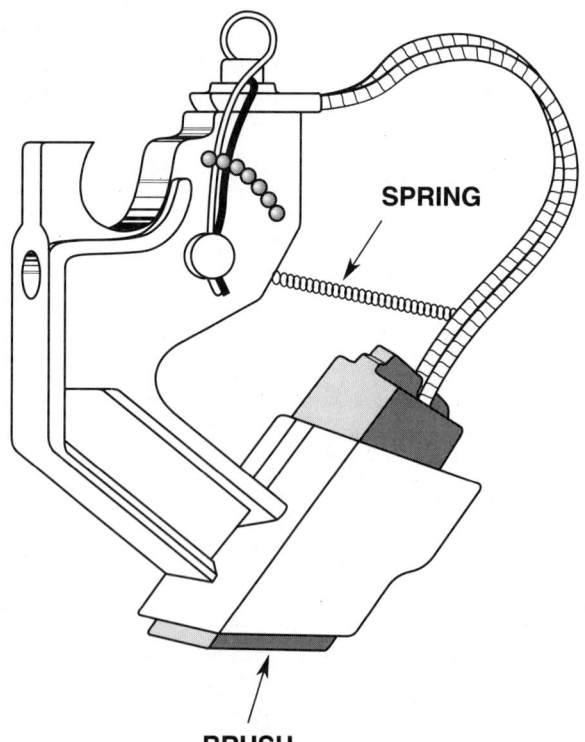

FIGURE 11–9
Brush and brush holder

DC Motor Theory and Construction

ative supply terminal. A cutaway view of a DC motor and its components is shown in Figure 11–10.

Ratings

Small and Medium-Sized Motors NEMA defines small and medium-sized motors as those that are smaller than 1.25 kw per rpm. NEMA defines the ratings for such motors in a number of areas, including form factor, horsepower, speed, voltage, time rating, ambient temperature, insulation class, and nameplate markings.

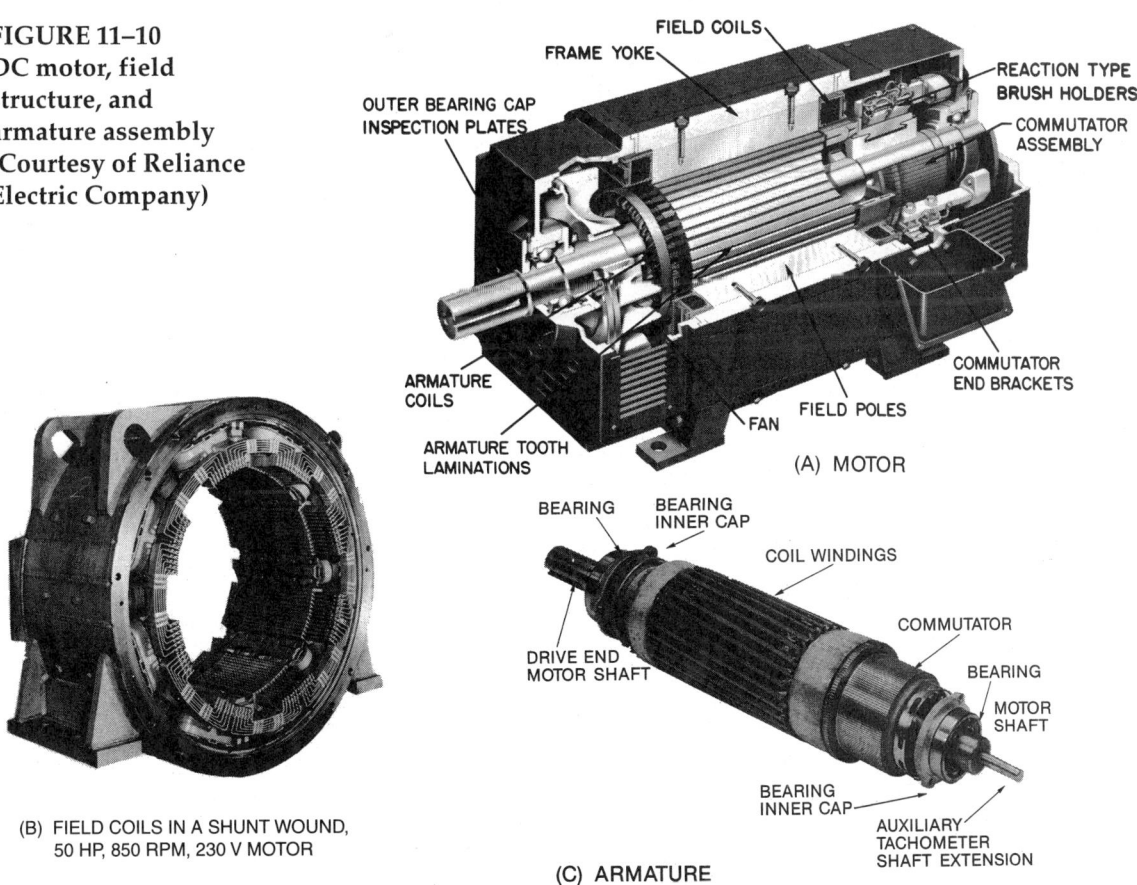

FIGURE 11–10 DC motor, field structure, and armature assembly (Courtesy of Reliance Electric Company)

(A) MOTOR

(B) FIELD COILS IN A SHUNT WOUND, 50 HP, 850 RPM, 230 V MOTOR

(C) ARMATURE

- Form Factor
  The basis of rating for DC small motors is the *form factor.* The form factor is a rating applied to the type of DC supply for the motor. A pure DC supply, with little or no ripple, will have a form factor of 1.0. As the ripple of the supply increases, the form factor increases.
- Kilowatt, Speed, and Voltage Ratings
  Tables 11–1 through 11–3 are representative samples of information available in the NEMA motor and generator standard MG-1. Notice that these tables specify horsepower, speed, voltage (field and armature), and form factor for small and medium-sized motors.
- Time Ratings
  Direct current motors can have either continuous or short time ratings. The marked short time ratings shall be 5, 15, 30, or 60 minutes.

Table 11–1 Horsepower and Speed Ratings for Small DC Motors Supplied by Low Ripple (Form Factor = 1.0) Supplies

(Courtesy of NEMA)

Hp	*Approximate Full Load, Rpm*			
1/20	3450	2500	1725	1140
1/12	3450	2500	1725	1140
1/8	3450	2500	1725	1140
1/6	3450	2500	1725	1140
1/4	3450	2500	1725	1140
1/3	3450	2500	1725	1140
1/2	3450	2500	1725	1140
3/4	3450	2500	1725	...
1	3450	2500	...	...

Table 11–2 Horsepower, Speed, Voltage Ratings for Small Motors Used with Adjustable Voltage Rectifier Power Supplies

(Courtesy of NEMA)

Hp	Approximate Rate-Load Speed, Rpm*				Rated Voltages, Average Direct-Current Values		Rated Form Factor
					Armature Voltages	Field Voltages	
Single-Phase Primary Power Source							
1/20	3450	2500	1725	1140			
1/15	3450	2500	1725	1140			
1/12	3450	2500	1725	1140			
1/8	3450	2500	1725	1140	75 volts	50 or 100 volts	See Notes I and II
1/6	3450	2500	1725	1140	90 volts	50 or 100 volts	
1/4	3450	2500	1725	1140	150 volts	100 volts	
1/3	3450	2500	1725	1140			
1/2	3450	2500	1725	1140			
3/4	3450	2500	1725	...	90 volts	50 or 100 volts	
1	3450	2500	...	...	180 volts	100 or 200 volts	
Three-Phase Primary Power Source							
1/4	3450	2500	1725	1140			
1/3	3450	2500	1725	1140			
1/2	3450	2500	1725	1140	240 volts	100, 150, or 240 volts	
3/4	3450	2500	1725	...			
1	3450	2500	...	...			

Note I – The rated form factor of a direct-current motor is the armature current form factor at rated load and rated speed and is an essential part of the motor rating.†

Note II – A direct-current motor is suitable for operation at the rated or lower form factor at rated load and rated speed. The rated form factor of a direct-current motor is determined by the motor manufacturer; see MG 1-10.66. Recommended rated for factors are given in Table 14-2 of MG 1-14.60.†

* Motors rated 1/20 to 1 horsepower, inclusive, are not suitable for speed control by field weakening.†
† Approved as Authorized Engineering Information.

Table 11-3 Horsepower, Speed, and Voltage for Industrial (Medium) DC Motors 240 Volt Armature

(Courtesy of NEMA)

Hp	Base Speed, Rpm									Field Voltage, Volts
	3500	2500	1750	1150	850	650	500	400	300	
	Speed by Field Control, Rpm									
½	...	...	...	...	1700	...	...	...	...	100, 150, or 240
¾	...	...	...	2000	1700	...	...	...	...	
1	...	...	2300	2000	1700	...	...	...	...	
1½	3850	3000	2300	2000	1700	...	...	...	...	
2	3850	3000	2300	2000	1700	...	...	...	...	
3	3850	3000	2300	2000	1700	...	...	...	...	
5	3850	3000	2300	2000	1700	...	...	...	...	
7½	...	3000	2300	2000	1700	1600	1500	1200	1200	150 or 240
10	...	3000	2300	2000	1700	1600	1500	1200	1200	
15	...	3000	2300	2000	1700	1600	1500	1200	1200	
20	...	3000	2300	2000	1700	1600	1500	1200	1200	
25	...	3000	2300	2000	1700	1600	1500	1200	1200	
30	...	3000	2300	2000	1700	1600	1500	1200	1200	
40	...	3000	2100	2000	1700	1600	1500	1200	1200	
50	...	...	2100	2000	1700	1600	1500	1200	1200	
60	...	...	2100	2000	1700	1600	1500	1200	1200	
75	...	...	2100	2000	1700	1600	1500	1200	1200	
100	...	...	2000	2000	1700	1600	1500	1200	1200	
125	...	...	2000	2000	1700	1600	1500	1200	1200	
150	...	...	2000	2000	1700	1600	1500	1200	1100	
200	...	...	1900	1800	1700	1600	1500	1200	1100	
250	...	...	1900	1700	1600	...	...	...	...	

- Rated Ambient Temperature
  For small and medium motors the rated ambient temperature is 40°C, unless otherwise marked. The allowable temperature rise for small and medium motors depends upon the motor size and its time rating Tables 11–4 through 11–7 show the ratings that NEMA assigns to small and medium motors.

Table 11–4 Allowable Temperature Rises for Small DC Motors
(Courtesy of NEMA)

	All Enclosures		
Class of Insulation System (See MG 1-1.65)	A	B	F
Time Rating (See MG 1-10.63)			
Temperature Rise, Degrees C			
1. Armature windings and all windings other than those given in item 2 – resistance	70	100	130
2. Shunt field windings – resistance	70	100	130
3. The temperatures attained by cores, commutators, and miscellaneous parts (such as brushholders, brushes, pole tips, etc.) shall not injure the insulation or the machine in any respect.			

Suggested Standard for Future Design 7-7-1965, reaffirmed 11-21-1968, NEMA Standard 11-12-1970, revised 3-6-1985.

Note I — Abnormal deterioration of insulation may be expected if the ambient temperature of 40°C is exceeded in regular operation. See Note III.

Note II — The foregoing values of temperature rise are based upon operation at altitudes of 3300 feet (1000 meters) or less. For temperature rises for motors intended for operation at altitudes above 3300 feet (1000 meters), see MG 1-14.04.

Note III — When the rated maximum ambient temperature is other than 40°C, the rated value should be 50°C or 65°C. The maximum temperature rise values for each machine part and for either method of measurement should be the value selected from the appropriate table with an adjustment as follows.

Rated Maximum Ambient Temperature Degrees C	Adjustment to Temperature Rise Values, Degrees C
50	Subtract 10
65	Subtract 25

Table 11-5 Allowable Temperature Rises, Medium Motors—Continuous Time Ratings
(Courtesy of NEMA)

	Totally-enclosed, Nonventilated and Totally-enclosed, Fan-cooled Motors, Including Variations Thereof				Motors with All Other Enclosures			
Class of Insulation System (see MG 1-1.65)	A	B	F	H	A	B	F	H
Time Rating	Continuous				Continuous			
Temperature Rise, Degrees C*								
1. Armature windings and all windings other than those given in items 2 and 3 – resistance	70	100	130	155	70	100	130	155
2. Multi-layer field windings – resistance	70	100	130	155	70	100	130	155
3. Single-layer field windings with exposed uninsulated surfaces and bare copper windings – resistance	70	100	130	155	70	100	130	155
4. The temperatures attained by cores, commutators, and miscellaneous parts (such as brushholders, brushes, pole tips, etc.) shall not injure the insulation or the machine in any respect.								

Note I — Abnormal deterioration of insulation may be expected if the ambient temperature of 40°C is exceeded in regular operation. See Note III.

Note II — The foregoing values of temperature rise are based upon operation at altitudes of 3300 feet (100 meters) or less. For temperature rises for motors intended for operation at altitudes above 3300 feet (1000 meters), see MG 1-14.04.

Note III — When the rated maximum ambient temperature is other than 40°C, the rated value should be 50°C or 65°C. The maximum temperture rise values for each machine part and for either method of measurement should be the value selected from the appropriate table with an adjustment as follows.

Rated Maximum Ambient Temperature Degrees C	Adjustment to Temperature Rise Values, Degrees C
50	Subtract 10
65	Subtract 25

Table 11-6 Allowable Temperature Rises, Medium Motors 30 and 60 Minute Ratings
(Courtesy of NEMA)

	Motors Rated 30 and 60 Minutes*							
	Dripproof, Forced-ventilated,† and Other Enclosures				*Totally-enclosed Nonventilated and Totally-enclosed Fan-cooled Motors, Including Variations Thereof*			
Class of Insulation System (see MG 1-1.65)	A	B	F	H	A	B	F	H
Temperature Rise, Degrees C*								
1. Armature windings and all windings other than those given in items 2 and 3 – resistance	70	100	130	155	80	110	140	165
2. Multi-layer field windings – resistance	70	100	130	155	80	110	140	165
3. Single-layer field windings with exposed uninsulated surfaces and bare copper windings – resistance	70	100	130	155	80	110	140	165
4. The temperatures attained by cores, commutators, and miscellaneous parts (such as brushholders, brushes, pole tips, etc.) shall not injure the insulation or the machine in any respect.								

* See MG 1-10.63

† Forced-ventilated motors are defined in 1.25.6, 1.25.7, and 1.26.6

Rated Maximum Ambient Temperature Degrees C	Adjustment to Temperature Rise Values, Degrees C
50	Subtract 10
65	Subtract 25

Table 11-7 Allowable Temperature Rises, Medium Motors 5 and 15 Minute Ratings
(Courtesy of NEMA)

	Motors Rated 5 and 15 Minutes*							
	Dripproof, Forced-ventilated,† and Other Enclosures				Totally-enclosed Nonventilated and Totally-enclosed Fan-cooled Motors, Including Variations Thereof			
Class of Insulation System (see MG 1-1.65)	A	B	F	H	A	B	F	H
Temperature Rise, Degrees C*								
1. Armature windings and all windings other than those given in items 2 and 3 – resistance	80	115	145	175	90	125	155	185
2. Multi-layer field windings – resistance	80	115	145	175	90	125	155	185
3. Single-layer field windings with exposed uninsulated surfaces and bare copper windings – resistance	80	115	145	175	90	125	155	185
4. The temperatures attained by cores, commutators, and miscellaneous parts (such as brush-holders, brushes, pole tips, etc.) shall not injure the insulation or the machine in any respect.								

* See MG 1-10.63
† Forced-ventilated motors are defined in 1.25.6, 1.25.7, and 1.26.6
1. See Referenced Standards, MG 1-1.01.
2. Notes approved as Authorized Engineering Information.

- Power Supplies
  As defined by NEMA for testing and checking DC motor characteristics, power supplies fall into five categories identified as A, C, D, E, and K. Figure 11–11 illustrates the diagrams of the five supplies. Note that Type A is designated as a low ripple supply with a form factor of 1.0. Other supplies may be designated as a low ripple supply by using a series inductor to obtain a ripple with less than a 6% peak-to-peak ripple.
- Nameplates
  NEMA requires slightly different information on the nameplate, depending upon what size and type of DC motor is being considered. The following information covers the information given in NEMA MG-1, Part 12:

DC Motor Theory and Construction

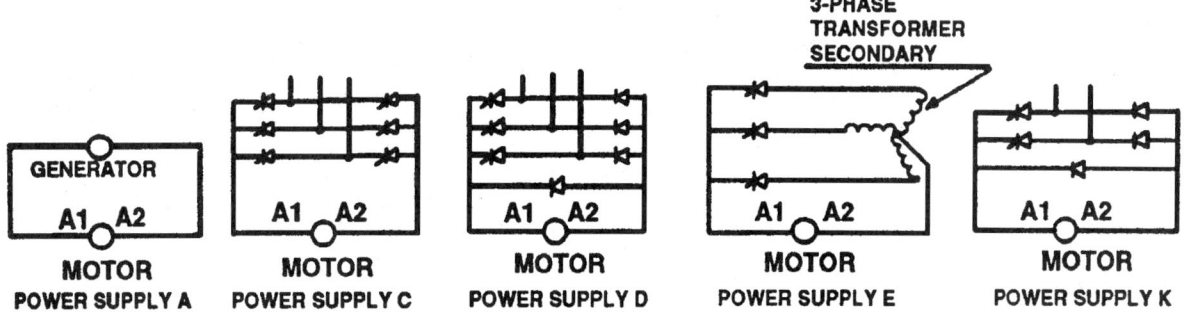

FIGURE 11-11
Test power supplies (Courtesy of NEMA)

Small Motors, Less than or Equal to 1/20 Horsepower

- Manufacturer's type designation and frame number
- Power output in millihorsepower (mhp)
- Full load speed
- Voltage rating
- The words "thermally protected" means that thermal protection is included in the motor itself
- The words "impedance protected" for motors with sufficient impedance to be protected from dangerous overheating due to overload or failure to start.

Small Motors Except Those Rated 1/20 Horsepower or Less

- Manufacturer's type designation
- Horsepower output at rated speed
- Time rating at rated speed
- Maximum ambient temperature
- Insulation system design (If field and armature have different systems, both must be marked.)
- Speed in rpm
- Rated armature voltage
- Rated field voltage or the letters "PM" if a permanent magnet design
- Armature rated load amperes at rated speed
- Rated form factor when operated from a rectifier power supply
- The words "thermally protected" for motors equipped with a thermal protector.

Medium Motors

- Manufacturer's type and frame designation
- Horsepower output at base speed

Time rating at rated speed
Maximum safe rpm for all series-wound motors, and for those compound-wound motors whose variation in speed from rated load to no load exceeds 35% with the windings at the constant temperature attained when operating at its rating
Maximum ambient temperature
Insulation system design (if field and armature are different, both must be shown)
Base speed at rated load
Rated armature voltage
Rated field voltage
Armature rated load current in amperes at base speed
Power supply identification
Winding type (shunt, compound, and so on)

Large Motors Many of the same ratings are provided for large motors. However, the ratings are somewhat more diverse than small and medium motors. For example, Table 11–8 is the horsepower, speed, and voltage rating for general industrial motors. Other NEMA tables are available for other types of large motors.

- Time ratings and temperature rises are determined by the manufacturer. The maximum ambient temperature is 40°C. The temperature rises based on insulation class are shown in Table 11–9.
- Nameplate information for large motors is very similar to the nameplate information for small motors:

Manufacturer's name and serial number
Horsepower output
Time rating
Temperature rise for rated continuous load
RPM at rated load
Rated voltage
Amperes at rated load
Winding: series, shunt, or compound

SELF-EXCITED DC MOTORS

Overview

Most DC motors obtain their field current by connecting the field to the DC supply in the same way that DC generators are connected. There are only a limited number of ways in which a motor may be connected. Each type of connections offers a slightly different combination of torque and speed

Table 11-8 Horsepower, Speed, and Voltage Ratings for Large, General Purpose, Industrial DC Motors
(Courtesy of NEMA)

Hp	850	650	500	450	400	350	300	250	225	200	175	150	125	110	100	90	80	70	65	60	55	50
250	…	…	…	…	…	…	…	…	…	…	…	A	A	A	A	…	…	…	…	…	…	…
300	…	…	…	…	…	…	…	…	…	…	A	A	A	A	A	…	…	…	…	…	…	…
400	…	…	…	…	…	…	A	A	A	A	A	A	A	A	A	…	…	…	…	…	…	…
500	…	…	…	…	…	A	A	A	B	B	B	B	B	B	B	C	…	…	…	…	…	…
600	…	…	…	…	A	A	A	B	B	B	B	B	B	B	B	C	C	…	…	…	…	…
700	…	…	A	A	A	B	B	B	B	B	B	B	B	B	B	C	C	C	C	C	…	…
800	…	…	A	A	A	B	B	B	B	B	B	B	B	B	C	C	C	C	C	C	…	…
900	…	A	A	A	A	B	B	B	B	B	B	C	C	C	C	C	C	C	C	C	C	…
1000	…	B	B	B	B	B	B	B	C	C	C	C	C	C	C	C	C	C	C	C	C	C
1250	C	C	C	C	C	C	C	C	C	C	C	C	C	C	C	C	C	C	C	C*	C*	C*
1500	C	C	C	C	C	C	C	C	C	C	C	C	C	C	C	C	C	C	C	C*	C*	C*
1750	C	C	C	C	C	C	C	C	C	C	C	C	C	C	C	C	C	C*	C*	C*	C*	C*
2000	…	C	C	C	C	C	C	C	C	C	C	C	C	C	C	C	C	C*	C*	C*	C*	C*
2250	…	C	C	C	C	C	C	C	C	C	C	C	C	C	C	C	C	C*	C*	C*	C*	C*
2500	…	…	C	C	C	C	C	C	C	C	C	C	C	C	C	C	C	C*	C*	C*	C*	C*
3000	…	…	…	…	C	C	C	C	C	C	C	C	C	C	C	C	C	C*	C*	C*	C*	C*
3500	…	…	…	D	D	D	D	D	D	D	D	D	D	D	D	D	D*	D*	D*	D*	D*	D*
4000	…	…	…	D	D	D	D	D	D	D	D	D	D	D	D	D	D*	D*	D*	D*	D*	D*
4500	…	…	…	…	…	D	D	D	D	D	D	D	D	D	D	D*	D*	D*	D*	D*	D*	D*
5000	…	…	…	…	…	…	D	D	D	D	D	D	D	D	D*	D*	D*	D*	D*	D*	D*	D*
6000	…	…	…	…	…	…	…	…	…	D	D	D	D	D*	D*	D*	D*	D*	D*	D*	D*	D*
7000	…	…	…	…	…	…	…	…	…	…	…	D	D*	D*	D*	D*	D*	D*	D*	D*	D*	D*
8000	…	…	…	…	…	…	…	…	…	…	…	…	D*	D*	D*	D*	D*	D*	D*	D*	D*	D*

Base Speed, Rpm

* These ratings are based on forced ventilation.

"A" indicates voltage rating at either 250 or 500 volts.
"B" indicates voltage ratings at either 250, 500, or 700 volts.
"C" indicates voltage rating at either 500 or 700 volts.
"D" indicates voltage rating at 700 volts only.

characteristics. The following paragraphs describe each connection and briefly explain how the motor is used.

Series-Connected Motors

The series-connected motor, Figure 11–12, is so named because the field winding is connected in series with the armature. Thus, the field current and the armature current are all the same. The field windings must be low resistance windings; therefore, it is wound with relatively few turns of large wire. The terminal designations for the armature are A_1 and A_2. The series field connections are S_1 and S_2.

The series motor has the highest amount of starting torque of all motors of equal horsepower rating. Unfortunately, the series motor has very poor speed regulation. In fact, if the load is suddenly removed from a series motor, the motor speed will increase rapidly. The speed will continue to rise, theoretically forever, until the motor literally flies apart.

Series motors are used to start large loads, such as internal combustion engines and are commonly used to drive cranes, hoists, winches, and electrical vehicles. In short, the series-connected motor may be used for any application that requires high starting torque, and has no particular requirement for speed regulation.

Shunt-Connected Motors

When the field windings are connected in parallel with the armature, the motor is shunt-connected, Figure 11–13. The shunt-connected motor has excellent speed regulation, but lower starting torque than the series-connected motor. Shunt motors are used extensively on lathes, milling machines, drills, planers, shapers, and other equipment that need a closely regulated speed. Note that the shunt motor can be speed-controlled by

FIGURE 11–12
Series-connected DC motor

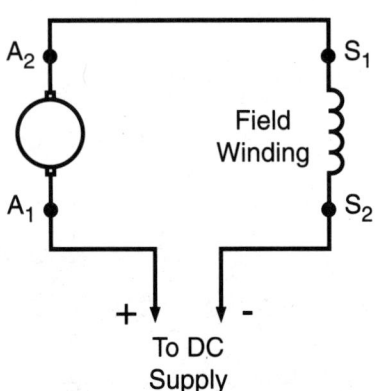

Table 11-9 Allowable Temperature Rises, Large DC Motors (Degrees C°)
(Courtesy of NEMA)

Item	Machine Part	Method of Temperature Determination†	General Industrial Service								Metal Rolling Mill Service							
			Semi-enclosed Continuous Rated 100% Load				Totally-enclosed Continuous Rated 100% Load				Metal Rolling Mills, Excluding Reversing Hot Mills Forced-ventilated or Totally-enclosed Water-air-cooled						Reversing Hot Mills Forced-ventilated or Totally-enclosed Water-air-cooled		
											Continuous Rated 100% Load			2 Hours‡ 125% Load			Continuous Rated 100% Load		
			Insulation Class				Insulation Class				Insulation Class			Insulation Class			Insulation Class		
			A	B	F	H	A	B	F	H	B	F	H	B	F	H	B	F	H
1	Armature windings and all other windings other than those given in items 2 and 3	Thermometer	50	70	90	110	55	75	95	115	40	60	75	55	75	95	50	70	90
		Resistance	70	100	130	155	70	100	130	155	60	90	110	80	110	135	70	100	130
2	Multilayer field windings	Resistance	70	100	130	155	70	100	130	155	70	100	120	80	110	135	70	100	130
3	Single-layer field windings with exposed uninsulated surfaces and bare copper windings	Thermometer	60	80	105	130	65	85	110	135	50	70	90	65	85	110	60	80	105
		Resistance	70	100	130	155	70	100	130	155	60	90	110	80	110	135	70	100	130
4	Commutator and collector rings	Thermometer	65	85	105	125	65	85	105	125	55	75	90	65	85	105	65	85	105
5	The temperatures attained by cores, commutators, and miscellaneous parts (such as brushholders, brushes, pole tips, etc.) shall not injure the insulation or the machine in any respect.																		

*For motors which operate under prevailing barometric pressure and which are designed not to exceed the specified temperature rise at altitudes from 3300 feet (1000 meters) to 13000 feet (4000 meters), the temperature rises, as checked by tests at low altitudes, shall be less than those listed in the foregoing table by 1 percent of the specified temperature rise for each 330 feet (100 meters) of altitude in excess of 3300 feet (1000 meters).
†Where two methods of temperature measurement are listed, a temperature rise within the values listed in the table measured by either method demonstrates conformity with the standard.
‡Temperature limits apply at end of 2-hour operation at 125-percent load following operation at rated load long enough to reach a stable temperature.

NOTE I—See MG 1-1.65 for description of classes of insulation.
NOTE II—Abnormal deterioration of insulation may be expected if the ambient temperature of 40°C is exceeded in regular operation.

[1] For totally-enclosed water-air-cooled machines, the temperature of the cooling air is the temperature of the air leaving the coolers. Totally-enclosed water-air-cooled machines are normally designed for the maximum cooling water temperature encountered at the location where each machine is to be installed. With a cooling water temperature not exceeding that for which the machine is designed:
  a. On machines designed for cooling water temperatures up to 30°C—the temperature of the air leaving the coolers shall not exceed 40°C.
  b. On machines designed for higher cooling water temperatures—the temperature of the air leaving the coolers shall be permitted to exceed 40°C provided the temperature rises of the machine parts are then limited to values less than those given in the table by the number of degrees that that temperature of the air leaving the coolers exceeds 40°C.
[2] See Referenced Standards, MG 1-1.01.

FIGURE 11–13
Shunt-connected
DC motor

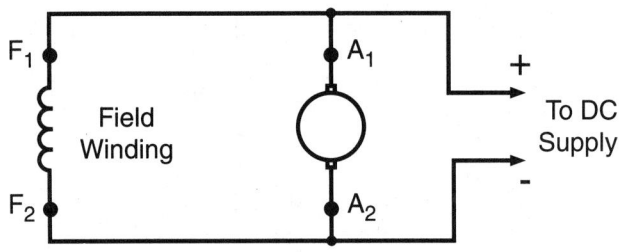

controlling the amount of field excitation. Also note that if the field is suddenly removed from a shunt motor, its speed will also increase indefinitely.

Compound-Connected Motors

Compound-connected motors, Figure 11–14, employ both a series and a shunt-connected field. The compound motor has speed and torque char acteristics that fall between shunt motors and series motors. They are used in applications which require higher starting torque than is available from a shunt motor, but better speed regulation than a series motor.

DC Motor Characteristics

DC motors have extremely high torque characteristics, can be readily speed-controlled, and have very high starting currents. Figure 11–15 is a comparison of the starting torque and speed regulation characteristics of

FIGURE 11–14
Compound-connected
DC motors

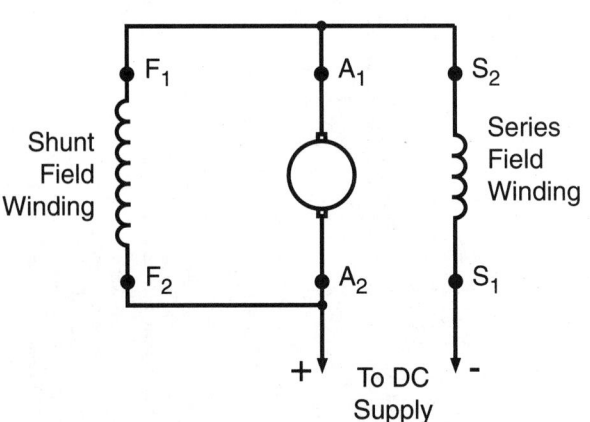

DC Motor Theory and Construction

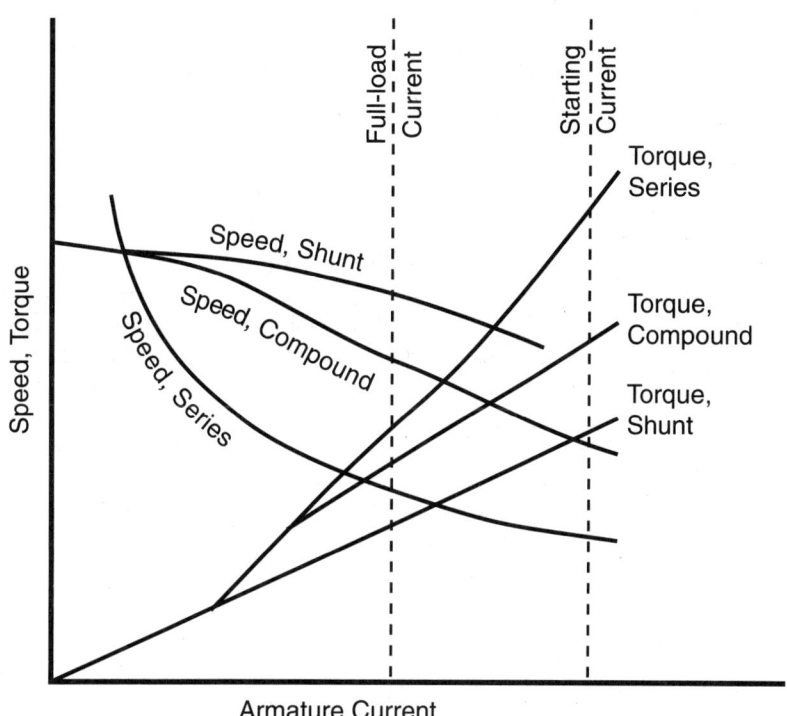

FIGURE 11-15
Comparative characteristics for shunt, series and compound DC motors

the three DC motor types. Table 11-10 is a tabular comparison of the various types of motors.

DC MOTOR SPEED CONTROL

DC motors can be operated below their normal speed by reducing the voltage that is applied to the armature. This can be accomplished by placing the resistor in series with the armature. Because of the high armature current, this method of speed control is rarely used, except with solid-state voltage supplies.

DC motors may be operated above rated speed by reducing the field flux. This is accomplished by putting a rheostat in series with the field winding.

Solid-state control circuits using silicon-controlled rectifiers can be used to vary the armature and/or the field current. This type of speed control is available in modern, solid-state units that provide extremely accurate speed regulation.

Table 11–10 DC Motor Characteristics Comparison

Speed Regulation	Speed Control	Starting Torque	Breakdown Torque	Applications
Series				
Varies inversely as load; races on light loads, full voltage.	Zero to maximum depending on control and on load.	Highest. Varies as square of voltage; limited by commutation and line capacity.	Highest. Limited by commutation, heating, and line capacity.	Where high starting torque is required and speed can be regulated. Traction, bridges, hoists, gates, car dumpers, car retarders.
Shunt				
Drops 3% to 5% from no load to full load.	Any desired range, depending on design, type of system.	High. With constant field it varies directly as voltage applied to armature.	High. Limited by commutation, heating and line capacity.	Where constant or adjustable speed is required and starting conditions are not severe. Fans, blowers, centrifugal pumps, conveyors, wood, and metal-working machines, elevators.
Compound				
Drops 7% to 20% from no load to full load depending on compounding.	Any desired range, depending on design, type of control.	Higher than for shunt, depending on amount of compounding.	Higher. Limited by commutation, heating and line capacity.	Where high starting torque and fairly constant speed is required. Plunger pumps, punch presses, shears, bending rolls, geared elevators, conveyors, hoists.

DIRECTION OF ROTATION

The direction of armature rotation of a DC motor depends upon the direction of the current in the field with respect to the current in the armature. To reverse the direction of rotation, the current direction in either the armature or the field may be reversed. Reversing the power supply leads *will not* reverse the direction of rotation.

SUMMARY

The construction of a basic DC motor is identical to a DC generator. Instead of connecting the armature to a load, however, the armature is connected to a DC voltage supply. Armature and field flux are created with such a relationship that the armature rotates. The DC motor also uses a commutator. As the armature rotates, the brushes are connected, alternately, to each side of the armature loop. The connection is made in such a way that the armature field reverses polarity so that the motor rotation is continued.

The commutator segments are positioned in such a way that the polarity is changed on the armature winding, just as the armature field is in exact opposition to the field flux. This prevents arcing and excessive brush and commutator wear. This point is called the neutral point.

"Real" DC motors have multiple windings on the armature, and a field created by an electromagnet. There are two types of field poles: the main poles and the interpoles. The main poles are used to turn the armature. The interpoles are used to help in the commutation process and minimize arcing at the brushes. The commutator is composed of copper segments insulated from each other with mica. The brushes are a carbon-based composition. There is a set of brushes for each main pole. Most DC motors have either two poles or four poles.

NEMA defines the standard ratings for DC motors including items such as kilowatt ratings, speed, and voltage. The standards provide a level playing field so that purchasers can compare features and characteristics.

Motors may be of the permanent magnet, series-connected, shunt-connected, or compound-connected type. Each type has its own specific features and capabilities. The shunt-connected motor is the most commonly used.

Motors have distinct speed and torque relationships, depending on the connection. The series motor has the highest torque and the worst speed regulation. The shunt motor has the lowest torque and a very flat speed characteristic. All DC motors have higher torque per horsepower than other motor types.

REVIEW QUESTIONS

1. If the brush rigging is not properly positioned at the neutral point, _____ will occur when the brushes momentarily connect one commutator segment to the adjacent segment.

2. Look at Figure 11–1. If the DC supply polarity is reversed, the motor will run counterclockwise. Why?

3. Why are the field pole cores laminated?

4. What is the difference between the interpoles on a DC generator and the interpoles on a DC motor?

5. What does the commutator do on a DC motor?

6. A certain motor rotates at 300 rpm and has a rated horsepower of 250. Does NEMA classify this as a large motor?

7. What is the allowable temperature rise for the armature of a totally enclosed, general industrial service motor with class F insulation?

8. What is form factor?

9. You have an application for a DC motor. The application requires very high torque and is relatively insensitive to speed variations. What type of DC motors might fit this application?

10. A certain DC, shunt-connected motor has a base speed of 1150 rpm. You want to speed control the motor from 850 rpm to 1250 rpm. Is this possible? How would you do it?

DC Machine Maintenance

12

OBJECTIVES

After studying this chapter, the student should be able to:

- *Describe the types of visual mechanical inspections to be performed on DC machines.*
- *Explain the symptoms that signify the need for maintenance on the brush and/or commutator system.*
- *Describe the electrical tests to be performed on a DC machine.*
- *Describe methods for troubleshooting various types of common problems encountered on DC machines.*

INTRODUCTION

The maintenance and troubleshooting procedures on DC motors and DC generators are very similar. This chapter will explain how to perform those procedures and will illustrate how to perform many of the more common ones. Differences between DC motor and DC generator procedures will be identified.

VISUAL MECHANICAL INSPECTIONS

Inspect the machine for signs of physical damage. Look for dents, burnt spots, frayed insulation, oil leaks, and other such problems. When possible turn the rotor slowly by hand to find obvious rough spots and/or rubs.

Anchorage and Mounting

Inspect the machine mounting equipment to make certain it is adequately installed. The normal operating vibration of some machines may loosen the mounting hardware over time. Check mountings with a torque wrench to make certain that they are properly installed.

Check the machine to make certain it is level. Even minor variations in horizontal or vertical alignment can cause long acceleration times and excessive wear on bearings.

Grounding

Verify that the machine frame is properly grounded and the grounding conductors and hardware are undamaged and properly installed. *This is an extremely important safety step!!* The ground obtained through the mounting hardware and assemblies is usually not sufficient for personnel protection. Machine frames should be solidly grounded and bonded to other electrical equipment, including the starter frame and/or any metallic conduits.

Connections

Carefully inspect the machine connection points. Inspections should involve two important steps:

- Infrared Scan
  With the machine running and fully loaded, all connections should be inspected using an infrared scan camera. Remake any connections that are hotter than the cables to which they are connected.
- Torque Wrench
  Use a torque wrench to tighten all connections, especially bad ones that have shown up in the infrared scan.

Hot connections account for a substantial number of failures—many of them catastrophic. Be certain that surge arrestor connections are never used for connections, Figure 12–1. The connection shown in Figure 12–1a uses the surge arrestor connection flange to carry machine current. The

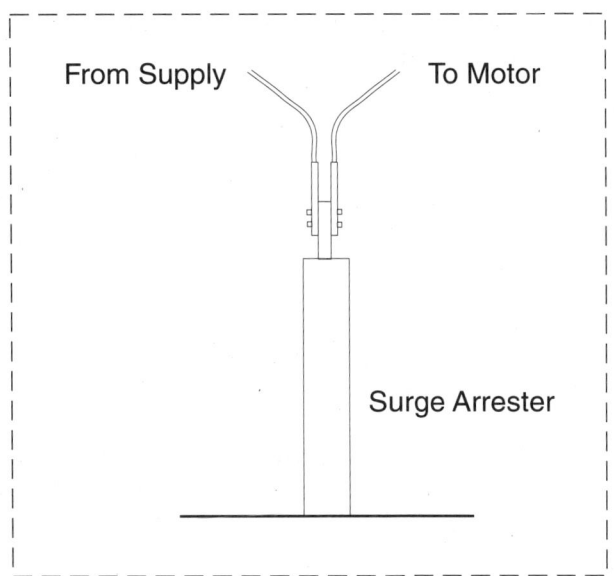

a. Wrong

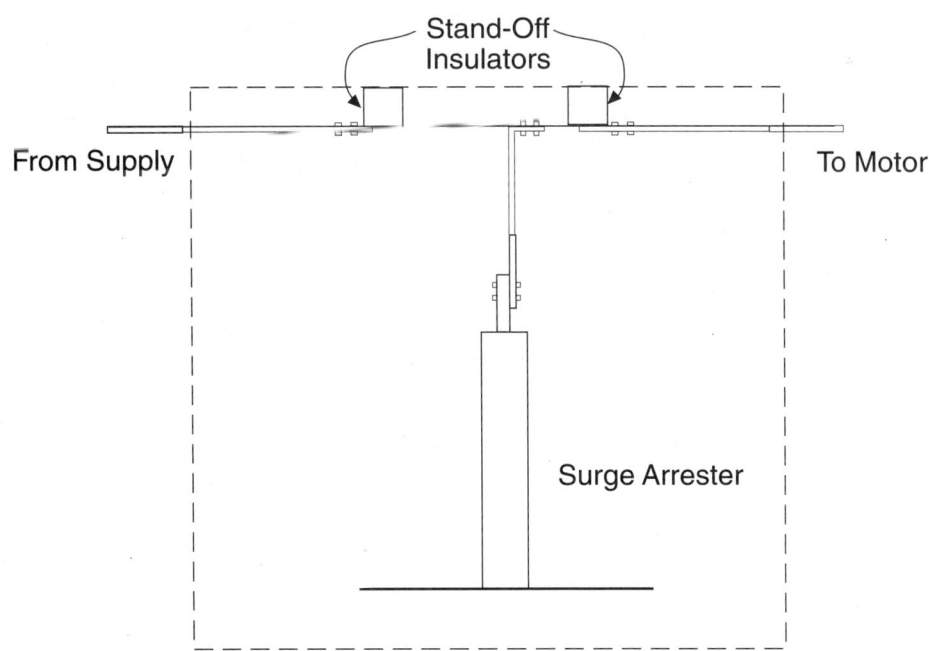

b. Correct

FIGURE 12–1
Surge arrester connections (Courtesy of Cadick Professional Services)

flange is not rated for such use and will overheat. Figure 12–1b shows the proper way to make the connection.

Lubrication

Check the machine bearings to ensure that the correct lubricant is present in the proper amounts. Some journal bearings are lubricated with oil rings, Figure 12–2. The oil rings are larger in diameter than the shaft, and suspended from the shaft into a reservoir of oil. As the shaft turns, the rings also rotate, carrying oil from the reservoir to the bearings. You may need to disassemble the bearing to inspect it and ensure that the oil ring is free to move.

Oil reservoirs should be filled to the proper level with the proper oil weight and grade. The shaft must be level to ensure proper lubrication.

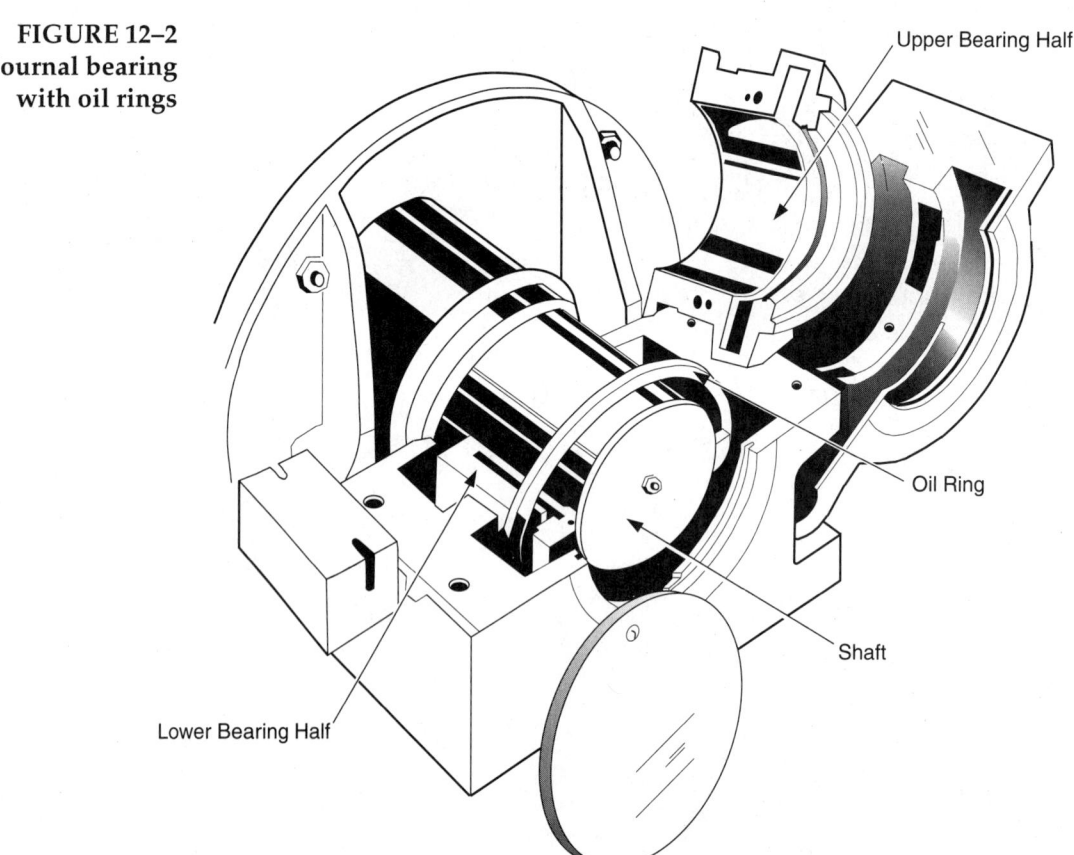

FIGURE 12–2 Journal bearing with oil rings

Ventilation

Make certain all ventilation ports are clear of foreign materials that may inhibit the movement of cooling air.

Rotor Air Gap

Check the rotor air gap to make certain the rotor is properly positioned in the stator, Figure 12–3. The air gap is defined as the radial, iron-to-iron clearance between the center of a pole face and the face of a core tooth. It is measured using a tapered air gap gauge or a long feeler gauge. Check the gap at a minimum of four locations (1, 2, 3, and 4 in Figure 12–3). The readings should fall within manufacturer's specifications.

Some machines are more difficult to check than others, and some machines may not be adjustable. Refer to manufacturer's instructions for detailed procedures.

BRUSHES AND COMMUTATOR MAINTENANCE

Note that while this section focuses on commutators, the bulk of the material applies equally to slip rings used on AC synchronous machines and wound rotor motors.

Normal Operation

When a brush and commutator system is operating properly, a very fine film will form on the surface of the commutator, Figure 12–4. This film

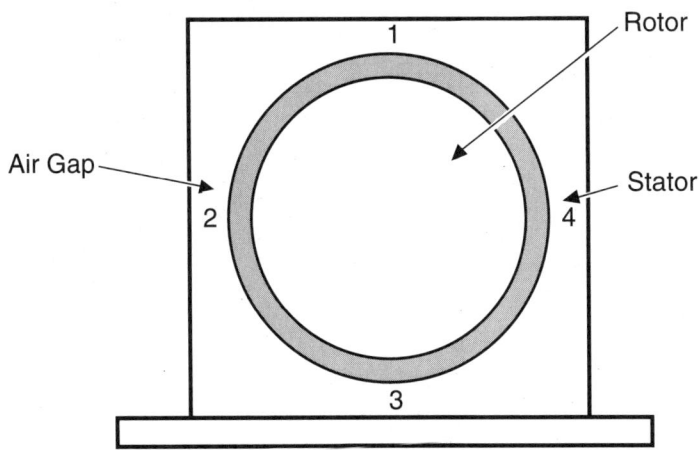

FIGURE 12–3
Measuring the rotor air gap

is composed of copper particles, carbon particles, and moisture. The film lubricates the commutator and provides an improved electrical connection.

The formation of this film is necessary for proper operation of the commutator. A good film is formed when the proper brush is used for the application and the commutator. The following sections will identify key information for proper installation and maintenance of brushes and commutators.

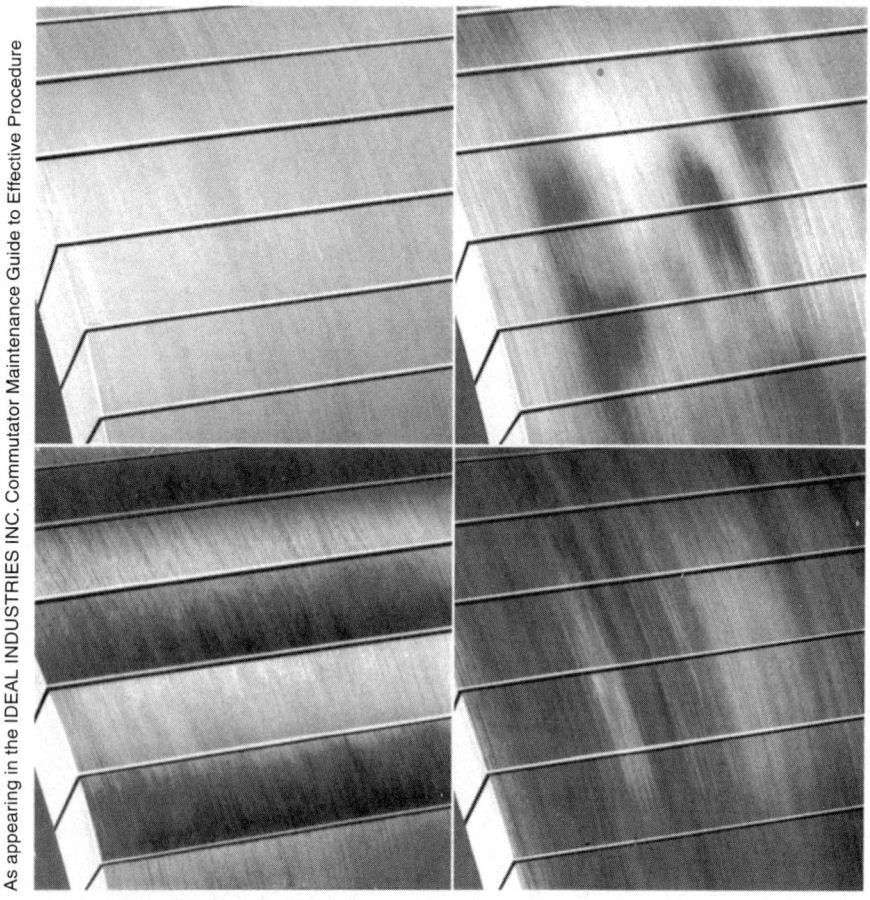

FIGURE 12–4
"Good" commutator film patterns (Courtesy of Ideal Industries Inc.)

Brushes

Brush Selection Brush size and material are selected based on the speed of rotation and the amount of current flow.

Brush cross section is defined in terms of the numbers of amperes per square inch of brush. Too little cross section will cause excessive wear on the brushes due to an excessively high current density. Too large a cross section will create an inadequate commutator film.

Incorrect brush material can also cause excessive wear. Too soft a brush will wear due to friction and too hard a brush will cause severe arcing and resultant wear.

NOTE: Be certain to replace worn brushes only with direct replacements. Know your brush. Do not assume that you have selected the right brush just because it looks the same. Wrong brushes can cause excessive wear and reduce the life of the commutator.

Brush Spring Tension Brush wear is caused by two different types of friction—mechanical and electrical. Both of these types of friction are affected by the spring tension, which is used to hold the brush in place, Figure 12–5.

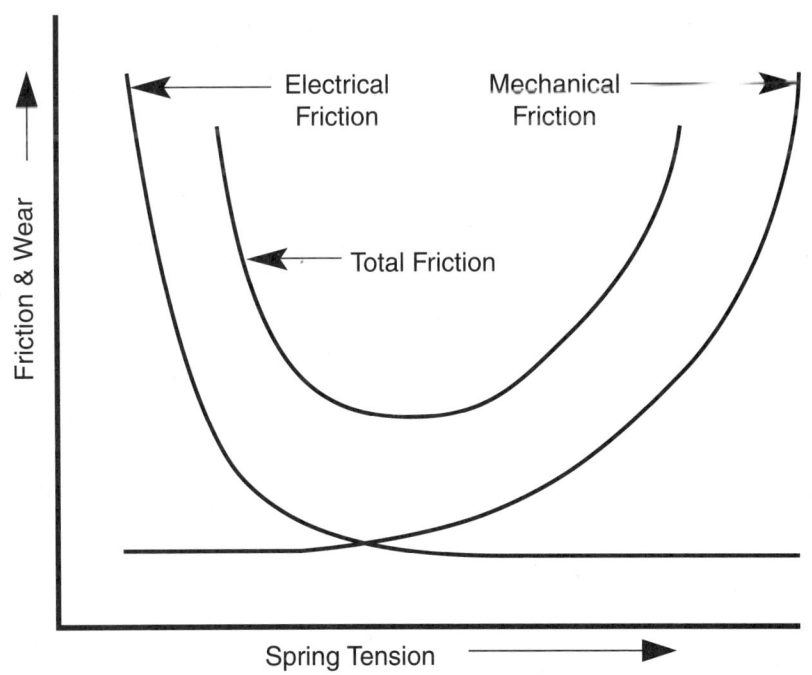

FIGURE 12–5 Mechanical and electrical friction (Courtesy of Cadick Professional Services)

Electrical friction is caused by heating and arcing at the brush/commutator interface. If the spring tension holding the brush down is too light, the interface resistance goes up and so does the heating. This causes the brush to arc and wear excessively. Mechanical friction is caused by the physical rubbing between the brush and the commutator. If the spring tension is too high, the brush will be worn excessively. The total friction will be a composite curve equal to the sum of the electrical and mechanical friction.

Setting proper brush tension varies depending on the type of brush rigging being used. Figure 12–6 shows one method that is commonly used on medium-sized to large machines. Place a piece of paper between the brush and the commutator, and attach a spring balance to the brush pigtail lead. Gently pull the spring balance until the paper is free to move. The reading on the spring balance is the spring tension. Adjust the spring tension using the method prescribed for the brush rigging on your machine.

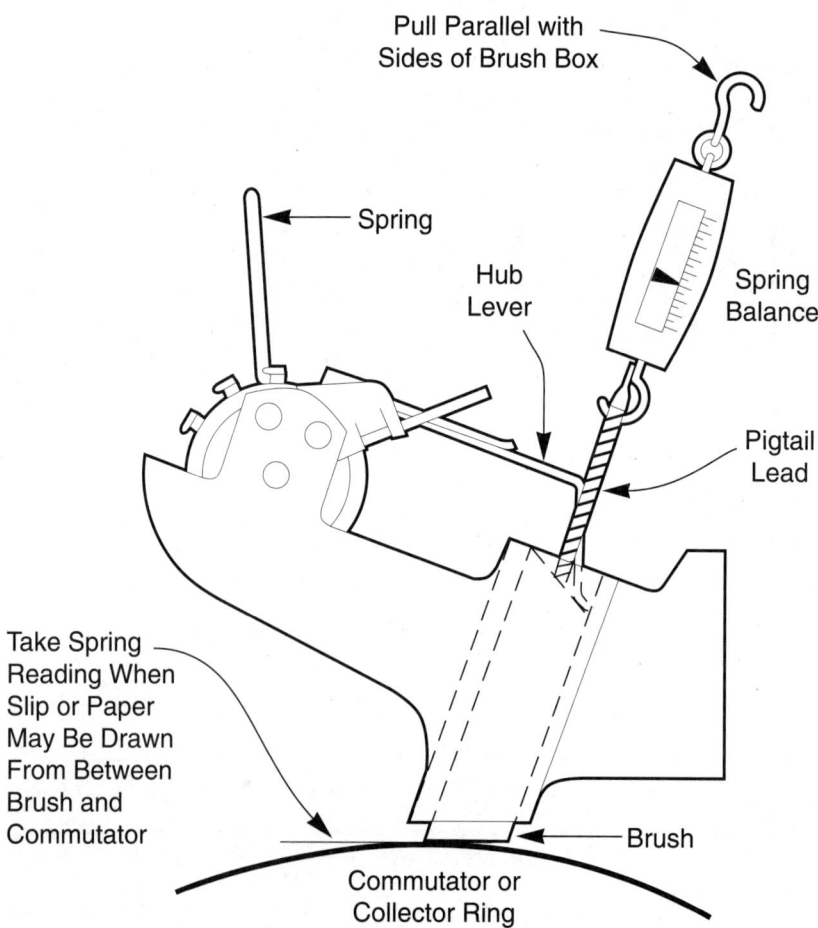

FIGURE 12–6 Setting/adjusting brush spring tension (Courtesy of Ideal Industries Inc.)

Some brush riggings have special assemblies that keep the brush tension constant as the brush wears. Even these types should be checked as part of a periodic maintenance procedure.

Finding the Neutral Point Several methods may be used to set the brush rigging at the neutral point. The following procedure applies in virtually all cases:

Mark one armature coil slot and its associated commutator segments with chalk. *CAUTION: Be certain to mark the commutator segments in such a way that you do not contaminate the film.*

Turn the rotor so the marked armature coil slot is directly under an interpole. If the machine does not have an interpole, position the coil slot so it is halfway between the main poles.

Adjust the brush rigging so that the brushes are directly on the marked commutator segments. This should be very close to the neutral point.

Operate the machine with load and slowly move the brush rigging to locate the point where minimum sparking occurs. This should be within one or two commutator segments of the one that was marked earlier.

> *CAUTION: Portions of the following procedure are performed while the machine is energized. Use extreme caution and all appropriate safety equipment while performing this procedure.*

Commutator

Commutator Problems Commutator problems fall into one of several areas. Table 12–1 identifies some of the common problems and possible causes.

Commutator Maintenance If a commutator is damaged (as indicated in Table 12–1), if the commutator bar segments are not level, or if the commutator or slip ring is out-of-round, the commutator will need to be repaired. Such repair falls into three basic categories: resurfacing, undercutting, and reseating the brushes.

Resurfacing:
One manufacturer recommends that the commutator be resurfaced under the following conditions:

- If individual commutator bar segments are more than .0002" different in height.
- If total out-of-round exceeds .003".

Several manufacturers make dressing stones (also called dressing blocks) for this purpose. A dressing stone is an abrasive mixture of materials that is used to grind and/or polish the commutator into a "like new" condition. Stones are available in a variety of grades, from extra coarse for heavy, fast cutting applications, down to polish by creating mirrorlike finishes. The stone to be used should be determined for the specific job at hand.

Table 12–1 Common Commutator Problems

Problem	Cause
Streaking: When the surface of a commutator begins to streak, metal is beginning to transfer to the carbon brush, Figure 12-7.	• Light electrical load • Light brush pressure • Abrasive or porous brush • Gas contamination • Abrasive dust contamination
Grooving: Grooves formed on the commutator surface will only increase in depth, Figure 12-8.	• Abrasive brush • Abrasive dust contamination
Threading: When fine lines appear on the commutator surface, there is excessive metal transfer resulting in rapid brush wear, Figure 12-9.	• Gas contamination • Light brush pressure • Light electrical load • Porous brush

As appearing in the IDEAL INDUSTRIES INC. Commutator Maintenance Guide to Effective Procedure

FIGURE 12–7 Commutator streaking (Courtesy of Ideal Industries Inc.)

DC Machine Maintenance

Table 12–1 Common Commutator Problems (Continued)

Problem	Cause
Pitch Bar Marking: Low or burned spots on the surface of the commutator. The number of these markings equals half of all the number of poles on the machine, Figure 12-10.	• Light brush pressure • Abrasive brush • Vibration • Faulty armature connection • Unbalanced shunt field
Copper Drag: Copper drag forms at the trailing edge of the bar. It is an abnormal buildup of commutator material that can cause flashover, Figure 12-11.	• Light brush pressure • Abrasive brush • Gas contamination • Vibration
Heavy Slot-Bar Marking: The trailing edge of the commutator bar is etched, rather than built up. The pattern is related to the number of conductors per slot, Figure 12-12.	• Gas contamination • Electrical overload • Electrical adjustment

FIGURE 12–8 Commutator grooving (Courtesy of Ideal Industies Inc.)

FIGURE 12–9
Commutator threading (Courtesy of Ideal Industries Inc.)

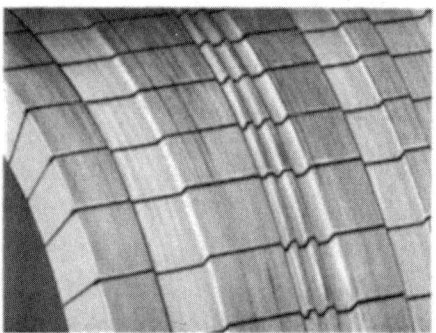

Most authorities agree that grinding the commutator in place, on the machine, is the correct approach. One manufacturer recommends the following precautions when preparing to dress a commutator in place:

- Be sure the commutator or slip rings are clean. If they are oily or greasy, wipe them dry and clean them with solvent.
- On machines, remove all brushes; on motors, remove as many brushes as possible. Run the machine at its normal speed with no load, and reduce voltage when possible. Note that the brush mounting bars can be used to mount the dressing equipment.
- Do not resurface under conditions of 440 V. Always use insulating materials such as floor mats and protective gloves.
- As a further safety precaution, the circuit breaker or fuse should be set to open at slightly above starting current for motors. Any short circuits will immediately open the circuit breaker.
- On high-voltage machines, or where the machine may not be removed from operation, the armature should be brought up to its highest speed, and the switch thrown. The grinding is done as the machine "free wheels" to a stop. Repeat as needed.

FIGURE 12–10
Commutator pitch-bar marking (Courtesy of Ideal Industries Inc.)

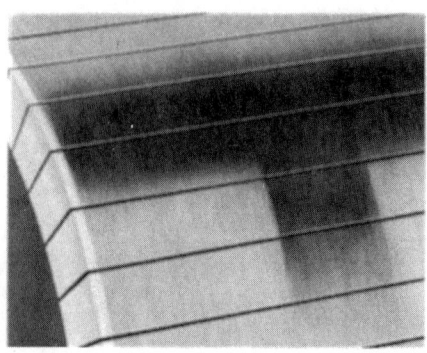

FIGURE 12–11
Commutator copper drag (Courtesy of Ideal Industries Inc.)

Whenever possible, grind on electrically-dead machines. Devise outside power sources to eliminate any possibility of accidental grounding.

- Even when grinding collectors on separately-driven machines or coasting machines, remember that the field rings of synchronous motors or machines should be short-circuited and operated with the brushes free from the rings. These field circuits must never be operated "open" because of the high induced voltages that build up.
- Always keep your resurfacer inventory clean and dry. Never permit resurfacers to become oil-soaked, and do not use them on an oil covered commutator. Oil ruins the cutting actions of a resurfacer. Remember that the resurfacer, although dielectric when dry, becomes a conductor if it absorbs a high content of moisture.
- Always protect your eyes with appropriate safety glasses or a face shield.
  Wear a dust mask.

Figures 12–13 through 12–15 show diagrams of three different rigs for resurfacing a commutator. Each of these three setups is using two resurfacing blocks to grind or resurface a commutator. The manufacturer of the rig

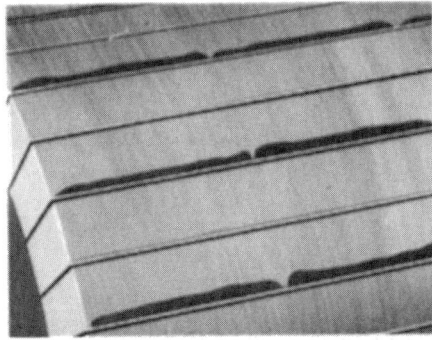

FIGURE 12–12
Commutator heavy slot-bar marking (Courtesy of Ideal Industries Inc.)

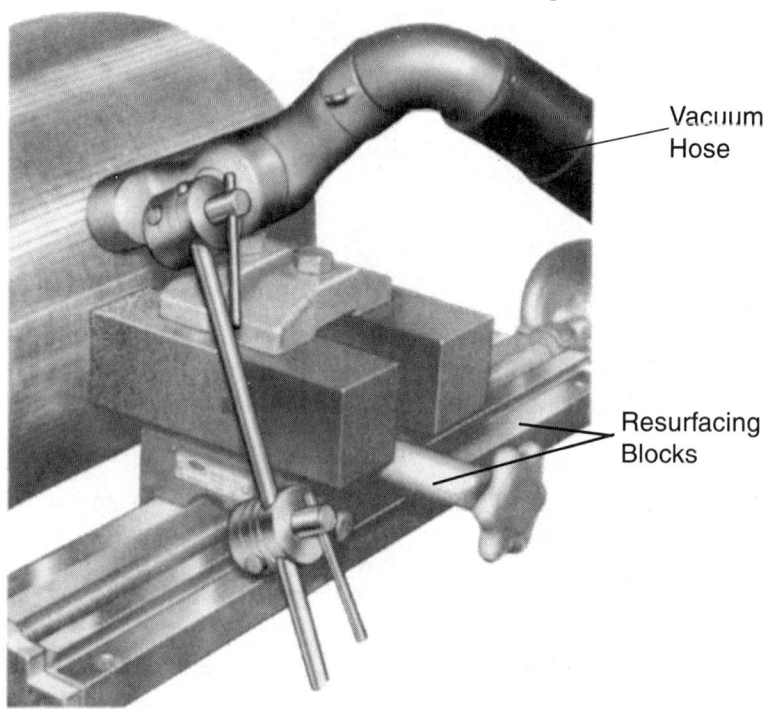

FIGURE 12–13 A resurfacing operation with a dust vacuum attached (Courtesy of Ideal Industries Inc.)

has provided different options to make the grinding project easier. In Figure 12–13, a commutator is being resurfaced using a rotating lathe. A vacuum hose is attached to draw off dust particles that are created by the grind.

In Figure 12–14, a commutator is being resurfaced in place. The resurfacing blocks are moved back and forth across the commutator, using an end drive adjustment wheel. Figure 12–15 is similar to Figure 12–14, except a right angle drive is employed to better position the controls for the worker.

Undercutting:

Mica is considerably harder than the copper. Because of this, the resurfacing will often leave the mica insulating layers too high with respect to the commutator segments. After resurfacing, the commutator will probably need to be *undercut*. Undercutting is the process wherein the mica is cut down to a reasonable depth. This procedure is best carried out using a tool called an undercutter, Figure 12–16. Figure 12–17 shows an undercutter being used.

Reseating the Brushes:

After the commutator is resurfaced, or when new brushes are installed, they must be seated, Figure 12–18. Seating the brush means simply starting

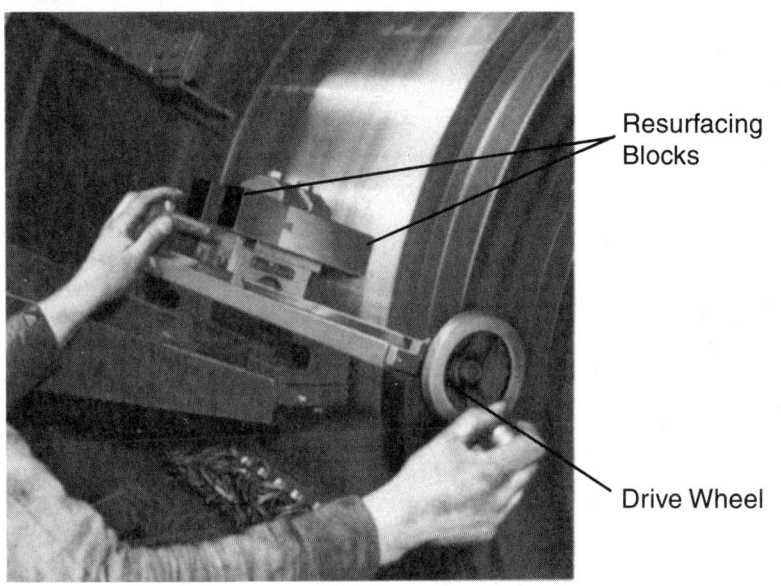

FIGURE 12–14
Resurfacing grinder with a rotary end drive (Courtesy of Ideal Industries Inc.)

the process of wearing the brush so that it conforms to the shape of the commutator.

To use a brush seater, the machine must be running at speed. Increase the spring tension or hold the brush firmly against the commutator. Press the seating stone firmly against the commutator, behind the brush to be seated. The stone will immediately start to wear and the powder will be carried under the brush. The powder wears the brush into the shape of the commutator. This procedure should take only a moment, and should be

FIGURE 12–15
Resurfacing grinder with a right angle drive (Courtesy of Ideal Industries Inc.)

FIGURE 12–16
An undercutter with blades (Courtesy of Ideal Industries Inc.)

repeated for each brush that needs to be seated. Avoid excessive use of the seating stone, because it will wear the commutator.

ELECTRICAL TESTS

DC machines should be electrically-tested periodically. Most of the tests are similar to the ones that have been discussed in previous chapters. At a minimum, the following tests should be performed periodically.

Measure Operating Voltages and Currents

When the machine is running at rated load, the field and armature voltages and currents should be measured and recorded. This information can be used for benchmarking purposes in later troubleshooting procedures.

Insulation Resistance Tests

The following insulation tests should be performed periodically:

FIGURE 12–17 Undercutting a commutator (Courtesy of Ideal Industries Inc.)

- One-minute insulation resistance test using a megohmmeter. (Refer to Chapter 4 for a complete description of insulation resistance measurements.)
- Perform polarization index tests (see Chapter 4).
- Optionally, perform an insulation power factor test (see Chapter 4).

Vibration Analysis

Perform vibration analysis tests. These tests can help identify worn bearings, unbalanced field flux, improper rotor positioning, and a number of other problems. The data collected from vibration analysis can also be put into databases for future analysis as part of a predictive maintenance program.

Protective Devices

Test protective devices. See Chapter 9 for a description of routine maintenance tests on common protective devices.

FIGURE 12-18
Seating the brushes
(Courtesy of Ideal
Industries Inc.)

TROUBLESHOOTING DC MACHINES

A number of problems can occur which will cause a machine to misoperate or to not operate at all. Some problems relating to the commutator have already been discussed. The following sections outline common problems for both generators and motors, Tables 12–2, 12–3 and 12–4.

DC Generators

Machine Does Not Generate

Table 12–2 Generator Does Not Generate: Causes and Explanations
(Courtesy of Ideal Industries Inc.)

Cause	Explanation
Loss of residual magnetism	A self-excited machine uses residual magnetism to "bootstrap" itself. If residual magnetism is lost, the machine will not start generating. To repair this problem, apply a source of direct current to the field for a few seconds.
Too much resistance in field circuit	If the field is open circuited, or has too high a resistance, the field will not be strong enough to start the generating process. Check for loose connections, poor brush contact, or broken brush pigtails.
Wrong field connection	If one or more field windings is connected backwards, the machine will not bootstrap, since the residual magnetism will "disagree" with the field magnetism. Reverse the field connections or reverse the machine rotation.
Wrong rotation	This problem is similar to the wrong field connection. To remedy, reverse the shunt field or the direction of rotation.
Shorted armature or field	Shorted windings will reduce the output, possibly to zero, depending on the degree of the short circuit.

Voltage Drops Excessively When Load is Applied, Table 12–3

Table 12–3 Generator Voltage Drops Off Excessively: Problems and Explanations
(Courtesy of Ideal Industries Inc.)

Cause	Explanation
Differential Connection	As explained previously, if the shunt field coil and the series field coil are connected backwards, the output voltage will drop off rapidly with application of load. To correct this problem, reverse the shunt field or series field connection.
Shorted Armature	A shorted armature will also cause overheating and smoke, if the load is allowed to continue.
Overload	If the load is excessive, the voltage drop will be very high. Reduce the machine load to within its capability.

DC Machine Maintenance

Voltage Does Not Build Up to Maximum, Table 12–4

Table 12–4 Generator Voltage Does Not Build to Maximum: Problems and Explanations

(Courtesy of Ideal Industries Inc.)

Cause	*Explanation*
Brushes not in neutral	If the brushes are not in the neutral position the DC voltage will drop off and the ripple will increase. To fix this problem, put the brushes in neutral as previously described.
Shorted armature or field coils	The degree of the short circuit determines what effect these problems have.
Resistance in the field circuit	A small resistance will cause the output voltage to drop off. Check connections.
Speed of machine too low	If the machine is not moving fast enough, the output will be low since the number of flux lines being cut per second will be too low. Run the machine at its design speed to resolve this problem.

DC Motors

Motor Fails to Run Failure to run may be caused by any or all of the following problems:

Open fuse or circuit breaker
Defective starter
Dirty or clogged brushes
Open armature circuit
Open field circuit
Shorted or grounded field
Shorted armature or commutator
Worn bearings
Grounded brush holder
Overload

Methods for identifying these types of problems have either already been covered or will be covered later in this chapter.

Motor Runs Slowly Slow motor operation can result from the following:

Shorted armature or commutator
Worn bearings
Open armature coils
Brushes off-neutral
Improper operating voltage
Excessive mechanical load

Motor Runs Too Fast Fast motor operation may be a result of:

Open shunt field
Series motor with no load connected
Shorted or grounded field
Differential connection in a compound motor

Motor Sparks This problem may be caused by the following:

Poor brush contact
Brushes off neutral
Dirty brushes or commutator
Wrong interpole polarity
Shorted or grounded field
Reversed armature leads
Open field circuit
High or low bars

High mica
Unbalanced armature

Motor is Noisy Noisy operation may be caused by one of the following:

Worn bearings
High or low commutator bars
Rough commutator
Unbalanced armature

Motor Runs Hot Overheating can be caused by:

Excessive mechanical load
Sparking
Tight or worn bearings
Shorted coils
Too much or too little brush pressure

In-Service and Troubleshooting Tests

Identifying the problems discussed in the previous sections requires that certain symptoms be diagnosed. Open coils, shorted coils, improper lead connections, and a number of other such difficulties can be detected using the procedures outlined in the following paragraphs. Note that many of these procedures should also be performed immediately prior to putting a new machine or repaired machine into service.

There are some differences between DC motors and DC generators in these procedures. Such differences have been noted where applicable.

Shorted Windings The way that shorted windings are found depends upon whether the short is in the field windings or the armature.

- Field winding shorts
  Identify field winding shorts by using an accurate ohmmeter. If no bench mark resistance for the entire winding is available, you must compare the resistance of each pole, Figure 12–19. Figure 12–19 shows a four-pole field system. To check for the short circuit, the following steps should be used:

  If an entire field winding resistance benchmark is available, measure the resistance between F_1 and F_2 (S_1 and S_2 for a series field).
  If the resistance is the same as the benchmark, the field is not short-circuited. If it is different, proceed to the next step.

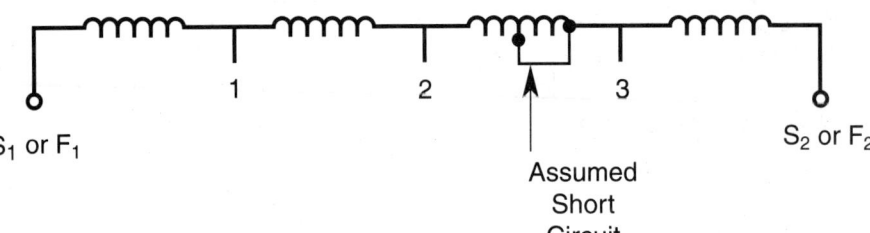

FIGURE 12–19 Measuring for field winding short circuits (Courtesy of Cadick Professional Services)

Disassemble the motor so that the field windings may be reached. Locate and remove the wires between the ends of the field coils and scrape the insulation off. (Points 1, 2, and 3.)

Using an accurate ohmmeter, measure the resistance between F_1 and 1, 1 and 2, 2 and 3, and 3 and F_2. (If this is a series field, substitute S_1 for F_1 and S_2 for F_2.)

Compare the four readings. If any of them is significantly different, that pole winding is probably short-circuited. In Figure 12–19 for example, the reading between 2 and 3 will be lower than the others.

After the shorted winding is repaired, take an entire field resistance reading and record it for a future benchmark reference.

- Armature shorts

An armature short circuit can be isolated using the growler, Figure 12–20.

Place the armature in the growler and activate the growler.

Place a hacksaw blade on the top of the armature, holding the blade so that it is directly on a slot.

If the coil is shorted, the magnetic field set up by the current will cause the hacksaw to vibrate and make a growling sound.

Continue this process until all of the slots have been checked.

NOTE: The growler will not work with all armatures. Use of meters to identify resistance differences in armature coils may be required.

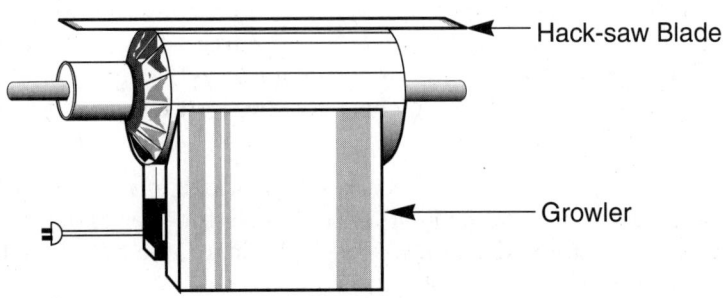

FIGURE 12–20 Using a growler to find a shorted turn in an armature (Courtesy of Cadick Professional Services)

DC Machine Maintenance

Open Windings

Field Windings
Open field windings may be determined using the circuit of Figure 12–19. Read the resistance of each field coil one at a time. Any infinite resistance readings are a sign of an open field coil.

Armature Windings
Locate opposite ends of each armature winding, one at a time. Use an ohmmeter to check for continuity.

Grounded Windings Grounded windings can be located using a megohmmeter.

Connect the earth lead to the stator frame, if field grounds are being located.
Connect the earth lead to the shaft, if a rotor is being checked.
Connect the line lead to the field winding, or the armature windings one at a time.
Energize the megohmmeter and take readings.
Readings should be at least 2 megohms. Lower values should be checked.

Lead Identification DC machine leads are normally identified as A_1, A_2, S_1, S_2, F_1, and F_2. Note that the shunt field leads, F_1 and F_2, are normally smaller wires than the others. This is not always true, however. Start by using an ohmmeter to determine which wires are part of the same winding. Also record the resistances for each pair of wires.

Shunt-Connected Motors:
The field winding will be the one with the higher resistance.
Series-Connected Motors:
Lift the brushes so they do not touch the armature. Take the measurements again. The armature winding will show open.
Compound-Connected Motors:
The field winding will be the one with the highest resistance. Lift the brushes. The armature winding will show open.
Five-Lead Compound Motors:
The motor must be disassembled and the leads traced.

Differential Connection Tests Although some generators are designed to operate with a differential compound connection, DC motors are rarely designed this way. The test for differential connections can be made as follows:

- DC motors

  After identifying the leads, compound connect the motor.
  Temporarily short out the series field winding.
  Connect the motor to a DC source and operate it. Take note of the direction of rotation.
  Remove the short on the series connection and again operate the motor. If the direction of rotation is the same with and without the series winding, the motor is not connected differentially.

- DC generators

  After identifying the leads, compound connect the generator.
  Run the generator and connect a variable load to the output.
  Gradually increase the load current while monitoring the output voltage.
  If the voltage drops off rapidly, the generator is differentially connected.

Correct Interpole Polarity The following test is used for a motor or a generator. A source of DC voltage is required; however, the rotor does not have to be removed. All directions are with respect to the commutator end of the motor.

Connect the DC supply leads to the armature and the interpole circuit. Disconnect all other leads.
Mark the brush positions and then rotate the brushes clockwise, or counterclockwise, so they are located halfway between the marks.
Momentarily apply the DC voltage to the motor, and note the direction of rotation.
If the motor rotates in the same direction as the brushes were moved (clockwise or counterclockwise), the interpoles have the correct polarity. Otherwise the interpole polarity must be reversed.
Restore the brushes to the correct neutral position, and reconnect the shunt field. Energize momentarily to ensure proper rotation. Reverse the shunt field if necessary.
Disconnect the shunt field and reconnect the series field. Repeat the previous step for correct rotation.
Reconnect all leads using the connections determined.
Verify that the brushes are in the correct neutral position by observing that little or no sparking is occurring.

Elevated or Depressed Rotor Bars Sometimes rotor bars can be disturbed and moved into high or low positions. This condition can be determined by feeling the commutator for raised or low bars.

To fix high bars, loosen the v-rings that hold the commutator segments in place. Gently tap the high bars with a hammer until the bar slips back into place. Be careful not to score the bar. Tighten the v-ring.

Low bars usually cannot be adjusted. Instead, the commutator must be resurfaced as described previously.

SUMMARY

DC machine maintenance falls into four basic categories: visual mechanical procedures, brushes and commutators, electrical tests, and troubleshooting procedures.

Visual mechanical procedures include the types of "common sense" techniques that are employed to make certain the machine is in proper physical condition. Procedures like infrared scan, checking mounting hardware, and lubrication are included in these techniques.

The brushes and commutators generally require a great deal of attention. In normal operation, the brushes form a film (on the commutator) composed of carbon, copper, and moisture. This film helps to lubricate and improve the electrical connection. Improper operation or electrical problems will often leave characteristic markings on the commutator. Proper diagnosis of these markings can lead the maintenance person to the proper maintenance procedure to repair the problem.

When commutators become severely damaged, they may need to be resurfaced. Resurfacing involves grinding and/or polishing the commutator surface so that it is smooth and round. The resurfacing procedure should be followed by undercutting, wherein the mica insulation is ground down to a level below the surface of the commutator segments. The final maintenance procedure is to seat the brushes using a seating stone.

The electrical tests for DC machines are very similar to the tests performed on all rotating equipment. Tests such as insulation evaluation, protective device tests, and vibration analysis are performed on a routine basis.

Troubleshooting DC machines is a relatively straightforward process. Operational problems for motors and generators are very similar, as are the tests to identify the problems.

REVIEW QUESTIONS

1. Electrical connection inspections should involve an _____ scan and a _____ wrench check.

2. The air gap on a DC machine should be checked at a minimum of _____ points around the circumference of the rotor.

3. What three materials combine to create the lubricating film on a properly operating DC machine commutator?

4. What two things determine the type and size of brushes to be used in a DC machine?

5. The specifications for a particular DC machine call for a brush spring tension of 4 lbs. The spring tension is set at 2 lbs, but the brushes are still wearing excessively. What is wrong?

6. A DC generator has excessive sparking at the brushes, and its output voltage drops off rapidly as the load current increases. What could the probable cause be?

7. A commutator inspection reveals that excessive copper is building up on the trailing edge of the commutator bars. List four things that could cause such a problem.

8. Inspection reveals that a commutator is .004" out-of-round. What should be done?

9. After resurfacing, you notice that the mica insulation is protruding above the commutator segment surfaces. What should be done?

10. What *external* causes could cause a DC motor's speed to be too low?

Part V

Transformers

Transformer Theory 13

OBJECTIVES

After studying this chapter, the student should be able to:

- *Explain the basic operating principles of transformers.*
- *Identify the various types of transformers.*
- *Describe the fundamental operating features of transformers.*

INTRODUCTION

Electrical transformers provide the backbone of modern electrical power systems. Transformer operation is based on Faraday's law. One winding, the primary winding, is connected to a source of alternating voltage. Another coil, the secondary, is wound in such a way that the magnetic flux lines from the primary cut through the secondary. You know, from Faraday's law, that this action causes a voltage to be established in the secondary winding. In fact, the magnitude of the voltage is determined by the magnitude of the primary voltage and the number of turns in the primary and the secondary.

This simple relationship was first put into use in electrical systems in the latter part of the 19th century. Prior to the advent of the transformer, direct voltage systems were competing for the electrical distribution market. DC systems are limited because the voltage level cannot be readily changed. Voltages that can be generated and used are too low to transmit any distance. Remember that the lower voltage requires higher current to transmit the same amount of power.

The transformer allows AC voltage to be readily converted from one voltage level to another. This way, the power can be generated at a relatively low voltage, say 13,800 V. A transformer is used to step the voltage up to a transmission level, 138,000 V Close to the use point of the power, the voltage is stepped back down to 13,800, and then routed to the transformer on the utility pole behind the house or close to the industrial plant. The final step down transformer changes the voltage to a utilization level such as 480 V or 120 V.

In this chapter, you will learn how transformers work. Subsequent chapters will explain how transformers are constructed, protected, and maintained.

PRINCIPLES

Figure 13–1 illustrates the basic operating principle of a transformer. Two windings, the primary and secondary, are placed close to each other. Often the coils are placed on top of each other on the same structure. A load, R, is connected to the secondary winding and a power supply, V_p, is connected to the primary.

The turns ratio, N, of the transformer is defined as the number of turns on the primary, N_p, divided by the number of turns on the secondary, N_s. The current, voltage, and impedance ratios are given as shown on Figure 13–1.

FIGURE 13–1 Basic operation of a power transformer

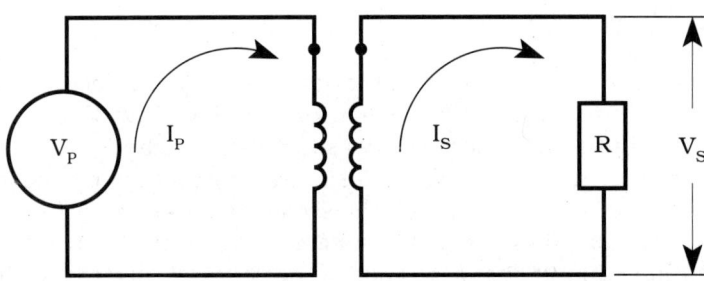

$$V_P = 480 \text{ V} \qquad V_S = 120 \text{ V} \qquad N = \frac{V_P}{V_S} = \frac{480}{120} = 4 \qquad R = 10 \text{ }\Omega$$

$$I_S = \frac{V_S}{R} = \frac{120}{10} = 12 \text{ A}$$

$$I_P = \frac{I_S}{N} = \frac{12}{4} = 3 \text{ A}$$

$$Z_P = \frac{V_P}{I_P} = \frac{480}{3} = 160 \text{ }\Omega \qquad Z_S = \frac{V_S}{I_S} = \frac{120}{12} = 10 \text{ }\Omega$$

$$\therefore Z_P = N^2 \times Z_S$$

$$W_P = V_P \times I_P = 480 \times 3 = 1440 \text{ W} \qquad W_S = V_S \times I_S = 120 \times 12 = 1440 \text{ W}$$

TRANSFORMER TYPES

Transformers are manufactured in a variety of types for different applications. In this section, you will learn about three of the common types of transformers: power, distribution, and instrument transformers. After the initial introduction of transformer types, the remaining parts of the chapter will cover power and distribution transformers.

Power Transformers

Power transformers are designed to handle large amounts of power. They are generally too large to mount on poles; therefore, they are usually mounted on concrete pads. Power transformers are available in kVA sizes from 500 and up and in voltage ranges from 13.8 kV and higher.

Power transformers are often equipped with external cooling apparatus such as fans or pumps. When so equipped, they often are given multiple kVA ratings. One, when the transformer is being self-cooled, and others for various stages of additional cooling.

Figure 13–2 is an example of a power transformer.

Distribution Transformers

Distribution transformers are smaller than power transformers. They are often used as the last transformer in the link between the generation source and the load, and are often pole-mounted units.

Distribution transformers are typically self-cooled, and have no fans or other auxiliary means of cooling. Figure 13–3 is a distribution transformer.

Instrument Transformers

Instrument transformers are relatively low power units that are used to reduce power system voltage and current values to safe levels usable by instruments. Some instrument transformers are used to isolate an electrical circuit from ground. Others are used to change a voltage to a value required for a particular device, such as a meter or relay.

Voltage Transformers Instrument transformers are available in two main types: voltage and current. Voltage transformers, Figure 13–4, are abbreviated VT, or sometimes PT (for potential transformers). PT is an obsolete term that has been replaced by VT. Voltage transformers are connected in parallel with the load as shown in Figure 13–5.

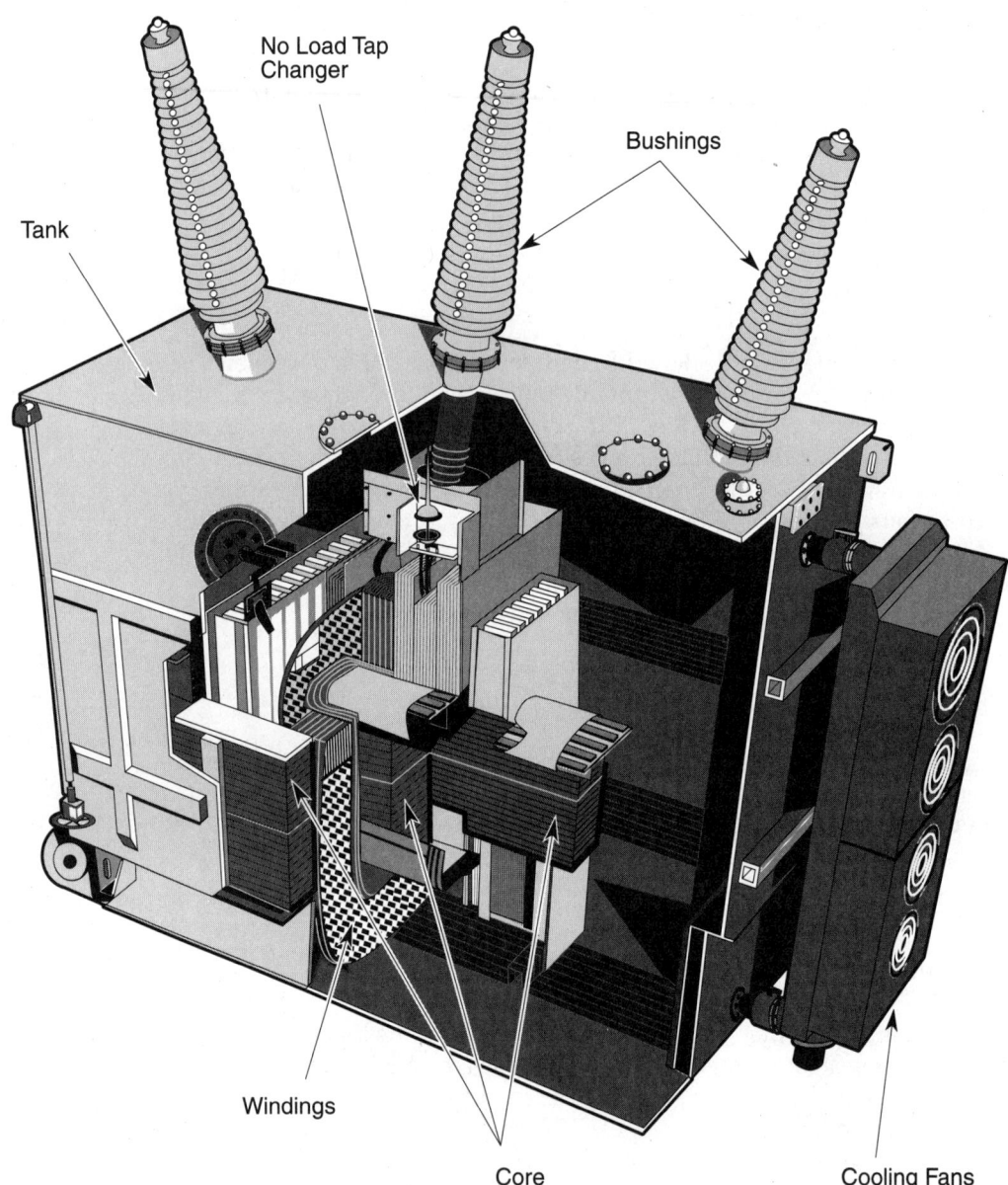

FIGURE 13–2
A modern power transformer

FIGURE 13–3 Pole-mounted distribution transformer (Courtesy of ABB Power T&D Company, Coral Spring, FL)

Voltage transformers are very much like power and distribution transformers. Some of them use liquid insulation and some have dry type insulation. Voltage transformers are usually single-phase units with 69-V or 120-V secondaries. Their primary voltage is determined by the voltage in the system in which they are applied.

Current Transformers Current transformers, Figure 13–6, are used to convert the system current from the relatively high values, to values that are more readily usable by relays, meters, and other instruments. Current transformers (CT) are available in two basic types: bar type and window type.

The bar type of CT bolts to the bus bars in the system. The bar in the bar type CT is the primary of the current transformer. The window CT, also called the doughnut CT, does not have the primary winding as part of the CT. Rather, the secondary winding is supplied with an opening in the middle to allow passage of a system conductor. The system conductor becomes the primary of the CT.

The primary of a CT is connected in series with the system, Figure 13–7. Because of this, the power system current determines that amount of magnetic flux that is present in the CT.

Current transformers are available in a variety of ratios. These ratios are usually expressed as the nominal primary current which produces a 5-A secondary current; 400/5, 1200/5, and 150/5 are all examples of current transformer ratios.

> *CAUTION: THE SECONDARY OF A CURRENT TRANSFORMER MUST ALWAYS BE CONNECTED TO A LOAD OR MUST BE SHORT-CIRCUITED. WHEN THE SECONDARY OF A CT IS OPEN-CIRCUITED, AN EXTREMELY HIGH, LETHAL VOLTAGE WILL BE DEVELOPED ACROSS THE SECONDARY CIRCUIT.*

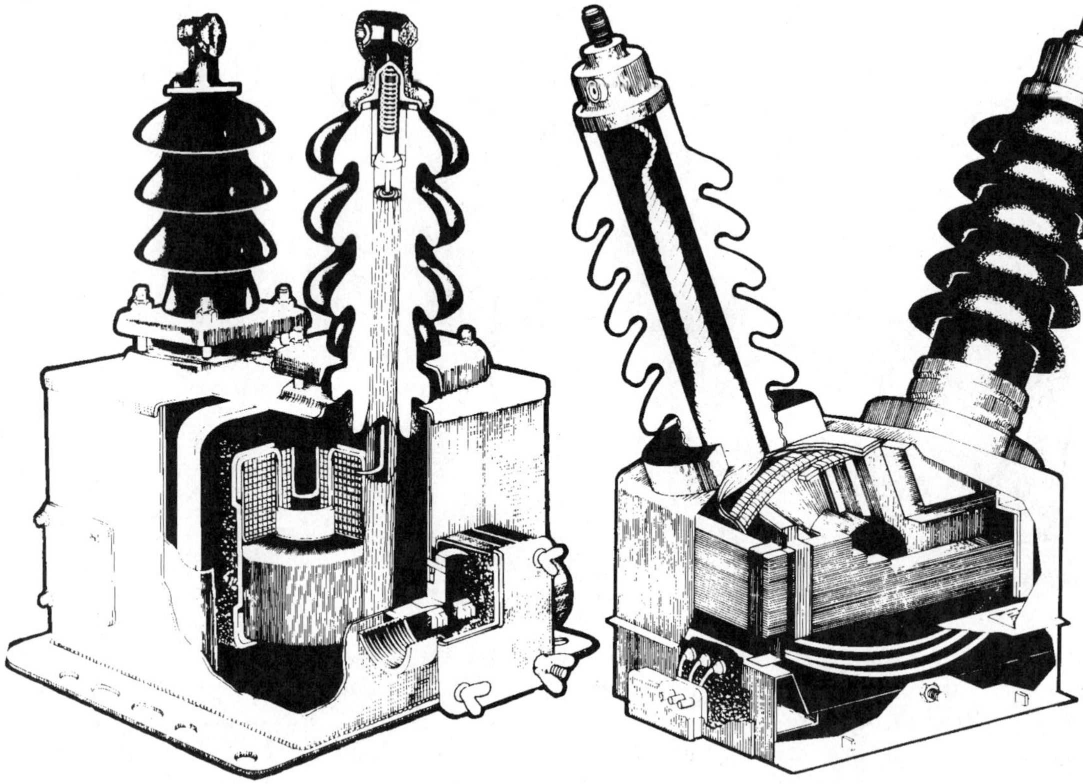

19,400 V
COMPOUND FILLED
69,000 V
OIL FILLED

FIGURE 13–4
Two types of instrument voltage transformers

A
B
C
±
±
Voltage Device

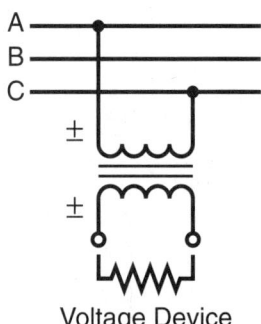

FIGURE 13–5
Voltage transformer connection

TRANSFORMER MAGNETIC STRUCTURES

Core Form

Figure 13–8a is a diagram of a core form transformer. In this type of magnetic structure, the primary and secondary windings are wound on separate legs of the same iron core. The core itself is made of square sheets of soft, high permeability steel, laminated into a solid center. The laminations are insulated from each other to minimize eddy current flow.

In the core form construction, the magnetic flux path is in the center of the windings. Core form design is normally used on smaller transformers such as distribution types and small power transformers.

Transformer Theory

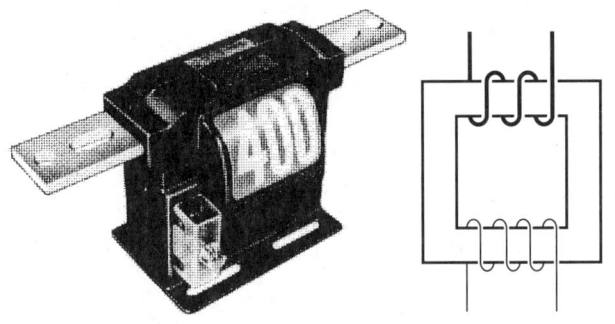
(a) Bar Type

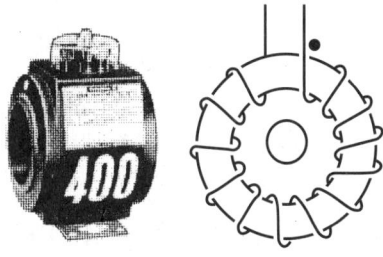
(b) Window Type

FIGURE 13–6
Two types of current instrument transformers

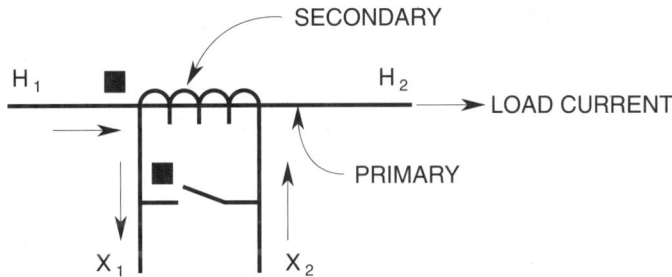

FIGURE 13–7
Current transformer connection

FIGURE 13–8
Shell and core form construction (Courtesy of Transformer Maintenance Institute)

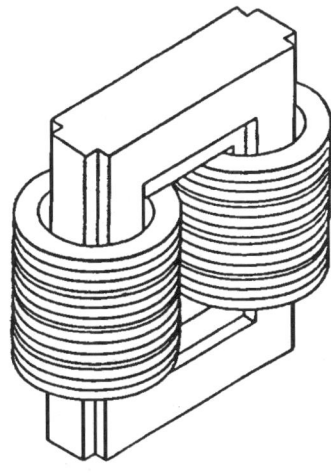
a. Core Form

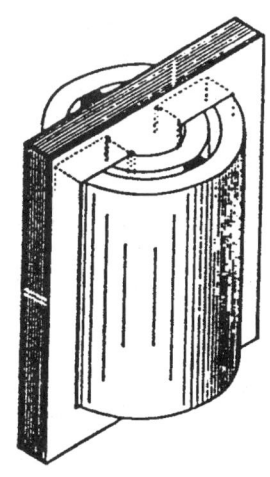
b. Shell Form

Shell Form

In the shell form construction, Figure 13-8b, the transformer magnetic flux forms a shell around the windings. The core of the transformer is made of flat steel laminations. In Figure 13-8b the core is a three-legged structure. The primary and secondary windings are wrapped around the center leg. The high permeability core channels the flux around the windings, surrounding the windings with magnetic flux. Shell form construction is used on larger transformers.

ELECTRICAL CONSTRUCTION

Terminal Markings

Transformer terminal connections are designed as $H_1, H_2 \ldots, X_1, X_2 \ldots,$ and $Y_1, Y_2 \ldots$ for the primary, secondary, and tertiary windings respectively. The tertiary winding is a third winding included in some transformers to provide a third output voltage. It works just like the secondary winding. For most single-phase transformers and all three-phase transformers, the H_1 connection will be on the right as you face the high-voltage side of the transformer. The other winding connections will be positioned as shown in Figure 13-9.

FIGURE 13-9 Transformer terminal markings

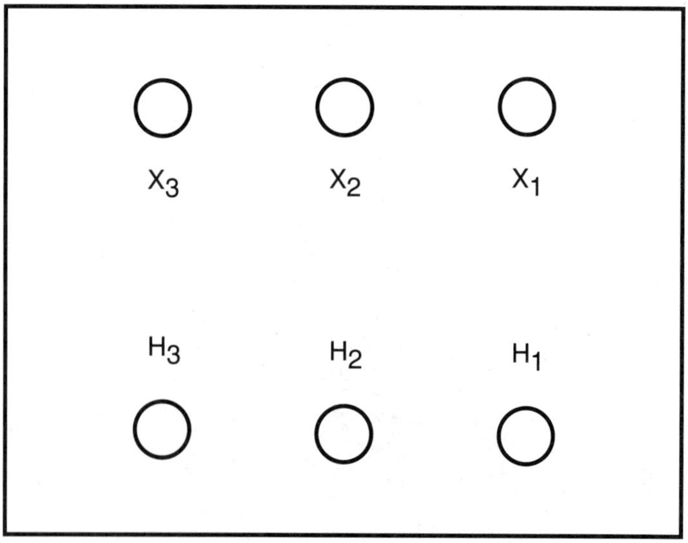

Polarity

The polarity marks on a transformer identify which terminals are in-phase with each other. The polarity terminals are usually marked by a ±, "a", or a · symbol. In Figure 13–5 for example, the ± symbol on the primary means the voltage at that point is in-phase with the voltage at the ± on the secondary.

The current phase angle relationships are also indicated by the polarity marks. The current flowing into the polarity mark on the primary is in-phase with the current flowing out of the polarity mark on the secondary.

Single-Phase Transformers

Connections Single-phase transformers are available in a variety of connection configurations. Figure 13–10 shows one of the most common. The transformer shown in this figure has one primary winding (H_1-H_2) and two secondary windings (X_1-X_2 and X_3-X_4). Each of these windings is rated individually for 115 V output. In Figure 13–10a, the two secondary windings have been connected in series to produce a total output voltage of 230 V. In Figure 13–10b, the windings are connected in parallel for a voltage output of 115 V.

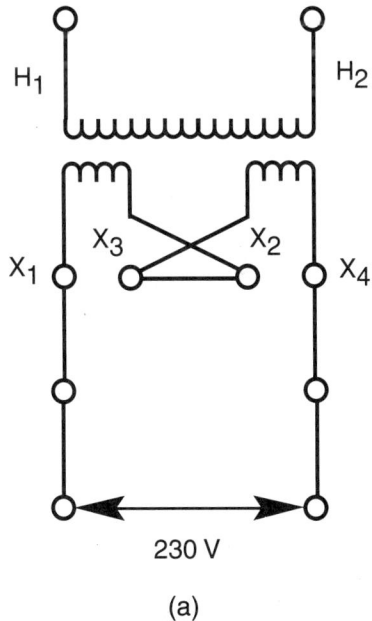
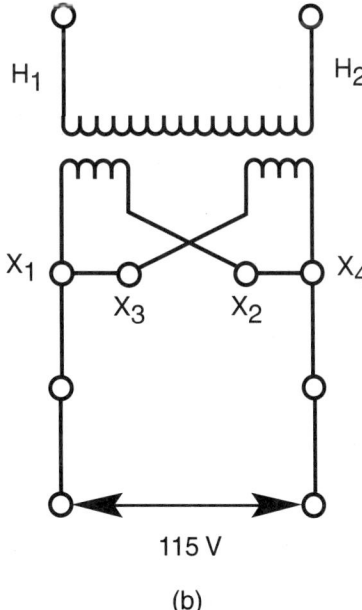

FIGURE 13–10 Single-phase transformer connections (Courtesy of Cadick Professional Services)

Note that in the parallel connection, the two currents from each winding add to each other, while the two currents are in series in the series connection. The total current capacity of the parallel connection is, therefore, twice as high as the series connection. Obviously then, the total kVA capacity of the transformer is the same in each case.

Polarity Single-phase transformers are available in either subtractive or additive polarity. In the subtractive polarity transformer, the H_1 terminal is brought out of the transformer directly across from the X_1 terminal. Subtractive polarity can be explained by using Figure 13–11. If the H_1 and X_1 terminals are connected, and the voltage is measured between H_2 and X_2, the result will be the difference between the primary voltage and the secondary voltage.

Additive polarity, Figure 13–12, is slightly different. In an additive polarity transformer, the H_1 terminal is brought out directly across from the X_2 terminal. When the H_1 and the X_2 terminals are connected, the voltage between the H_2 and X_1 terminals will be equal to the sum of the two voltages.

The terms additive and subtractive polarity predate the use of the polarity markings on transformers. Obviously, in each case, the H_1 and the X_1 terminals are the polarity terminals.

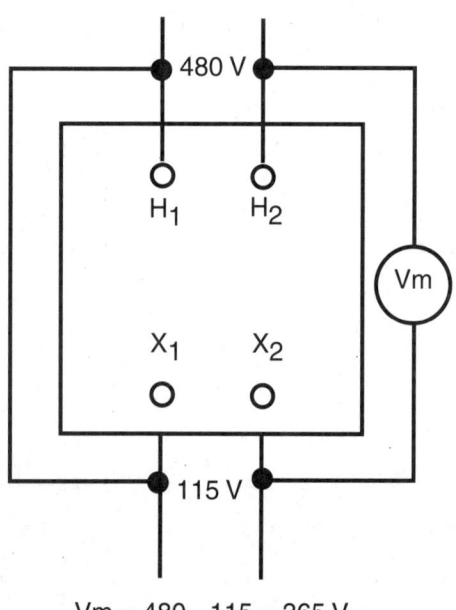

FIGURE 13–11 Subtractive polarity transformer connections

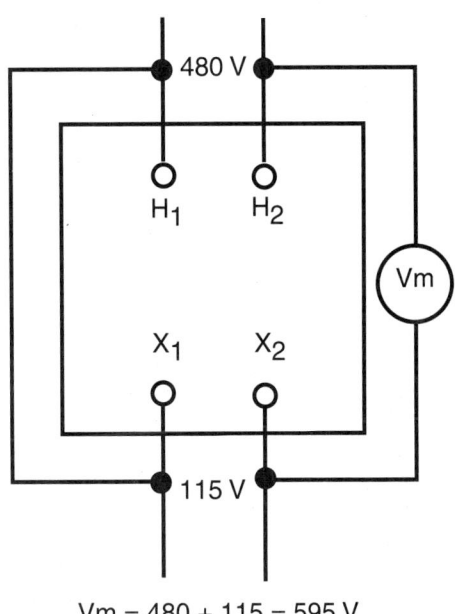

FIGURE 13–12
Additive polarity transformer connections (Courtesy of Cadick Professional Services)

Vm = 480 + 115 = 595 V

Three-Phase

Basic Connections The three windings for a three-phase transformer may be connected in either a delta or a wye (pronounced *wi*) configuration. Figure 13–13 shows an example of each connection. Notice that the wye connection has three more voltages that the delta connection. The voltage between any two terminals is called the *phase-to-phase* voltage or the *delta* voltage. The voltage between any terminal and the common point of the

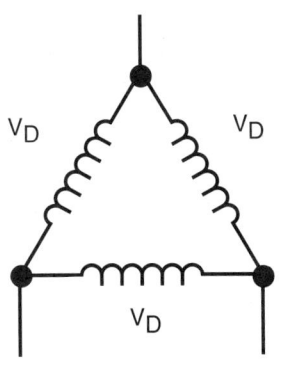

a. Delta

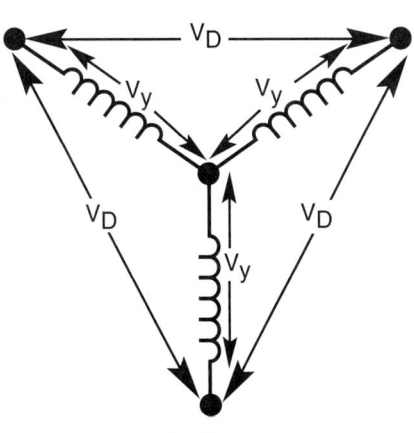

b. Wye

FIGURE 13–13
Delta and wye connections (Courtesy of Cadick Professional Services)

wye connection is called the *phase-to-neutral* voltage. The common point of the wye connection is called the *neutral* point.

Notice that the voltages on a three-phase transformer have the relationship:

$$V_D = V_Y \times \sqrt{3}$$

Where V_D is the phase-to-phase voltage of a wye winding and V_Y is the phase-to-neutral voltage of the wye winding.

Terminal Markings Figure 13–14 shows the ANSI standard configuration for a three-phase transformer. Note the X_0 terminal. This is the neutral terminal for the X windings. This means the transformer secondary is a wye connection. If the transformer primary is also a wye, the H_0 terminal will also be brought out.

ANSI Configurations The American National Standards Institute standardizes the connections for three-phase transformers. Although many standards exist, the four most common are shown in Figure 13–15.

FIGURE 13–14
Three-phase transformer terminal markings

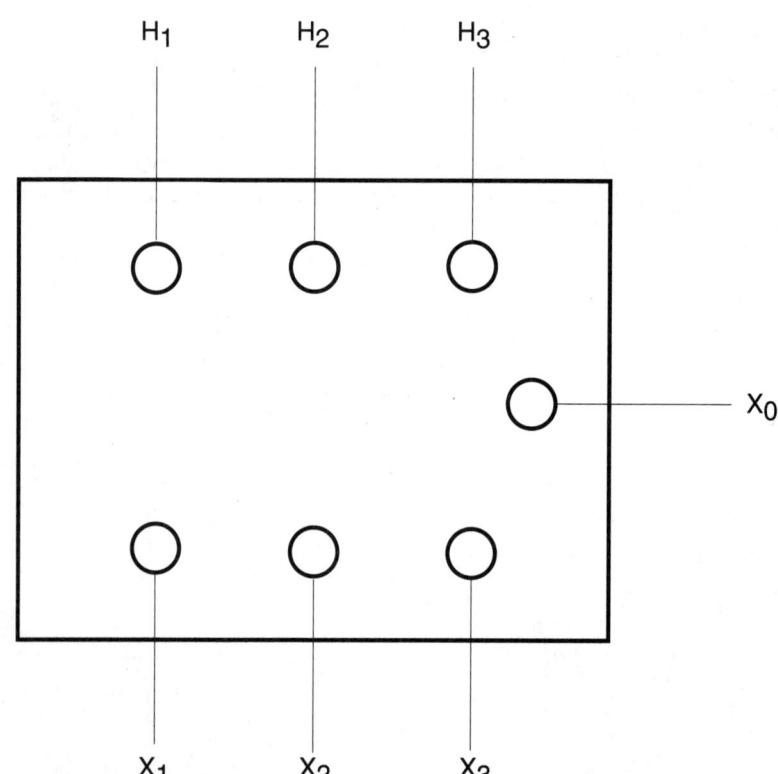

Transformer Theory

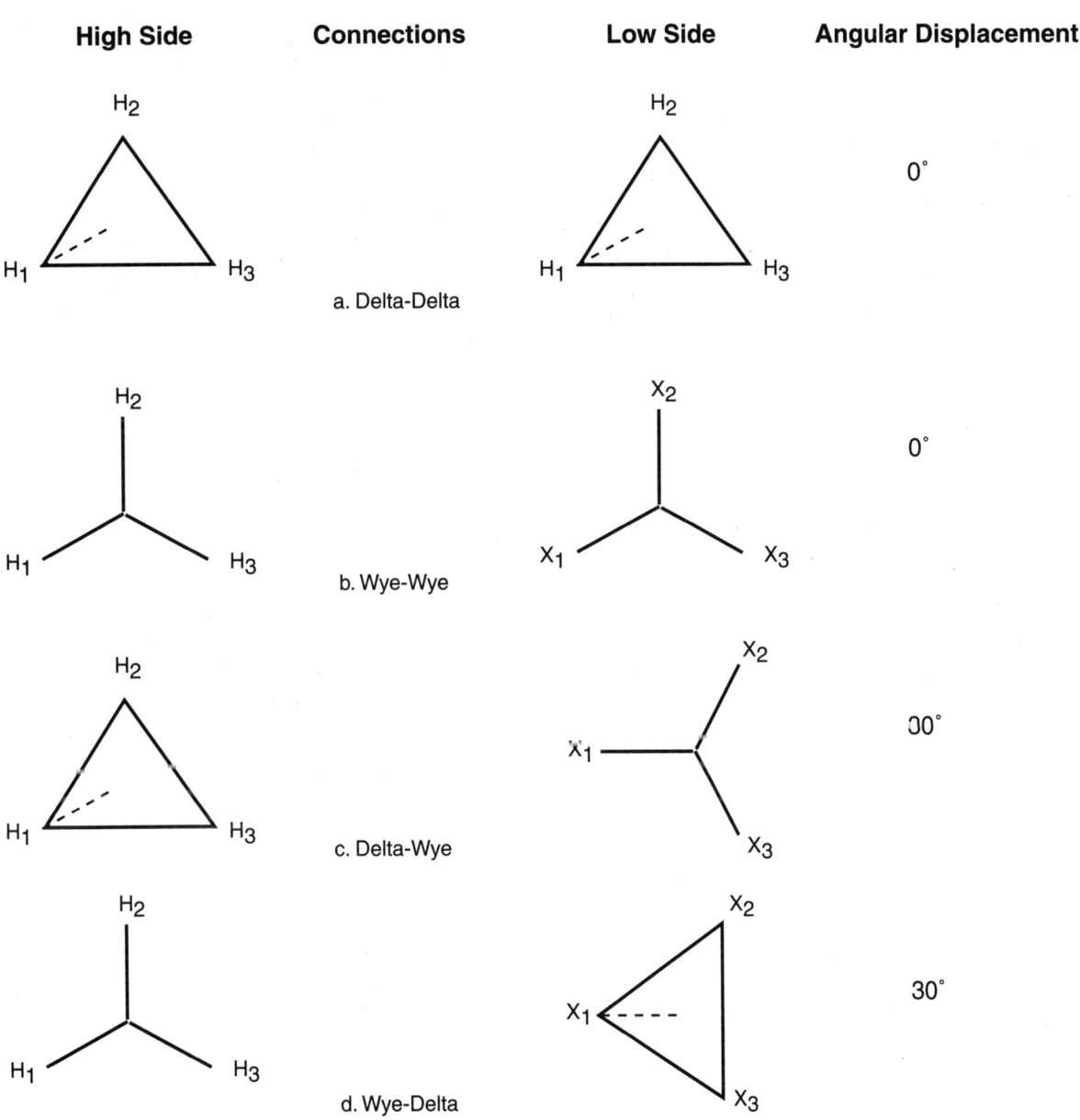

FIGURE 13–15
ANSI standard connections for three-phase transformers

These connection diagrams identify not only the way the various terminals are connected but also the angular relationship between the primary and the secondary currents and voltages. In connections *a* and *b*, the primary voltages and currents will be in-phase with each other; thus, these connections are 0° connections.

The second connections, *c* and *d*, are called 30° connections. Note the dashed lines in each of the delta connections. These lines represent the phase-to-neutral voltage for the delta. Of course, since the delta does not have a neutral, these voltages are hypothetical. However, the dashed lines also represent the phase angle of the delta line voltage.

Figure 13–16 shows a little more detail. Note that with the 30° connections, the secondary current lags the primary current by 30°. Other connections are allowed by ANSI; however, the 30° lag is the most common.

FIGURE 13–16
Current relationships for transformer connection of Figures 13–15d (Courtesy of Cadick Professional Services)

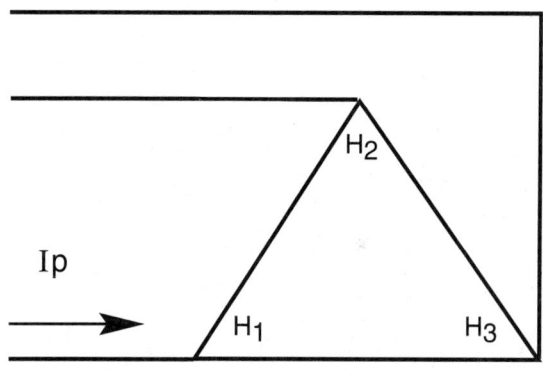

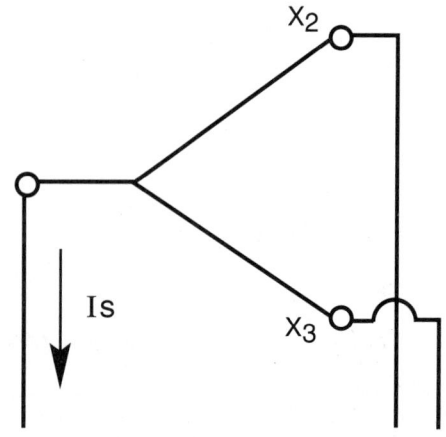

Is lags Ip by 30°

Voltage Relationships The voltage relationships for three-phase transformers are a little more complex than the single-phase. Note that when a transformer voltage is given, it is always expressed as the phase-to-phase value. Thus, a 138 kV to 13.8 kV transformer has 138 kV phase-to-phase on the primary and 13.8 kV phase-to-phase on the secondary.

Delta-delta and wye-wye connections are very straightforward. The phase-to-phase voltages on the primary and the secondary are in-phase and have the same ratio as the turns. Therefore, if the transformer primary has 10 times as many turns on the primary as the secondary, the voltage on the primary will be 10 times as large as the secondary voltage.

The delta-wye and wye-delta transformers are slightly more complex, Figure 13–17.

In these transformers, because of the square root of three relationship, the transformer is wound with windings that have a ratio that is 1.732 times bigger than the voltage ratio.

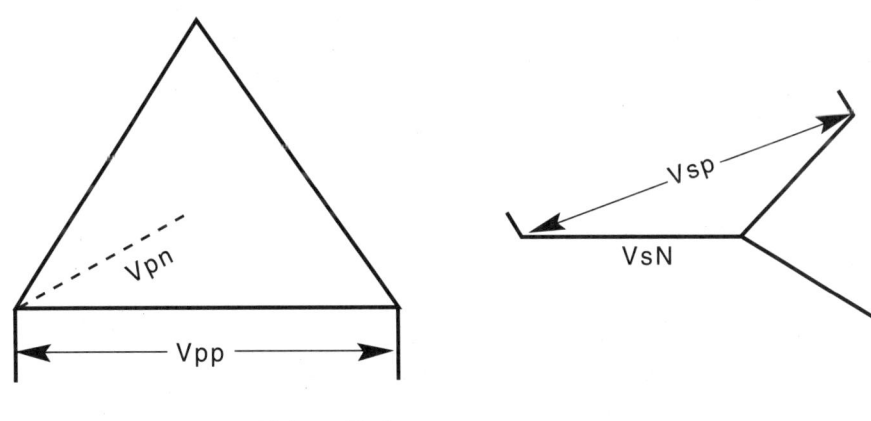

FIGURE 13–17 Voltage and turns ratio for delta-wye transformers (Courtesy of Cadick Professional Services)

Voltage Ratio =

$V_{pp} = 138KV$, $V_{pN} = \dfrac{138KV}{\sqrt{3}} = 79.7KV$

$V_{sp} = 13.8KV$, $V_{sN} = \dfrac{13.8}{\sqrt{3}} = 7.97KV$

Voltage Ratio $= \dfrac{138KV}{13.8KV} = \dfrac{10}{1}$

Turns Ratio $= \dfrac{138KV}{7.97KV} = \dfrac{17.32}{1}$

FIGURE 13–18 Autotransformer connections

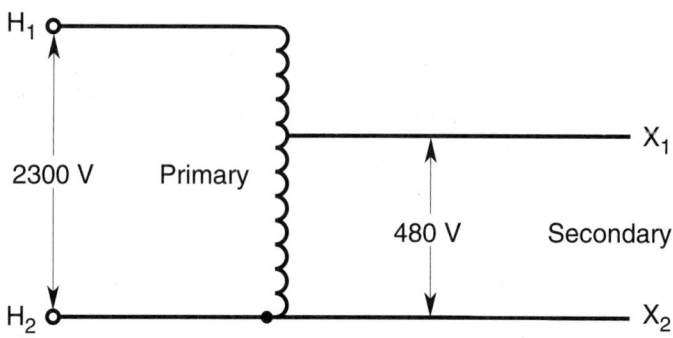

Autotransformers

Autotransformers are constructed in such a way that their primary and secondary windings are connected both electrically and mechanically, Figure 13–18. In these transformers, the transformation ratio is determined by the number of turns between H_1 and H_2, divided by the number of turns between X_1 and X_2. In the example shown, the turns ratio 2300/480 = 4.8. Autotransformers are somewhat less expensive than similar models made with separate coils.

Sometimes the autotransformer secondary will be adjustable to change the number of windings available. Such a device is called a variable autotransformer and is commonly used in 115-V sizes as a variable voltage power supply.

Autotransformers have one hazard that is unique, Figure 13–19. If the secondary winding opens as shown, the secondary voltage will be equal to the primary voltage. Because of this problem, autotransformers are not often used for step down service from medium-voltage and above to low-voltage.

FIGURE 13–19 Autotransformer hazard

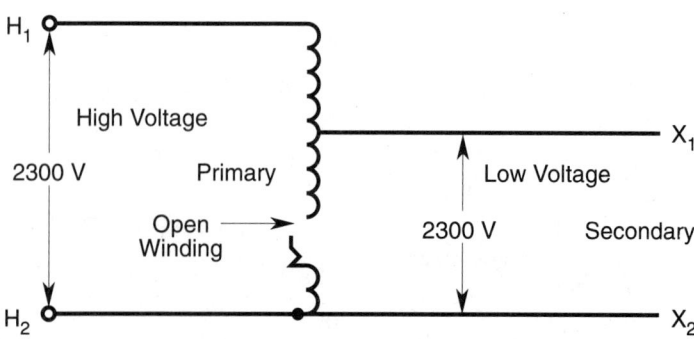

Transformer Theory

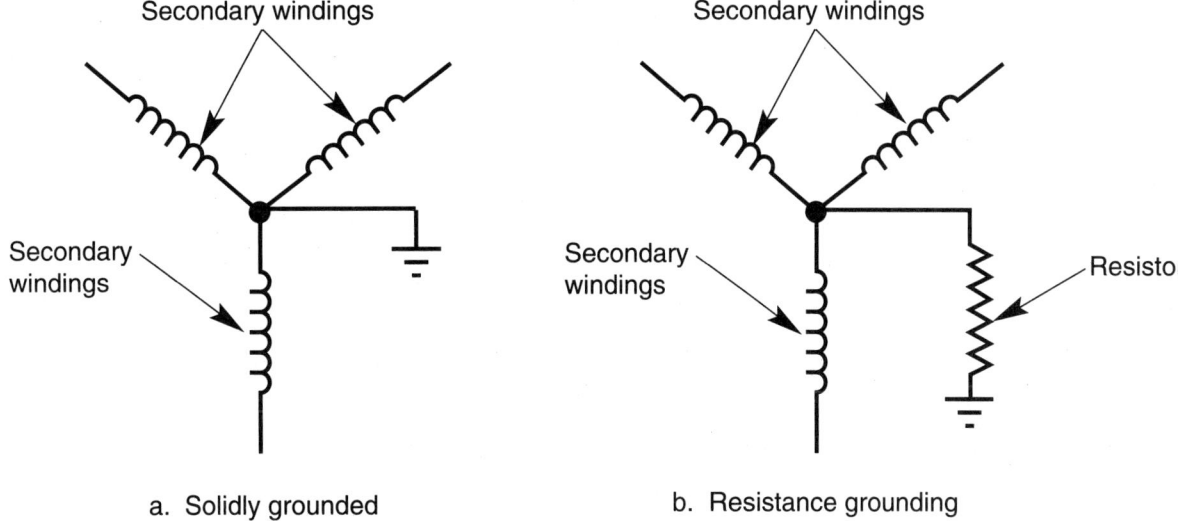

FIGURE 13-20
Various grounding methods for three-phase power and distribution transformers (Courtesy of Cadick Professional Services)

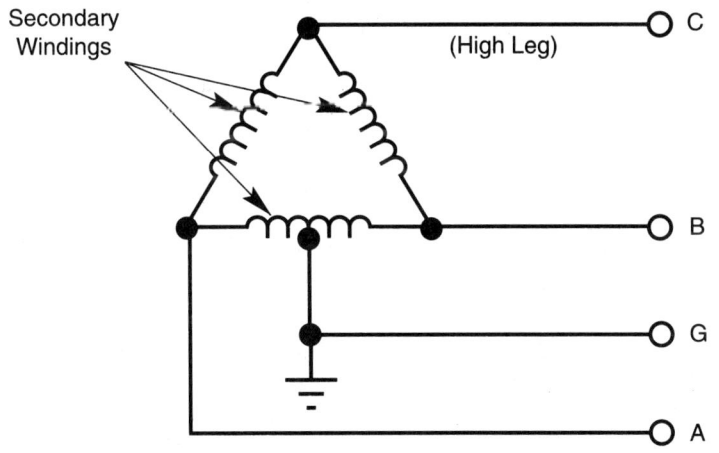

FIGURE 13-21
Common lighting transformer connections

$V_{AB} = V_{BC} = V_{CA}$ = 240 Volts

$V_{BG} = V_{AG}$ = 120 Volts

V_{CG} = 208 Volts

Transformer Grounding

For a variety of safety and operational reasons, transformer windings are often grounded. The point chosen to ground the transformer depends upon the connection. Figures 13–20 and 13–21 show two different grounding methods used for various purposes.

Figure 13–20 shows two common connections for three-phase power and distribution transformers. In this connection, the neutral point of a wye winding is grounded. This stabilizes the system, minimizes voltage transients, and provides a reference for the transformer. The resistance grounded connection minimizes the amount of ground fault current in the event that a phase wire falls to the earth. This type of connection is limited by the National Electrical Code.

Figure 13–21 is a very low-voltage connection used to obtain three very common voltages. Three 240-V windings are connected in a delta and one of them is center-tapped and grounded. As shown, this makes 240 V, 208 V, and 120 V available for use.

SUMMARY

Transformer operation is based on Faraday's law. Two coiled conductors are placed close to each other, and a voltage or current is applied to one of them. The current flow in this "primary" winding creates magnetic lines of flux which cut through the "secondary" winding. By Faraday's law, the flux lines cutting through the secondary create a voltage on the secondary. The magnitude and phase angle of this voltage are determined by the magnitude of the primary voltage and current, the phase angle of the primary voltage and current, the ratio of the number of turns in the primary versus the number of turns in the secondary, and the relative position of the primary to the secondary.

Transformers are available in many fundamental types: power, distribution, and instrument. The principal difference between power and dis tribution transformers is size. Power transformers are typically 500 kVA or larger with primary and secondary voltages from 13.8 kV and higher. Distribution transformers are generally smaller than power transformers; however, there is some overlap. Distribution transformers are frequently the last transformer between the generation source and the load.

Instrument transformers are used to change the voltage or current from high, unusable system values, to lower values suitable for use in system relays, meters, and other such instruments. Voltage transformers usually have 120-V or 69-V secondary levels. Current transformers ratios are expressed in terms of the nominal primary current which will produce a secondary current of 5 A.

The principal *working* components of a transformer are the core and coils. The two common varieties of core and coil construction are the core form and the shell form. Most large transformers use shell form construction while the smaller ones are of the core form variety.

REVIEW QUESTIONS

1. A 13,800-V to 480-V delta-wye transformer has a secondary current flow of 500 A. How much current is flowing in the primary?

2. What is the voltage ratio of the transformer discussed in question 1? What is the turns ratio of coils in the primary windings to the number of coils in the secondary winding?

3. How much current is flowing in the secondary circuit of a 400/5 CT when 800 A are flowing in the primary?

4. The secondary of an energized CT must always be connected to its load or must be _____ .

5. Which of the two types of core construction features coils wound on separate legs of the same iron core?

6. Draw a terminal diagram of a three-phase, delta-wye power transformer. Label the terminals properly.

7. What is the phase-to-neutral voltage of a 138,000-V primary, 4160-V secondary, delta-wye transformer? Give the answer for both the primary and secondary windings.

8. A low voltage service in a commercial building requires 240, 208, and 120-V service. Draw a three-phase transformer supply that might be used to provide such a service.

9. An ammeter in the secondary of a 480 to 120 V transformer shows 100 A. How much current is flowing in the primary?

10. A certain autotransformer has a 230-V primary and a 120-V secondary. What is its turns ratio?

Transformer Construction   14

OBJECTIVES

After studying this chapter, the student should be able to:

- *Describe the magnetic, electrical, and mechanical construction features of electrical transformers.*
- *Describe the construction and operations of the cooling systems used on typical transformers.*
- *Identify the information included on transformer nameplates.*

INTRODUCTION

The transformer cooling system is used to dissipate the large amounts of heat that are generated in the core and coils. The use of air, oil, fans, pumps, and other such devices allows relatively small transformers to carry large amounts of load.

The modern transformer is a sophisticated assembly of component parts. In addition to the core, coils, and cooling system, a modern transformer has a variety of valves, gauges, and other such equipment.

This chapter will discuss the more common aspects of transformer construction, including the cooling system and other component parts. It will finish with a description of a large power transformer nameplate.

TRANSFORMER INSULATION SYSTEMS

Transformer insulation is composed of several component parts. The core and coils themselves are insulated using a variety of insulation including plastic, wood, phenolic, paper, varnish, transformer oil, and, more recently, epoxy. The core and coils are immersed in some other medium to insulate and cool. In small transformers the medium may be a solid epoxy or air; larger transformers typically use liquid insulation such as transformer oil. The following paragraphs describe the construction of the more popular types of insulating and cooling materials—air and transformer oil.

Dry-Type Transformers

The term *dry-type* implies that no liquid is used for insulating or cooling the transformer. Although most dry-type transformers use air as the principal cooling and insulating medium, nitrogen and fluorocarbon gases have also been used. With increased concerns about the environment, fluorocarbon gases are no longer used.

Figure 14–1 shows the core and coils of a modern, dry-type transformer. In this assembly, the core and coils are cast in an epoxy resin. The epoxy provides a good electrical insulation as well as physical strength.

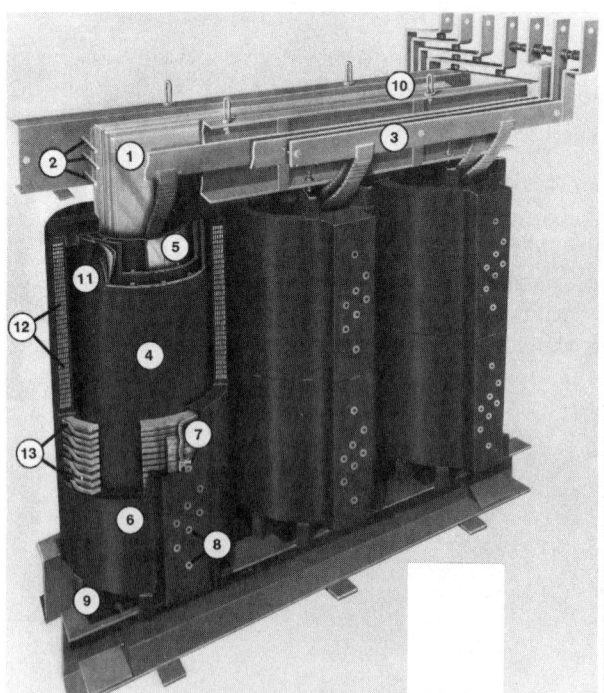

FIGURE 14–1 Dry-type transformer (Courtesy of ABB National Industrial Transformers, Hampton, VA)

Legend
1. Core
2. Framing Bolts
3. Low Voltage Bus
4. Low Voltage Coil
5. Sheet Conductor
6. High Voltage Coil
7. Fiber Glass Blankets
8. High Voltage Taps
9. Support Blocks With Resilient Pads
10. Lifting Lugs
11. Air Ducts
12. Disk Winding Section
13. Upset Crossover Connections

Transformer Construction

Liquid-Filled Transformers

Most power transformers, and many large distribution transformers, use oil as the insulating/cooling medium. Such transformers are available in five basic designs. Each design is actually a different liquid preservation system, isolating the oil from air and moisture, both of which will cause deterioration of the transformer oil. Each of these five designs is discussed in the following paragraphs.

Open Tank   Open tank, also called free breathing transformers are vented to the atmosphere through either an open pipe or a dehydrating breather. The dehydrating breather uses a silica gel to remove moisture from the air. The breather is bidirectional and will allow air to flow in and out of the transformer.

The open tank design is found on older transformers. This design costs less when new than other designs. The disadvantage of this design is that the dehydrating breather must be inspected at regular intervals. The free breather is the least effective in protecting the transformer oil. Open tank transformers are rarely used in new designs.

Sealed Tank   The sealed tank design, Figure 14–2, was developed as an alternative to the open tank design. The core, coils, and oil are completely enclosed with no vent. An inert gas blanket (usually nitrogen) is sometimes

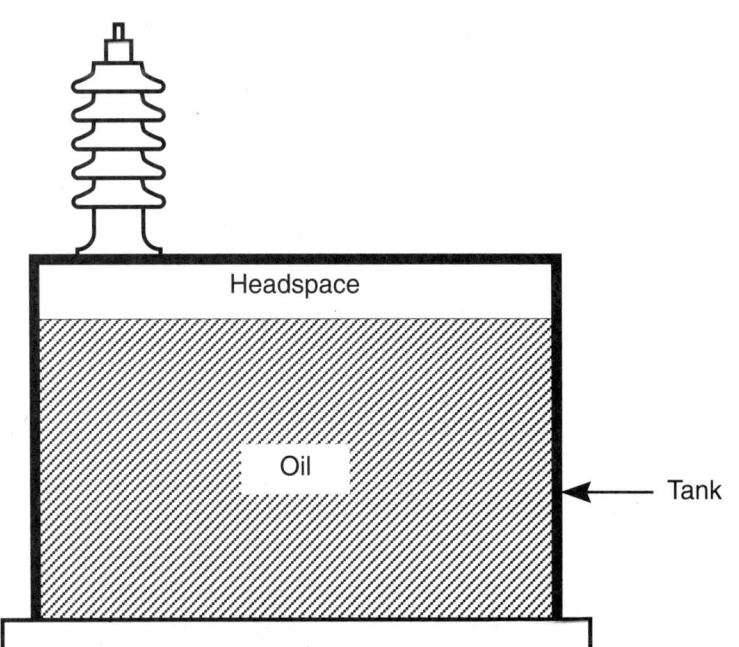

**FIGURE 14–2
Sealed tank transformer**
(Courtesy of IEEE Std 242-1986, IEEE Recommended Practices for Protection and Coordination of Industrial and Commercial Power Systems)

added on larger transformers. This design prevents the outside air from coming into contact with the oil and deteriorating it. Tank covers are often welded instead of being bolted. This is one additional step to prevent leakage of air into the system.

Many transformers, particularly the lower voltage types, still use the sealed tank design. Pole-mounted transformers are sealed tank design, but they do not have a gas blanket.

Conservator/Expansion Tank The conservator type design, Figure 14-3, uses an auxiliary tank placed above the main transformer tank. The main transformer tank is completely filled with oil. The expansion tank and pipe are filled to approximately one-half their volume. The expansion tank itself can be quite large, holding from 3%-10% of the total oil in the transformer. Conservator transformers breathe through a dehydrating breather installed in the expansion tank. The main transformer tank is sealed.

As the oil in the transformer expands and contracts, outside air is drawn into the conservator tank. Air pollutants are drawn into the conservator tank and settle into the sump where they can be drawn off. Some of the newer

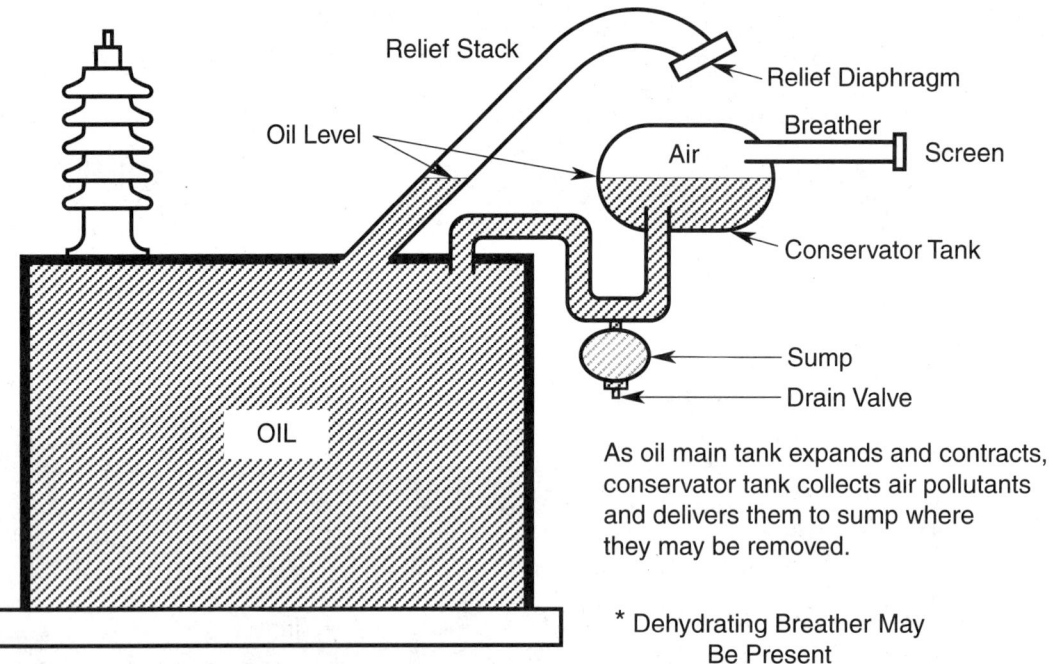

FIGURE 14-3
Conservator/expansion tank transformer
(Courtesy of IEEE Std 242-1986, IEEE Recommended Practices for Protection and Coordination of Industrial and Commercial Power Systems)

conservator designs use a synthetic bag installed in the expansion tank. All the oil is kept completely isolated from the air in this style of expansion tank.

Gas-Oil Seal This type of transformer, Figure 14–4, uses an auxiliary tank also. The auxiliary tank in this design is divided into two sections. One section is connected to the transformer's main tank headspace and serves as an expansion tank. The second section is vented to the atmosphere through a breather. The two sections are separated from each other by a gas-oil seal.

The gas-oil seal operates on the manometer principle. As the temperature of the transformer increases, the oil expands. This expansion pushes the nitrogen blanket into the expansion tank. As the gas pressure increases, the oil in the expansion tank is transferred to the vented compartment. As the transformer cools, the nitrogen is forced into the main transformer tank. The oil in the main tank never comes into contact with the oil in the expansion tank.

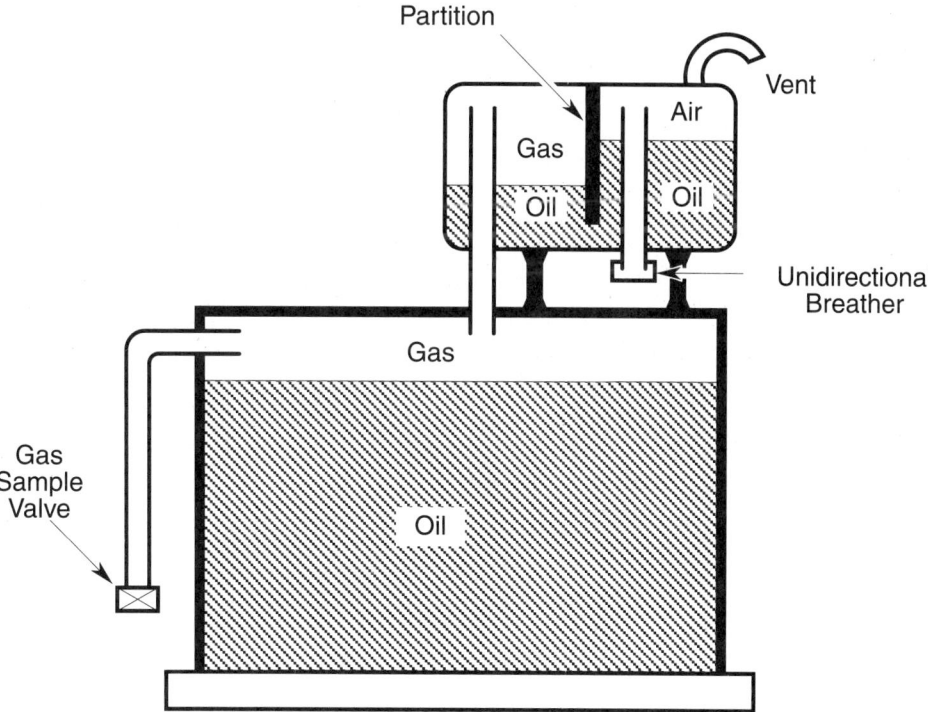

FIGURE 14–4
Gas-oil seal transformer
(Courtesy of IEEE Std 242 1986, IEEE Recommended Practices for Protection and Coordination of Industrial and Commercial Power Systems)

**FIGURE 14–5
Automatic gas
pressure transformer**
(Courtesy of IEEE Std 242-1986,
IEEE Recommended Practices for
Protection and Coordination
of Industrial and Commercial
Power Systems)

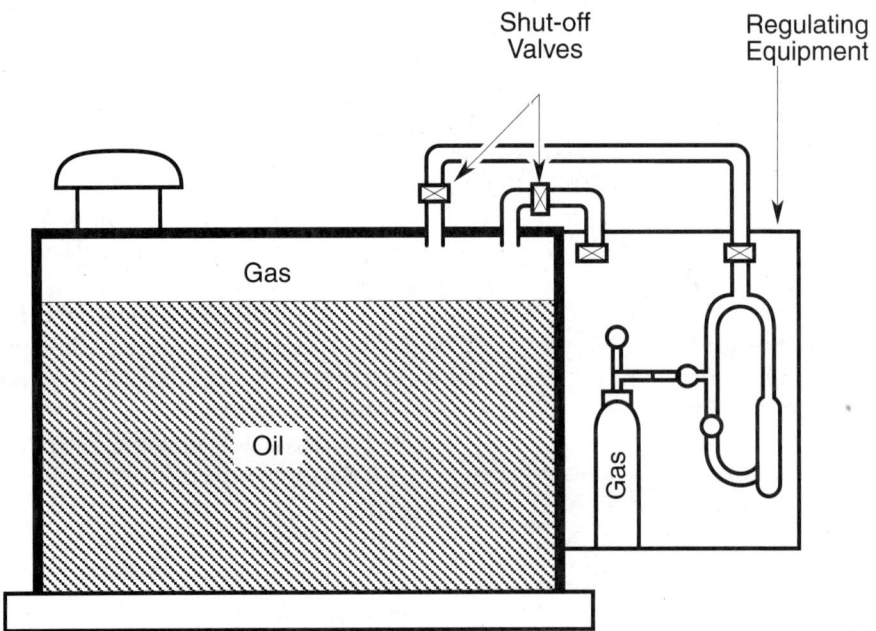

Automatic Gas Pressure In the automatic gas pressure design, Figure 14–5, the transformer tank is sealed. A gas blanket of inert gas, usually nitrogen, is maintained above the oil at a positive pressure. A cylinder of gas is placed with the transformer and connected to the headspace through regulating equipment. The regulating equipment maintains the gas pressure through all temperature variations. An alarm is often connected to the regulator so that the bottle of gas can be changed before its pressure falls too low. As long as the bottle maintains a positive pressure in the transformer tank, no air can reach the oil. Should the bottle run out of nitrogen, the nitrogen in the tank will be exhausted as the oil expands. When the oil contracts, air would be drawn into the main tank.

Automatic gas pressure transformers are one of the most popular designs for large transformer designs.

Cooling Classifications

Transformers are given a cooling letter code depending on the type of cooling employed. Table 14–1 defines the most common cooling types:

TRANSFORMER COMPONENTS AND INDICATORS

The construction and components used in transformers depend upon the size of the transformer. Small units may have only the tank, core, coils, and

Table 14-1 Transformer Cooling Categories
(Courtesy of Cadick Professional Services)

Transformer Type	Cooling Designation	Description
Dry-Type	AA	Self-Cooled
	FA	Forced Air-Cooled (Fans)
	AA/FA	Self-Cooled and Forced-Air (Fans energize when temperature rises).
Liquid-Filled	OA	Liquid-Filled, Self-Cooled
	OA/FA	Self-Cooled and Forced-Air, (Fans blow across self-circulating cooling fins).
	FOA	Forced-Oil, Fan-Cooled. Oil pumps circulate the oil through cooling fans. This type usually has three stages based on temperature. As the temperature goes up, the fans energize. The next temperature plateau energizes the oil pumps.
	OW	Water-cooled heat exchangers jacket the transformer instead of air.
	FOW	Pumps circulate both oil and water for extra cooling.

terminal connections. Larger units will have a more sophisticated array of auxiliary equipment such as protective devices, instruments, tap changers, and the like. The following paragraphs explain many of the components and transformers.

Tank

The transformer tank, Figure 14-2, is the outer container of the transformer. On liquid-filled transformers, the tank holds the insulating liquid. The tank also performs a number of other important functions; it protects the windings and core from damage, seals the insulating liquid and solid insulation from moisture and oxygen, and provides a mounting structure for auxiliary components such as bushings and cooling equipment.

The tank itself is constructed of high quality steel with a fiber liner. Sometimes the inside is coated with a high quality, insulating paint to prevent rusting. The outside is typically covered with a highly weather resistant paint for protection against weathering.

Dry-type transformers are often enclosed with a sheet metal enclosure rather than a tank. This enclosure will have panels that may be removed for easy access to the core and coils.

Small transformers may use the core itself as part of the external protection. The transformer coils are wound on the core, and end caps are placed on the lamination bolts to protect the coils. This construction is used only on very small transformers.

Core and Coils

The core is the magnetic heart of the transformer. Virtually all power, distribution, and instrument transformers have a core made of high permeability, steel laminations. The two types of core designs are the shell and the core form type. These two types were discussed in the last chapter and are illustrated in Figure 13–8.

Transformer cores are usually made of cold-rolled silicon steel. Regular steel is not used because its internal crystals are arranged in random. The crystals are arranged in all directions which causes the magnetic lines of force to wander and bunch up, creating losses in the form of heat, Figure 14–6.

By using cold-rolled, crystal-oriented silicon steel, the magnetic lines of force flow more smoothly with less reluctance, because the crystals are arranged in the same direction. This crystal arrangement produces less loss in the core, which increases efficiency, Figure 14–7.

Bushings

On small, low voltage transformers, the voltage leads pass directly through openings in the transformer case or tank. On higher voltage transformers, the voltage stress would be too great for direct entry of this type.

Bushings, Figure 14–8, provide terminal connections to the windings and also insulate the connections from the tank and each other. To do their job properly, bushings must perform four basic functions:

1. The outer shell of a bushing must be a good insulator.
2. A bushing must be water tight.
3. A bushing must be gas tight.
4. A bushing must be oil tight.

Transformer Construction

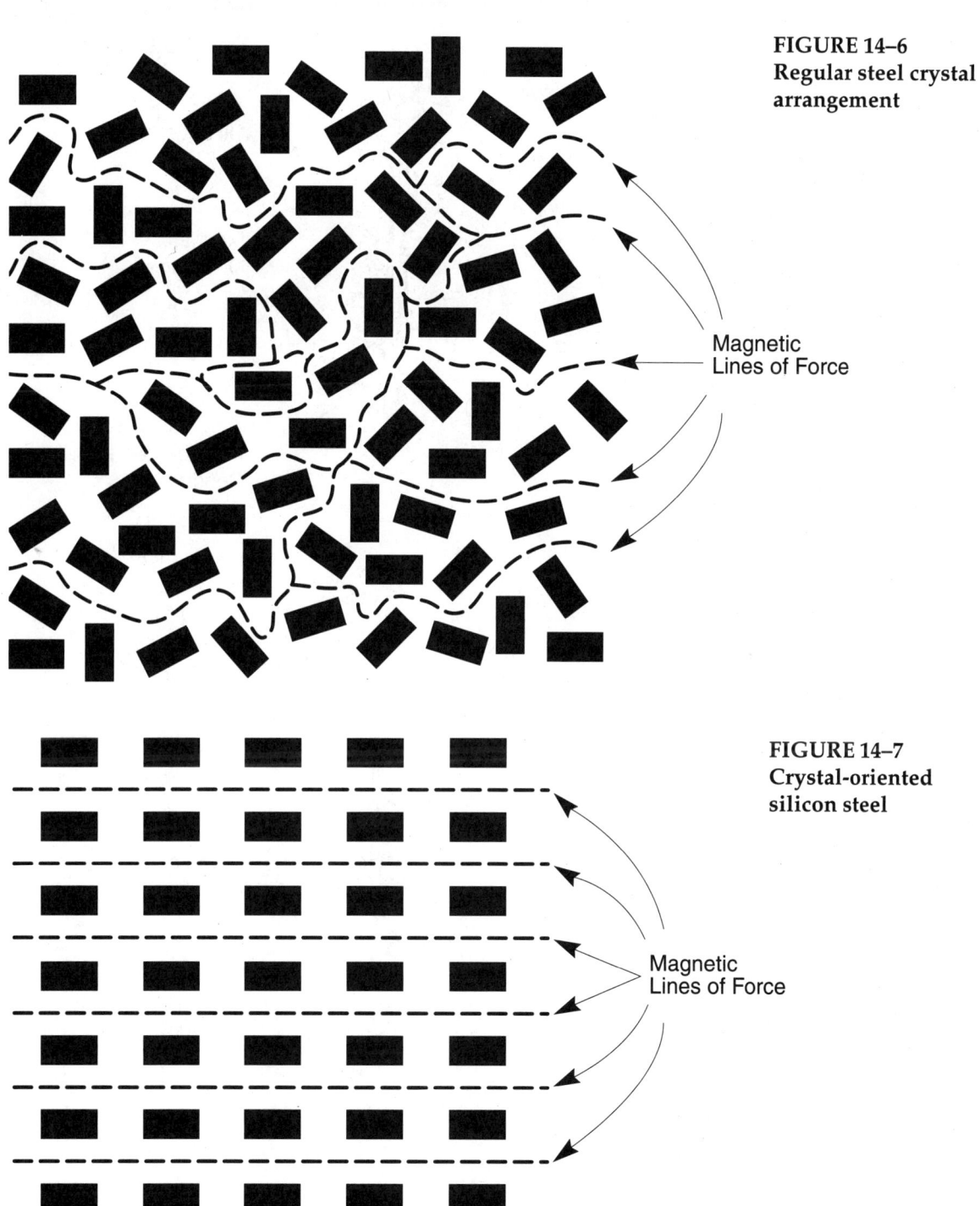

FIGURE 14–6
Regular steel crystal arrangement

FIGURE 14–7
Crystal-oriented silicon steel

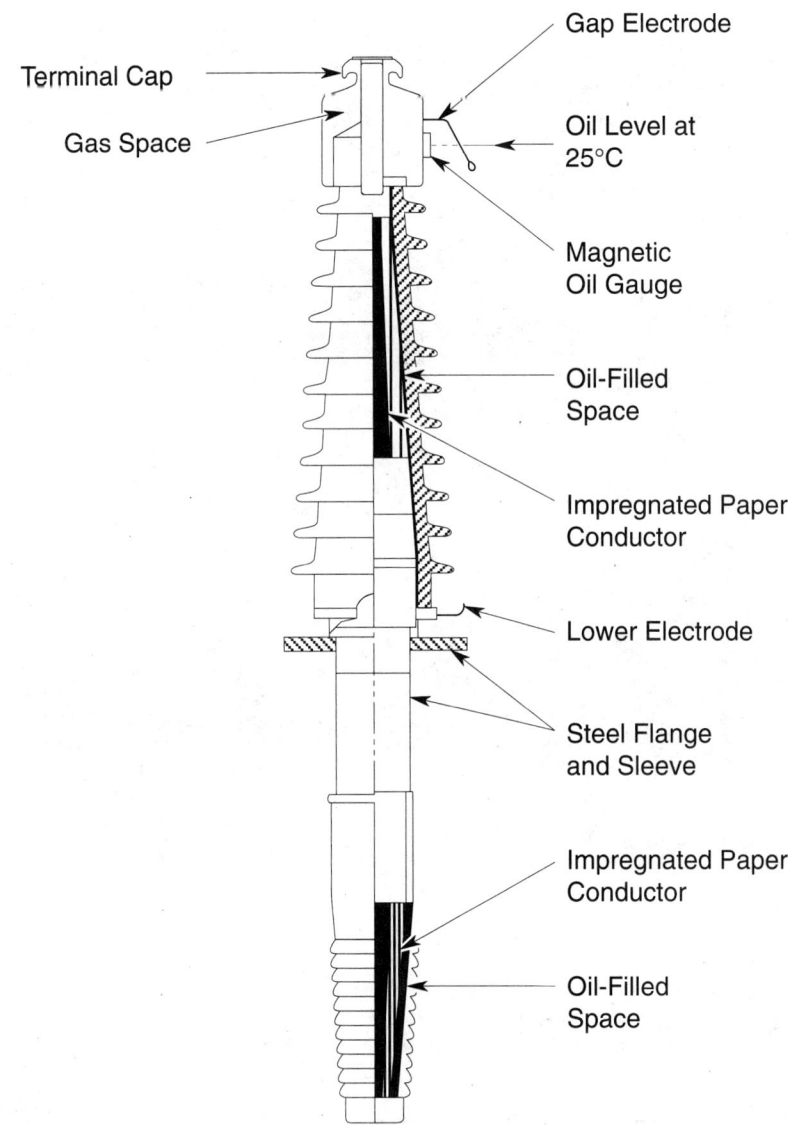

FIGURE 14–8
Transformer bushing type "O", 69 kV (Courtesy of ABB Power T&D Company Inc.)

Bushings may be constructed of a variety of materials depending on the voltage level. Almost all bushings have an outer covering of porcelain. Porcelain offers excellent mechanical and electrical strength.

Bushings for use up to 25 kV are most often made of ceramic clad with porcelain. Bushings from 13.8 kV to 115 kV (note the overlap) have a porcelain exterior shell with a synthetic resin-bonded paper or oil-impregnated paper inside.

Petticoats (or watersheds) are often used on bushings for outdoor applications. These petticoats increase the distance from the line terminal to the tank. This reduces the possibility of flashover due to moisture or contamination. The bushing shown in Figure 14–8 is a large, high voltage, oil-filled unit. A bushing of this type may be as much as 15 ft long.

Tap Changers

Many transformers are equipped with tap changers. A tap changer allows the transformer turns ratio to be varied to allow for supply deviations or secondary load changes. No-load tap changers allow a small adjustment in voltage ratio to allow for supply line deviations. Typical positions are nominal voltage, two steps above nominal, and two steps below nominal. Each step in such a tap changer represents a 2½% adjustment. Figure 14–9 shows a typical, primary winding, no-load tap changer.

Each progressive tap from A-B to E-F reduces the number of windings on the primary. This means that for a given input voltage, the output voltage goes up as the tap is moved from A-B towards E-F. The primary tap changer is used to adjust the transformer output for differences in input voltage.

Example:

The transformer of Figure 14–9 is designed for 13,800–480 V operation on tap C-D. In a particular application the input voltage is 13,110 V. The desired output is still 480 V. What tap should be used?

Solution:

13,110 V is 95% of the nominal primary voltage. In this instance the primary tap should be set to the E-F setting. This puts the primary turns 5% below nominal and boosts the output to 480 V.

Some large transformers have under-load tap changers or ULTC. These tap changers are set to operate automatically to increase or decrease the voltage level as the load decreases or increases.

PERCENT VOLTS	DIAL POSITION	TAP CHANGER CONTACTS
105	1	A TO B
102.5	2	B TO C
100	3	C TO D
97.5	4	D TO E
95	5	E TO F

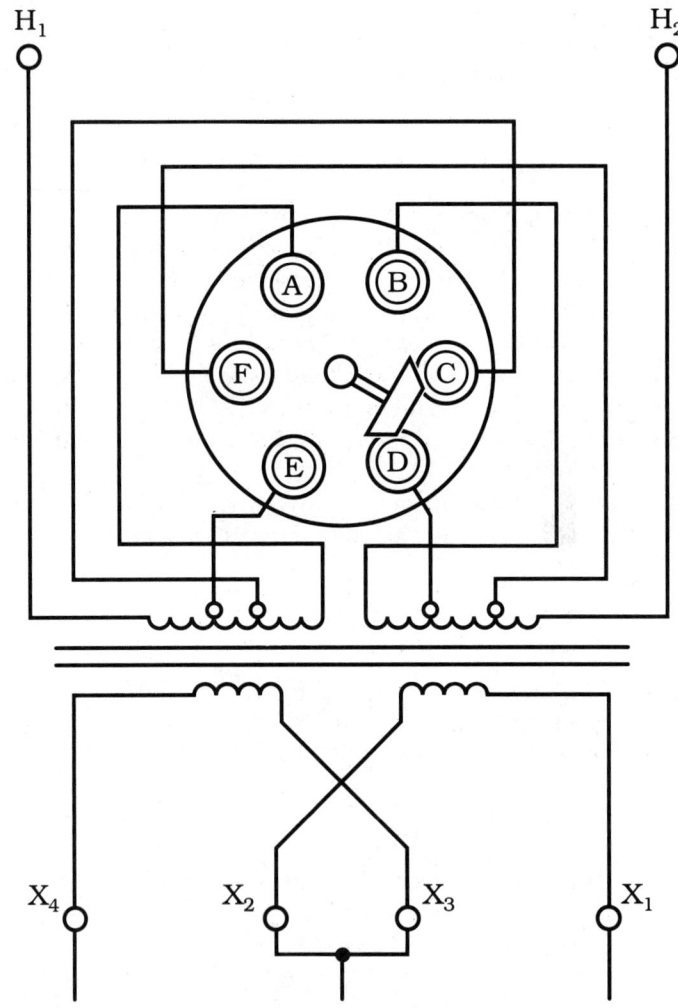

FIGURE 14–9
Typical primary, no-load tap changer diagram (Courtesy of ABB Power Tip Company Inc.)

Sudden Pressure Devices

The sudden pressure device, Figure 14–10, is also called a sudden pressure relay. This device is a small diaphragm switch that is activated by pressure from within the transformer. Sudden pressure increases in a transformer can be caused by internal arcing or other such major problems inside the transformer. The sudden pressure device will respond very quickly to sudden pressure changes and close a set of contacts to trip circuit breakers that

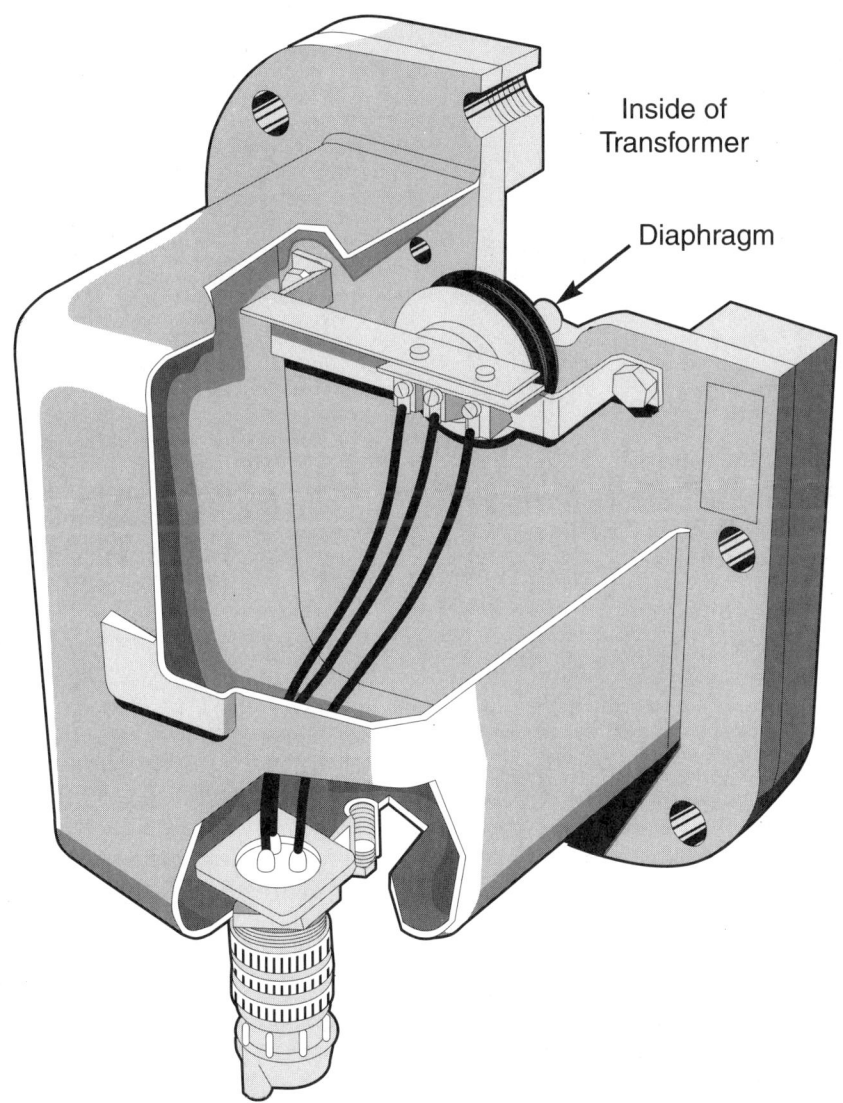

FIGURE 14–10
Sudden pressure relay (Courtesy of ABB Power T&D Company Inc.)

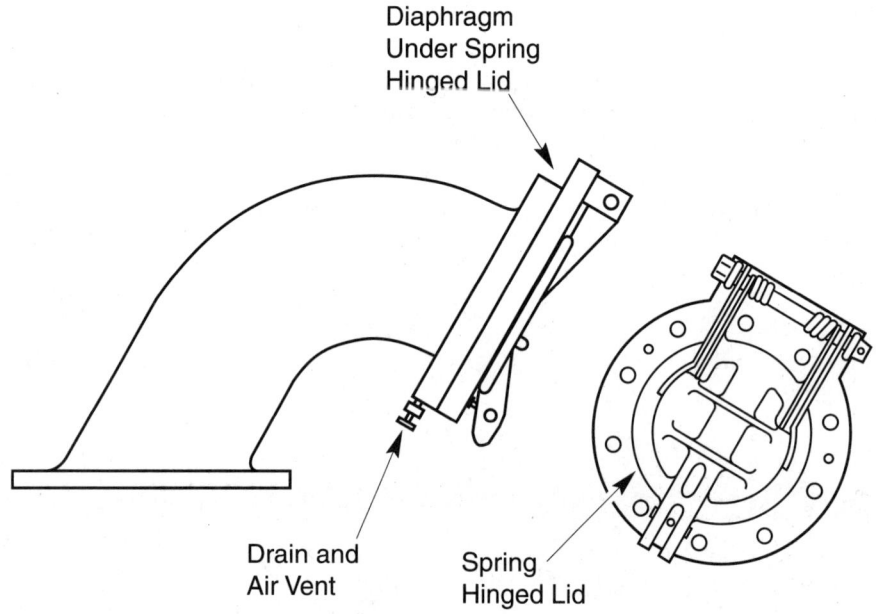

FIGURE 14–11
Diaphragm pressure relief device

feed the transformer. Slow pressure changes will not trigger the device. Sudden pressure relays are among the fastest protection available for power transformers.

Small transformers and dry-type transformers are usually not equipped with sudden pressure devices.

Pressure Relief Devices

Sudden pressure devices are available in two basic types. The diaphragm type, Figure 14–11, consists of a large diameter pipe, a diaphragm, and a hinged cover. The diaphragm ruptures at an internal pressure of 5–6 lbs per square inch. After the diaphragm ruptures, it must be replaced. This is the type of device used on the conservator/expansion tank transformer.

The valve pressure relief device is a self-resetting, nondestructive unit, Figure 14–12. When the internal pressure rises above its setting, typically 5 to 6 psi, the diaphragm compresses the springs and opens the gaskets, thus allowing pressure relief inside the transformer. The semaphore pops up at the same time, indicating that the unit has operated. This type of device is normally used on large transformers and rarely on smaller distribution units.

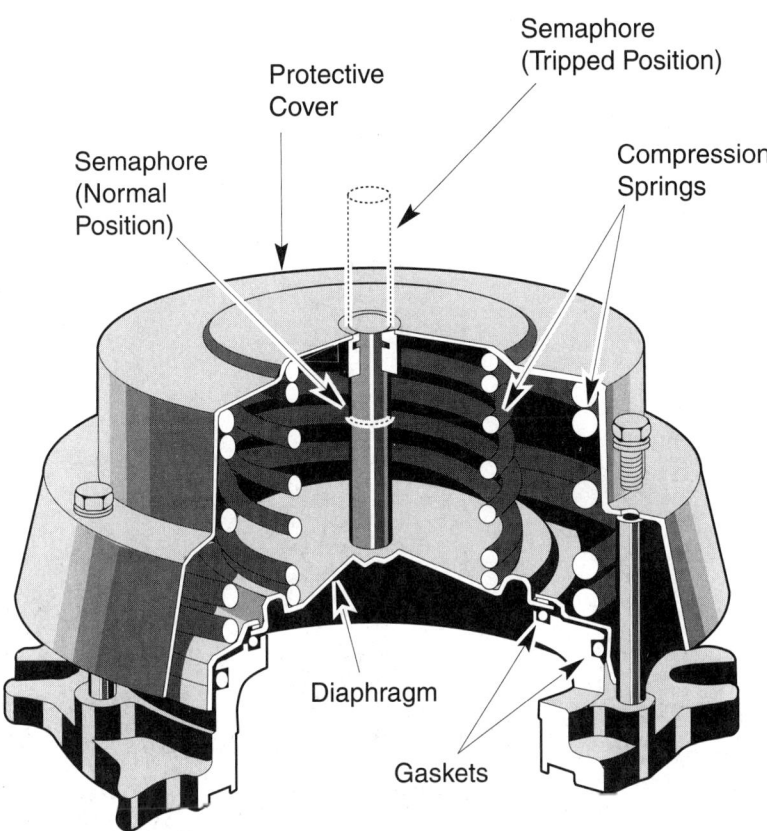

FIGURE 14–12 Valve pressure relief device (Courtesy of ABB Power T&D Company Inc.)

Cooling Fans

Cooling fans, when present, are externally mounted at various locations on the transformer. Fans are normally operated in groups by a temperature sensor; they allow the transformer to carry more load without overheating. Cooling fans are only one way of increasing the load carrying capabilities of transformers (Figure 14–13). Radiators, cooling fins, and pumps may also be used.

Liquid Level Indicators

Oil level indicators show the level of oil in the tank. Since the oil expands and contracts as the temperature changes, gauges are equipped with calibration points which show the full mark for a 25°C liquid temperature. Indicators are available in several types, including the sight glass type, Figure 14–14a, and the dial type, Figure 14–14b.

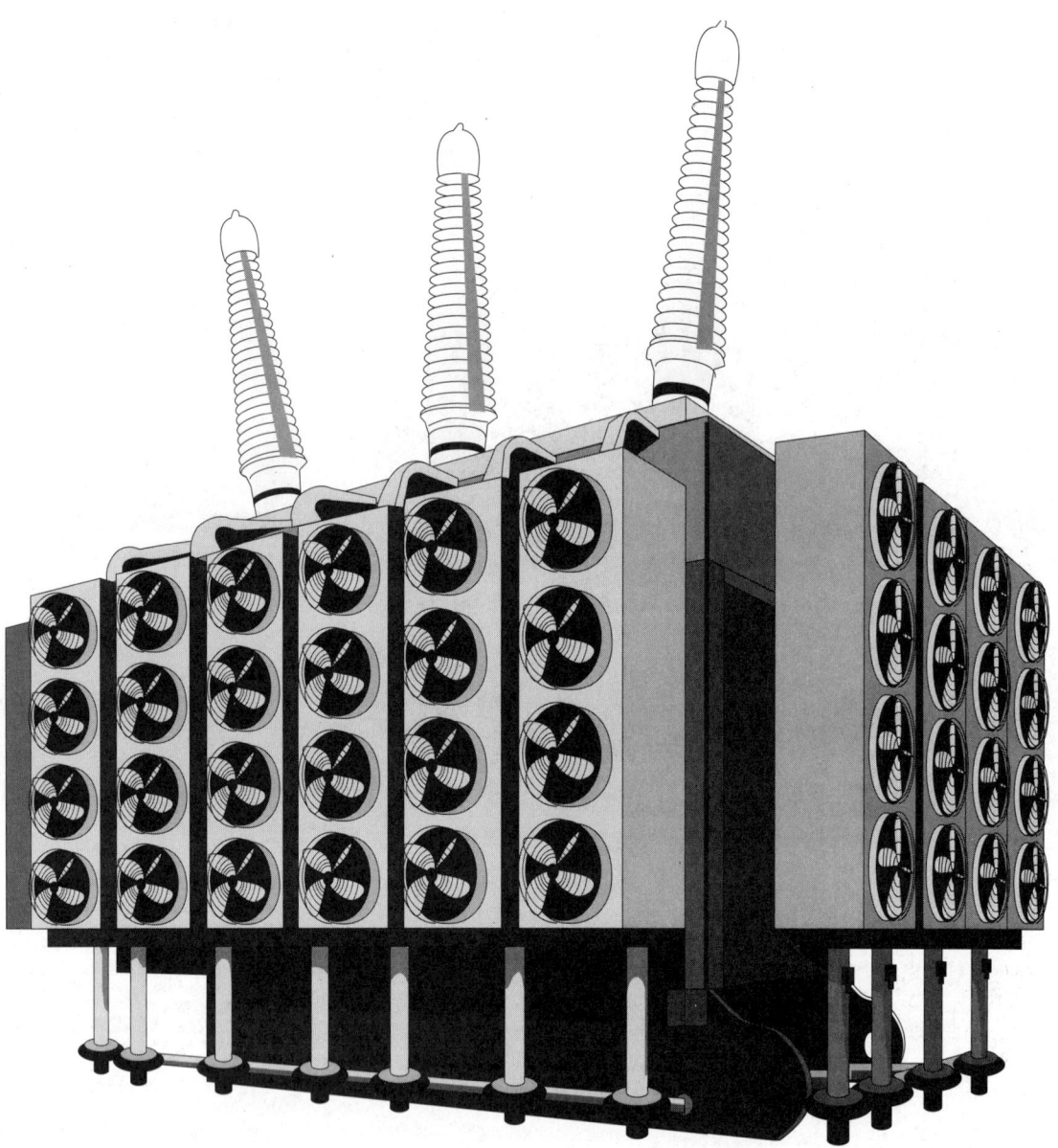

FIGURE 14–13
Large OA/FOA transformer (Note the large banks of cooling fans)

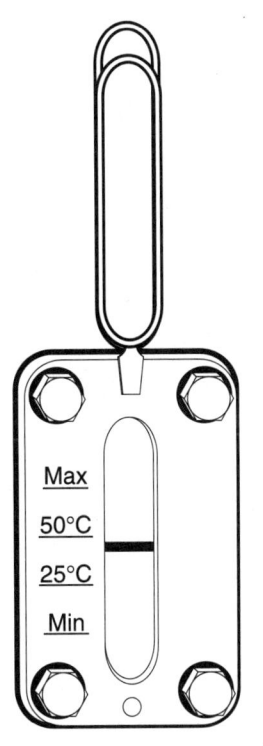 

FIGURE 14–14
Two types of liquid level indicator (Courtesy of ABB Power T&D Company Inc.)

(a) Flush Type, Sight Glass Level Indicator

(b) Indirect reading Indicator

Temperature Indicators

Liquid temperature indicators, Figure 14–15, read the oil temperature directly from sensors placed in the tank.

Hot-spot indicators, Figure 14–16, give a reading of the oil temperature in a well of oil heated by a coil. A current transformer is connected to the coil and supplies current proportional to load current. As load current varies, the heating effect also varies. The additional heating supplied by the heating coil is calibrated to match the characteristics of the transformer's hottest spot. The transformer's hottest spot is usually considered to be two-thirds of the way up the coil stack and one-third of the way into the high-voltage winding.

Liquid temperature indicators and hot-spot indicators may be combined into one unit with two pointers operating independently of each other.

FIGURE 14–15
Liquid temperature indicator (Courtesy of ABB Power T&D Company Inc.)

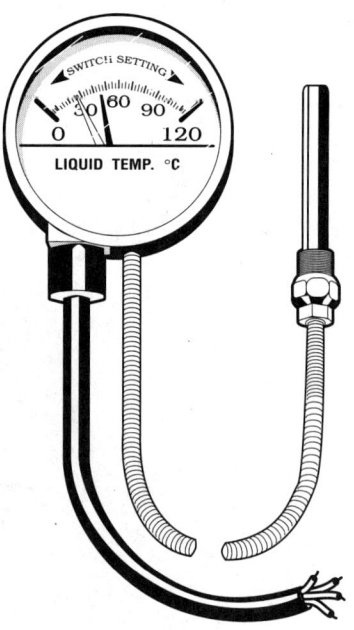

FIGURE 14–16
Winding temperature indicator (Courtesy of ABB Power T&D Company Inc.)

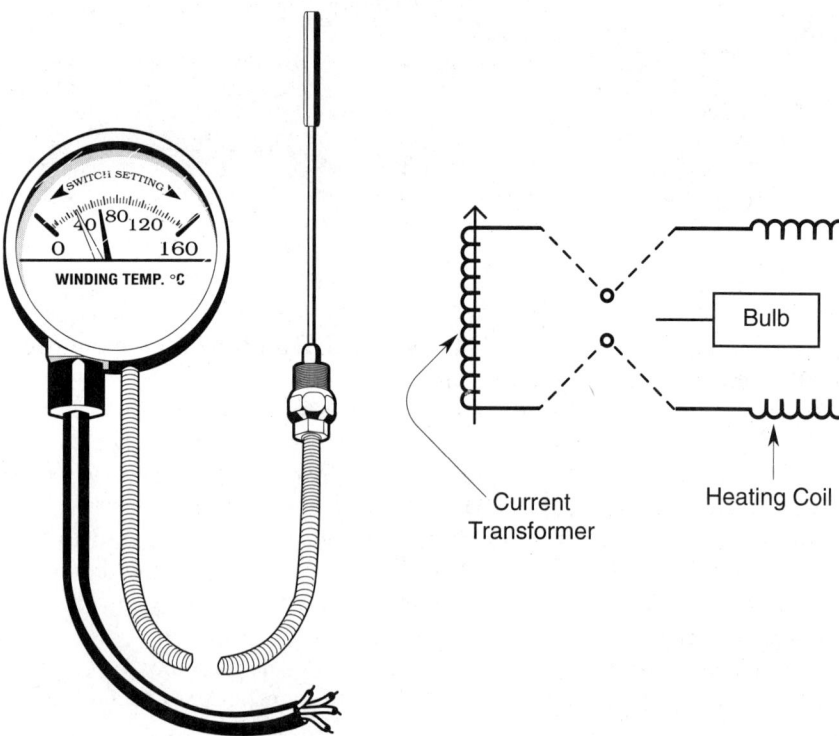

Fault Gas Detector

When electrical faults occur inside transformers, they form gases that are not normally present. Fault gas monitors, Figure 14–17, are used to sample the gas in the head space of the transformer, and to analyze that gas for the presence of fault gases. These detectors can be used on-line, or they may be portable units used for preventive maintenance procedures.

Pressure/Vacuum Indicators

Large, enclosed transformers should be operated with a slight positive pressure. Pressure/vacuum gauges, Figure 14–18, show the pressure inside the transformer.

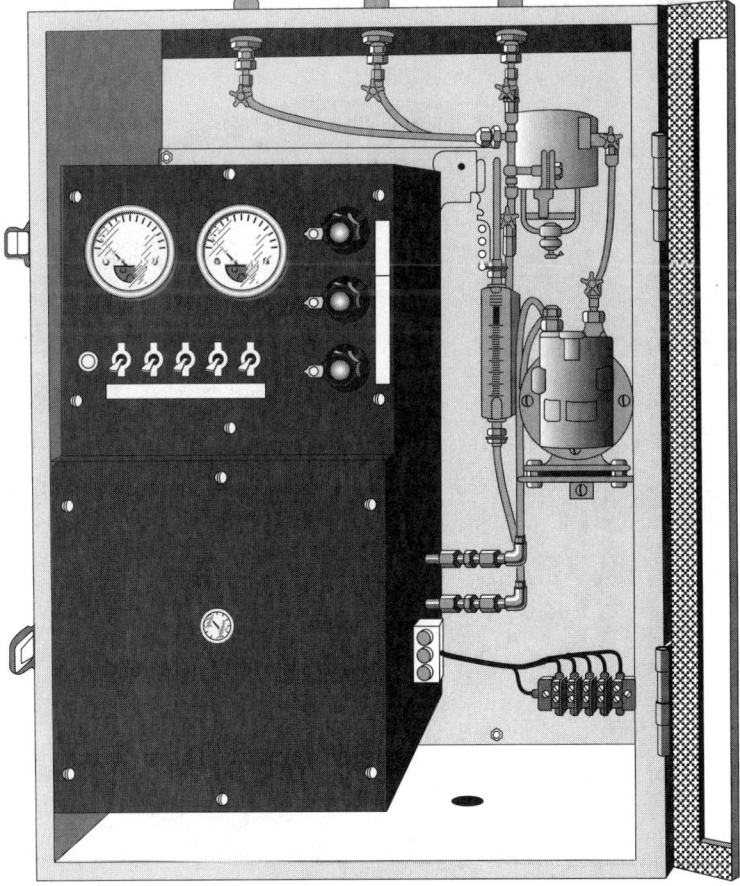

FIGURE 14–17
Fault gas monitor

TRANSFORMER CHARACTERISTICS AND RATINGS

Industrial transformers have a number of industry standard ratings that are assigned to them.

- *kVA rating*—kVA rating is the maximum kVA that the transformer will carry on a continuous basis. ANSI defines standard kVA ratings. Table 14–2 shows ANSI standard kVA ratings for liquid-immersed power and distribution transformers. Table 14–3 provides the same information for dry-type transformers.
- *voltage ratings*—Two voltage ratings apply to transformers. The *insulation class* identifies the insulation type and categorizes the transformer accordingly. The *winding voltages* are given phase-to-phase and identify the turns ratio of the transformer. Remember that in delta-wye

FIGURE 14–18 Pressure/vacuum indicator (Courtesy of ABB Power Tip Company Inc.)

Table 14-2 ANSI Standard kVA Ratings for Liquid Immersed Distribution and Power Transformers

(Courtesy of IEEE Std C 57.12.00-1993, IEEE Standard General Requirements for Liquid-Immersed Distribution, Power and Regulating Transformers)

Single-Phase Transformers	Three-Phase Transformers
5	15
10	30
15	45
25	75
37.5	112.5
50	150
75	225
100	300
167	500
250	750
333	1000
500	1500
	2000
833	2500
1250	3750
1667	5000
2500	7500
3333	10,000
	12,000
5000	15,000
6667	20,000
8333	25,000
10,000	30,000
12,500	37,500
16,667	50,000
20,000	60,000
25,000	75,000
33,333	100,000

transformers, the voltage ratio and the turns ratio vary by the square root of three. Table 14–4 shows ANSI standard voltage ratings.
- *cooling*—Each transformer is rated according to the type of cooling. See Table 14–1 for a listing of standard cooling ratings and their meanings.

Table 14-3 ANSI Standard kVA Ratings for Dry-Type Power and Distribution Transformers

(Courtesy of IEEE Std C 57.12.01-1989, IEEE Standard General Requirements for Dry-Type Distribution and Power Transformers Including Those with Solid Cast and/or Resin Encapsulated Windings)

Single-Phase Transformers	Three-Phase Transformers
1	15
3	30
5	45
7.5	75
10	112.5
15	150
25	225
37.5	300
50	500
75	750
100	1000
167	1500
250	2000
333	2500
500	3750
833	5000
1250	7500
1667	10,000
2500	12,000
3333	15,000
5000	20,000
6667	
8333	
10,000	

- *impedance*—Transformer impedance level is given in percent and is usually based on the transformer's full load capacity. For example, assume a 1500-kVA transformer with a primary voltage of 13,800 V and a secondary voltage of 480 V. Assume the transformer has a percent impedance of 5.75%. The transformer's full load current is calculated from:

$$I_{fl} = \frac{1500}{0.48 \times \sqrt{3}} \; 1804 \text{ A} \tag{1}$$

Table 14-4 Standard Voltage Ratings for Three Phase Transformers
(Courtesy of Cadick Professional Services)

Secondary Substation	
High-Side Voltage	Low-Side Voltage
15-kV-class insulation	600-V-class insulation
13,800	600
13,200	480
12,000	480Y/277
7,200	240
6,900	208Y/120

Primary Substation		
69-kV-class insulation	15-kV-class insulation	5-kV-class insulation
67,000	14,400	4,800
	13,800	4,360
	13,200	4,160
46-kV-class insulation	13,090	2,520
43,800	12,600	2,400
	12,470	
34.5-kV-class insulation	12,000	
34,400	8,720	
26,400	8,320	
	7,560	
	7,200	
25-kV-class insulation	6,900	
22,900	5,040	

The maximum fault current for a short circuit on the secondary of the transformer is calculated from:

$$I_{sc} = \frac{I_{fl}}{\%Z} \times 100 = \frac{1804}{5.75} \times 100 = 31{,}373 \text{ A} \qquad (2)$$

The 100 is added because the impedance is in percent. The use of percent impedance greatly simplifies the calculation of short circuits in transformers.

- *voltage taps*—Many transformers are equipped with primary tap changers. This was discussed earlier.
- *basic impulse level (BIL)*—The basic impulse level is the maximum or crest value of a voltage impulse. It is a measure of how much of a voltage surge the transformer will withstand. The manufacturer tests transformers for BIL as part of the manufacturing process. Table 14–5 identifies standard BIL ratings for liquid immersed power and distribution transformers.

TRANSFORMER NAMEPLATES

Virtually all power and distribution transformers are equipped with a nameplate that contains all of the critical information for the transformer. The following paragraphs discuss the information on a very comprehensive nameplate, Figure 14–19.

Frequency

The frequency rating of a transformer is the normal system frequency. Operation at any value other than the rated frequency requires that the transformer ratings be reduced. This transformer has a frequency rating of 60 Hz.

Phase

This rating indicates the number of phases for which the transformer is designed. In this case, the transformer is a three-phase unit.

Class

This is the cooling class of the transformer. This particular transformer has three stages of cooling, OA/FA/FOA.

Table 14-5 Transformer BIL Levels. Liquid Immersed Power and Distribution Transformers

(Courtesy of IEEE Std C 57.12.00-1993, IEEE Standard General Requirements for Liquid-Immersed Distribution, Power and Regulating Transformers)

Application	Nominal System Voltage (kV rms)	Maximum System Voltage (from ANSI C84.1-1982 [16] and ANSI C92.2-1981 [18]) (kV rms)	Basic Lightening Impulse Insulation Levels (BIL) in Common Use (kV crest)			
Distribution	1.2		30			
	2.5		45			
	5.0		60			
	8.7		75			
	15.0		95			
	25.0		150	125		
	34.5		200	150	125	
	46.0	48.3	250	200		
	69.0	72.5	350	250		
Power	1.2		45	30		
	2.5		60	45		
	5.0		75	60		
	8.7		95	75		
	15.0		110	95		
	25.0		150			
	34.5		200			
	46.0	48.3	250	200		
	69.0	72.5	350	250		
	115.5	121.0	550	450	350	
	138.0	145.0	650	550	450	
	161.0	169.0	750	650	550	
	230.0	242.0	900	825	750	650
	345.0	362.0	1175	1050	900	
	500.0	550.0	1675	1550	1425	1300
	765.0	800.0	2050	1925	1800	

NOTES: (1) BIL values in bold type-face are listed as standard in one or more of ANSI C57.12.10-1977 [2], ANSI C57.12.20-1981 [4], ANSI C57.12.21-1980 [5], ANSI C57.12.22-1980 [6], ANSI C57.12.23-185 [7], ANSI C57.12.24-1982 [8], ANSI C57.12.25-1981 [9], and ANSI C57.12.26-1986 [10].

(2) Single-phase distribution and power transformers and regulating transformers for voltage ratings between terminals of 8.7 kV and below are designed for both Y and Δ connection and are insulated for the test voltages corresponding to the Y connection, so that a single line of transformers serves for the Y and Δ applications. The test voltages for such transformers when operated Δ connected are, therefore, higher than needed for their voltage rating.

(3) For series windings in transformers, such as regulating transformers, the test values to ground shall be determined by the BIL of the series windings rather than by the rated voltage between terminals.

(4) Values listed as *nominal system voltage* in some cases (particularly voltages 34.5 kV and below) are applicable to other lesser voltages of approximately the same value. For example, 15 kV encompasses nominal system voltages of 14,440, 13,800, 13,200, 13,090, 12,600, 12,470, 12,000, 11,950, etc.

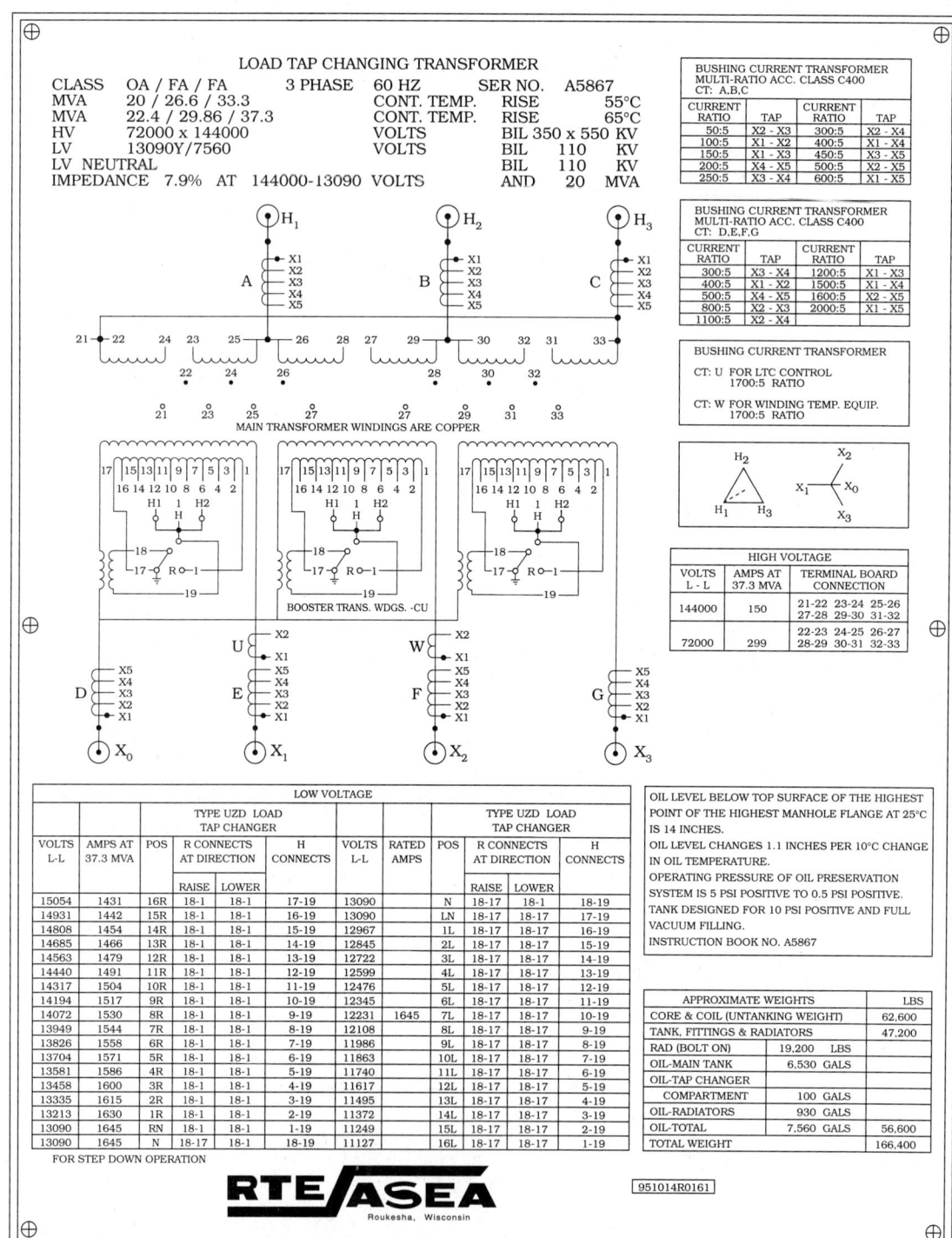

FIGURE 14–19
Typical transformer nameplate

Insulating Medium

This rating tells the reader what kind of insulating material is to be used in the transformer. In the example nameplate, the transformer is identified as being an oil-filled transformer.

Serial Number

The manufacturer will put a serial number on larger units. This number is used to identify the transformer for record keeping purposes.

Voltage Ratings

This identifies the winding voltage ratings for the transformer. The voltages are given in phase-to-phase volts. This particular transformer has a dual voltage winding delta primary and a wye-connected secondary. The primary can be connected for operation either on 144,000 V or 72,000 V. The secondary output voltage is 13,090 V delta and 7560 V wye.

In this case, since the transformer has a secondary automatic tap changer, the secondary voltage is the midpoint voltage for the tap changer.

Temperature Rise

The temperature rise indicates the maximum allowable temperature *above ambient* that the transformer may be operated. Ambient temperature for transformer rating purposes is 30°C. A transformer with a 55°C rise may be operated at a total temperature of 55°C above 30. Therefore, this transformer should never be operated at a temperature in excess of 85°C.

The sample transformer, Figure 14–19, has a dual temperature rise capability of 55° or 65°.

kVA Capacity

The kVA rating is the maximum load the transformer can handle at its rated voltage without exceeding its rated temperature rise. The kVA is always stated for the self-cooled condition and for each additional state of cooling. In the transformer for this example, the triple rating is 20/26.6/33.3 for the 55° rise. The 65° rise rating is 22.4/29.86/37.4. Note that these values are in mega-volt-amperes or mVA.

Instruction Book Number

This information tells the user where to look for more detailed information on the transformer.

Fluid Capacities

The fluid capacities indicate the amount of insulating liquid in each of the major components of the transformer. Typical components that may be included are the main tank, the tap changer compartment, the conservator tank, and the radiators. In the sample transformer, the following capacities are given:

Main Tank	6530 gallons
Tap Changer	100 gallons
Radiators	930 gallons

Impedance

This rating was described previously. For this particular transformer, the impedance is given as 7.9% at 20 mVA.

Basic Impulse Level

The BIL is given for both the primary and the secondary windings. The primary winding has a dual BIL since it has a dual voltage. In this case, the BIL is 350 × 550 for the 72,000-V connection and the 144,000-V connection, respectively. The BIL for the secondary is 110,000 V.

Weights

The transformer weights with and without liquid are also given on the nameplate.

Connection Diagram

This diagram indicates how the windings may be connected. All internal connections are shown, including those that are made by tap changers. If disconnect links are used, their location is shown. All bushings are shown and identified along with polarity, electrical location, and ratios of internal current transformers. The connection diagram can be an aid in troubleshooting and testing as well as initial connections.

Vector Diagram

This shows the phase relationships between the primary voltages and currents and the secondary voltages and currents. The sample transformer is an ANSI standard delta-wye connection with the secondary lagging the primary by 30°. This is the same as Figure 13–15c.

Tap Changer Voltage Charts

These charts show the voltages at certain tap changer positions above and below nominal or neutral position. Tap changer voltage charts show all available positions of the tap changer indicator, all internal connections at the indicated position, and the voltage and current relationships at the indicated position. They also show any internal manual connections required, and the voltage/current relationships at those connections.

Liquid Levels

The liquid level information states the proper liquid level in relation to its temperature. The liquid level at 25°C is given with respect to a fixed part of the transformer. In this example, the liquid level is given as 14 inches below the surface of the highest point of the highest manhole flange.

Vacuum Filling Information

This gives the maximum vacuum that may be drawn on the transformer. This transformer is designed for full vacuum. This area also indicates that this transformer is designed for a maximum positive pressure of 10 lbs per square inch.

Operating Pressure

The normal operating pressures for the sample transformer are given as 5 psi to 0.5 psi positive pressure.

Winding Conductor Material

When the windings are copper, there may be nothing noted in this area. Aluminum windings are always indicated. In the sample transformer, the windings are copper.

Instrument Transformers

If a transformer has internal current transformers, their location, polarity, and ratios are given. Current transformers used for the operation of the transformer (tap changer and hot spot temperature), and those brought out for external connections, are also identified. There may be additional information pertaining to the current transformers such as individual identification, types and tap ratios of multi-ratio current transformers. Note that these blocks do not the actual ratios for the transformer, but rather the ratios that are available.

Special Notes and Warnings

Any notes or information of importance to the safe operation of the transformer are given. If the transformer is unusual in any respect, it will appear in the notes. These notes take precedence over the instruction book.

SUMMARY

Transformer operation is based on Faraday's law. Two coiled conductors are placed close to each other and a voltage or current is applied to one of them. The current flow in this "primary" winding creates magnetic lines of flux which cut through the "secondary" winding. By Faraday's law, the flux lines cutting through the secondary create a voltage on the secondary. The magnitude and phase angle of this voltage are determined by the magnitude of the primary voltage and current, the phase angle of the primary voltage and current, the ratio of the number of turns in the primary versus the number of turns in the secondary, and the relative position of the primary to the secondary.

Transformers are available in many fundamental types including power, distribution, and instrument transformers. The principal difference between power and distribution transformers is size. Power transformers are typically 500 kVA or larger with primary and secondary voltages from 13.8 kV and higher. Distribution transformers are generally smaller than power transformers; however, there is some overlap. Distribution transformers are frequently the last transformer between the generation source and the load.

Instrument transformers are used to change the voltage or current from high, unusable system values to lower values suitable for use in system relays, meters, and other such instruments. Voltage transformers usually have 120-V or 69-V secondary levels. Current transformers ratios are expressed in terms of the nominal primary current which will produce a secondary current of 5 A.

The principal *working* components of a transformer are the core and coils. The two common varieties of core and coil construction are the core form and the shell form. Most large transformers use shell form construction while the smaller ones are of the core form variety.

The mechanical and electrical construction of transformers depends on the size and application of the transformer. Electrical connections are usually either single-phase or three-phase design. The mechanical and cooling features are usually more sophisticated on larger transformers, with very elaborate methods being used on large transformers.

Transformer nameplates are used to tell the user all about the transformer. KVA capacity, voltages, connection diagram, impedance, and so on are included on the nameplate.

REVIEW QUESTIONS

1. What keeps atmospheric contaminants out of the main tank of a conservator tank design transformer?

2. What is the weight of the bolt on radiators for the transformer whose nameplate appears in Figure 14–19?

3. What is the temperature of the liquid indicated in the gauge shown in Figure 14–15? What was the maximum temperature?

4. You need to select a transformer for a 65-kVA load. What dry-type kVA size would you use?

5. What is the basic impulse level (BIL) for a 15-kV liquid immersed power transformer?

6. What is the maximum, fan-cooled kVA rating for the transformer whose nameplate is shown in Figure 14–19? Assume that the maximum, allowable temperature rise is 55°C.

7. What is the purpose of the oil in the upper chamber of the transformer shown in Figure 14–4?

8. Explain the terms OA, FA, and FOA.

Transformer Protection   15

OBJECTIVES

After studying this chapter, the student should be able to:

- Explain the types and intervals of visual and diagnostic inspections to be performed on transformers.
- Describe the methods used to sample insulating liquid.
- Describe the types of tests to be performed on transformer insulating liquid.
- Describe the types of tests that may be performed on transformers using direct voltage test equipment.
- Describe the types of tests that may be performed on transformers using alternating voltage test equipment.

INTRODUCTION

Like all pieces of electrical power equipment, transformers must be maintained. This chapter will describe the various types of visual and electrical procedures and tests that should be performed on transformers. It will also provide a suggested listing of the tests and test intervals for transformers, based on their size, insulating medium, and voltage.

Note carefully that not all tests identified in this chapter are recommended for all transformers. Be certain to perform only those procedures that are consistent with the particular transformer upon which you are working. Also note that the electrical tests referred to in the early part of the chapter are described in detail later in the chapter.

INSPECTIONS

Visual (Routine) Inspections

Routine inspections require the transformer indicators to be checked and the readings recorded. Some of the more common checks are:

Load Current Check once a week. If the load current exceeds the full load current rating of the transformer, overheating can occur, shortening the life of the transformer. In case of an overload, take steps to reduce the current demand of the transformer.

Load Voltage Check load voltages once a week. Overvoltages can injure the transformer and its load by stressing the insulation. Undervoltages can damage a transformer's load. Generally, transformers are designed to operate within a range of ± 5% of their nameplate rated voltage.

If there appears to be a problem with either the voltage or current, a recording volt/amp meter should be connected to the secondary of the transformer. This will show all fluctuations of load voltage and current and help simplify troubleshooting. Recording instruments will also diagnose problems that may be occurring between inspections.

Liquid Level Indicators Check liquid level indicators once a week, especially after a period of low load at low ambient temperatures. The oil is at its lowest point under these conditions. Liquid level indicators come in two types: flush or sight glass and dial. New insulating liquid should be added before the level falls below the sight glass or low indicator mark. Transformers are at their optimum level at the 25°C mark when their oil temperature is 25°C (Figures 15–1 and 15–2). Check the nameplate or manufacturer's instruction book to determine the rise and fall of the liquid with the temperature.

Pressure/Vacuum Gauges Take readings weekly. These gauges are normally found on sealed transformers, Figure 15–3. By comparing readings to the recommended values given by the manufacturer, a great deal of information may be obtained. Excessive pressures may be an indication of an overload or internal problem and should be investigated immediately. If readings are at the neutral point (either pressure or vacuum), then the transformer may be breathing outside air.

Liquid Temperature Indicator Check all temperature gauges weekly, Figures 15–4 and 15–5. Transformers are rated to carry their nameplate rated load, with a given heat rise, when the ambient temperature is at 30°C. Another way of saying this is when fully loaded, the indicated temperature should not be more than the rated temperature rise plus the ambient tem-

FIGURE 15–1 Flush type liquid level gauge (Reprinted with permission of ABB Power T&D Company, Inc.)

FIGURE 15–2 Liquid level indicator (Reprinted with permission of ABB Power T&D Company, Inc.)

perature. If the temperature is higher than normal, an overload or a reduction in the cooling ability of the transformer is possible. If a transformer is operated at a higher than normal temperature, the life expectancy of the liquid and solid insulation is reduced.

Special Inspections Water-in and water-out temperatures need to be checked weekly on forced-water-cooled (FOW) transformers. On forced-oil

FIGURE 15–3 Pressure-vacuum indicator (Reprinted with permission of ABB Power T&D Company, Inc.)

FIGURE 15–4
Liquid temperature indicator (Reprinted with permission of ABB Power T&D Company, Inc.)

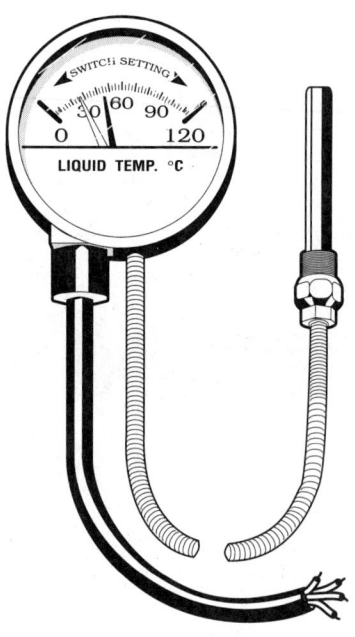

FIGURE 15–5
Hottest spot/winding temperature indicator (Reprinted with permission of ABB Power T&D Company, Inc.)

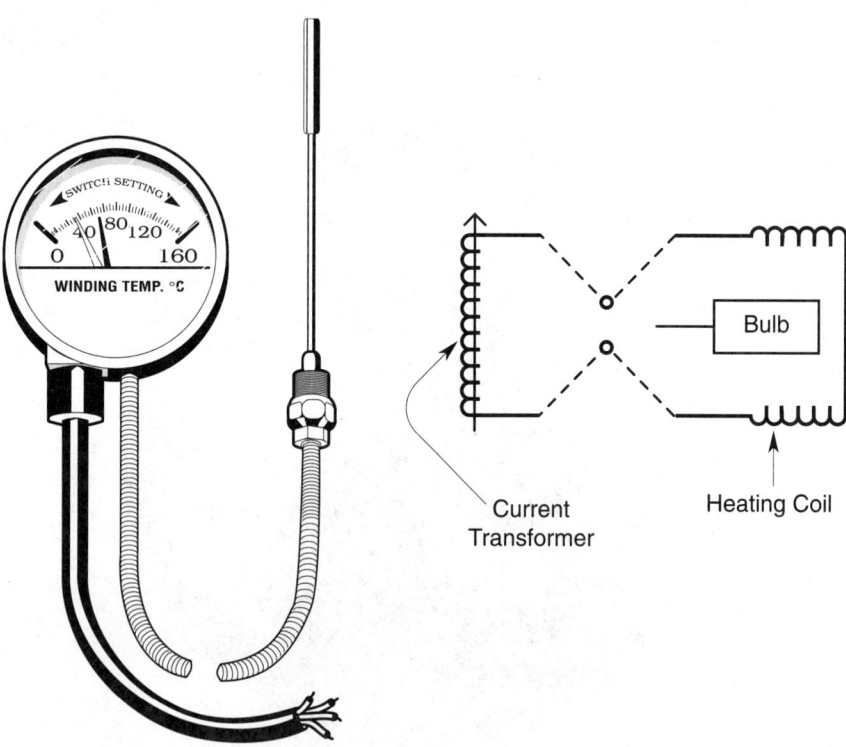

(FOA) transformers, check the oil-in and oil-out temperature weekly. If the transformer is enclosed in a vault, take monthly ambient temperature readings. Temperature plays a large role in extending or shortening the life of a transformer. If any of the above-mentioned temperature readings are above what the manufacturer calls for, the cause should be investigated immediately and corrected.

Diagnostic Inspections

In addition to routine inspections performed to check on the day-to-day condition of a transformer, a more thorough inspection should be done in accordance with the guidelines given at the end of this chapter. A diagnostic inspection is more investigative than routine inspections. If an abnormal condition can be found before it becomes a problem, much money and down time can be saved.

Precautions

- Condensation—Moisture has always been a problem in transformers. Moisture, in the form of condensation, can collect in transformers when they are opened. For example, when a cool transformer is opened on a warm morning, moisture will condense on any surface that is cooler than the surrounding air. If the surrounding air is warmer than the transformer, allow it to stand until all signs of external condensation disappear.

    Try not to deenergize a transformer too far ahead of the time the work is to be performed. A good rule of thumb is to keep the transformer 10°C warmer than the outside air or measured dewpoint.

    Moisture reduces the overall insulation resistance. In some cases, insulation resistance may be affected by relative humidity even when the temperature is above the dewpoint. A general rule-of-thumb is not to open a transformer unless absolutely necessary, and if it is necessary, try not to open it if the relative humidity is above 65%.

    Relative humidity can be determined with a hygrometer or psychrometer and psychrometric chart, Figure 15–6. A sling psychrometer is a simple device used to measure relative humidity. It contains two thermometers, one wet bulb and one dry bulb. The wet bulb thermometer has a moistened wick over the bulb, which causes it to read a lower temperature than the dry bulb (due to evaporation). The psychrometer is twirled around by its handle, and the temperature is then read from both thermometers. The wet bulb and dry bulb temperatures are located on the psychrometric chart. By drawing intersecting lines from each, the relative humidity may be found. There are also

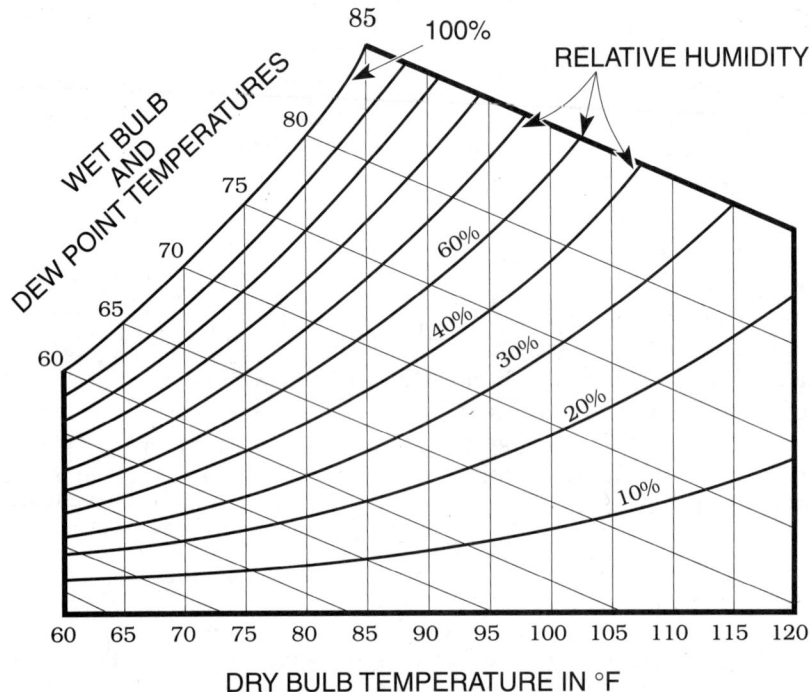

FIGURE 15–6
Psychrometric chart

digital hygrometers available that give not only relative humidity, but also temperature. These units simplify taking readings greatly.

Dew point is the temperature at which condensation occurs at a given pressure. At the same pressure, air will hold less moisture at a lower temperature before condensing than it will at a higher temperature. In recent years, dewpoint has become more of a tool for determining dryness, than for determining whether to open a transformer or not.

If a transformer must be opened during inclement weather or high humidity because of the results of diagnostic tests, remove the nitrogen or oil under vacuum. Connect a source of clean, dry air to the transformer. Aerate the transformer with dry air before and during maintenance or inspection, being sure to keep the transformer as sealed as possible to prevent moisture from entering. Upon completion of maintenance, remove the dry air, refill the transformer with nitrogen and allow it to sit for 24 hours. This allows the moisture in the transformer to reach a state of equilibrium with the insulation and gas. After 24 hours, a dewpoint measurement may be taken to determine dryness.

- Instruction Leaflets—*Caution:* In all cases of maintenance and repair, have on hand the manufacturer's instruction book for the device that

is being worked on. This is very important because each device has different tolerances and capabilities.
- Safety—Never work on a transformer alone, and be sure to restrict access to the testing site. Is someone there who is not doing anything? There is a good chance that person should not be there. The larger the number of people present, the greater the chance of an accident.

The first thing to do when taking a transformer out of service is to get an electrical clearance. Be certain everyone who needs to know is notified. Make sure all possible sources of voltage are disconnected. Remember, transformers can be back-fed. Apply locks and tags to power switches.

Apply ground straps to all bushings and terminals to discharge static electricity. Leave all lines grounded whenever possible. Lines to the transformer can be energized by induced voltage from adjacent power lines.

Is this the right transformer? If the transformer that is to be tested is not deenergized, someone could get hurt. Test to make sure that the right transformer was deenergized.

If a ladder is to be used, tie it to ensure its stability. People can and do fall off shaky, unsecured ladders. Restrict access to the test area with safety tape or other restrictions and signs as required.

If a bridge test is being performed on a transformer, keep everyone off and away from it. Bridge testing can cause an inductive kick. Anytime a DC test is performed, ground the winding long enough to dissipate the stored energy.

General   Check all transformer auxiliaries in accordance with the manufacturer's instructions. Run all pumps and fans, checking for noisy bearings, leaks or a device that will not operate. Manually operate level and temperature switches and sudden pressure relays. Make certain they alarm or operate. Nitrogen regulators need to be checked for leaks, low pressure alarms, and proper headspace pressure.

Make an external visual examination of the transformer. Check all gauges, valves, and fittings for proper tightness, using the factory torque specifications. Look for leaks, not only around fittings, but also at any gasket and welded joints. An askarel transformer will seep or, as it is sometimes called, weep from any pinhole in gaskets.

Look inside pressure relief valves for obstructions, since dirt, trash, debris or even birds nests may be hidden inside. Inspect for cracking, chipping, or peeling paint, or rust since transformers will rust through. Check the packing in fill and drain valves. Inspect air-cooled transformers for dust and lint buildup or anything that prevents free air circulation.

Bushings   Since transformer bushings are so important to the functioning of a transformer, extra care should be given to their inspection. Do *not* hang on or climb on transformer bushings. They are not designed to carry the extra weight and stress.

Hand wipe bushings when the transformer is deenergized. In dirty or corrosive climates, bushings may need to be periodically washed with demineralized water or dry-cleaned with limestone powder or ground corncobs. After cleaning, when possible, cover the bushing with automotive wax or a silicone compound. This will help keep contaminants from damaging the porcelain and, in the case of silicone compounds, prevent water filming, leakage currents, and contamination paths from forming.

In addition to cleaning and waxing bushings, there are a number of items that should be checked. Look for cracks or chips. Moisture will penetrate the bushing if cracks in the porcelain are left unattended. Seal minor chips with epoxy: large chips or cracks may render a bushing non-serviceable. If the bushing has an oil level indicator, check it for proper oil level. Look for leaks of oil or compound. Even in small amounts, a leak is a potential hazard. Moisture could be drawn into the bushing through the leak. To check for cracks that are not visible, perform an ultrasonic test.

Once every two years, a power factor test should be performed on bushings. Results should be compared with previous results, factory test data, and test results from other similar units.

Tap Changers Check the oil in the load tap changer compartment. On windings of 15 kV and below, the oil should test to at least 22 kV before breakdown. On windings of 15 kV and above, the oil should test to at least 26 kV before breakdown. If the oil tests to a value lower than these limits on the dielectric breakdown voltage test, it should be filtered or replaced. Before refilling the load tap changer, clean the tap changer compartment to remove carbon and other contaminants. If the tap changer is rated at 69 kV or above, or if it contains vacuum interrupters, fill the tap charger compartment under vacuum.

After draining the oil and before filling, inspect tap changer contacts for excessive wear, positive alignment, and freedom of movement. Check all connections for tightness. After inspecting and refilling the load tap changer, double check the oil level.

Lightning Arresters Lightning arresters generally require very little in the way of maintenance. Clean all types of lightning arresters in the same manner as bushings are cleaned. If an arrester is excessively dirty, flashover may occur at the wrong voltage. Or, an arc started by a high surge may fail to clear after the high voltage condition returns to normal. If the arc fails to clear in an expulsion arrester, it may burst or burn. Valve arresters may puncture or crack the resistor element.

In addition to cleaning lightning arresters, check for blown fuses on elementary arresters, inspect the porcelain on valve arresters, and the thickness of the fiber tubes on expulsion arresters. An insulation resistance test can be performed at this time also. This test will indicate the condition of the arrester housing and elements. Moisture, corrosion, broken resistors, and punctured discs may also be located by the insulation resistance test.

Testing Take a sample of insulating fluid from the transformer and test it. Remove askarel and other synthetics from the top of the transformer and mineral oil from the bottom. Use standard sampling procedures when taking a sample.

If the transformer has a gas headspace, check a gas sample for the presence of combustible gases using a portable fault-gas detector.

When checking the gas in the head space, also check for the presence of excessive oxygen, which accelerates decay and the formation of sludge. If the oxygen content exceeds 1%–5%, (depending on manufacturer) it will need to be blown out three or four times with nitrogen. After the blow out is completed, check the pressure of the system (headspace and, if included, nitrogen bottle).

An infrared scan performed while the transformer is operating is of great value. This test will show hot-spots, loose connections, and plugged tubes or radiators.

At each diagnostic inspection the transformer should be electrically tested. Each manufacturer and testing organization recommends a variety of tests to be performed.

Recommendations for Testing and Maintenance

Maintenance (All Transformers)

- Weekly:
    1. Check all indicators and record readings. On level and temperature indicators, record minimum, maximum, and present readings.
    2. Reset all indicators. Zero or negative readings should be reported to supervisor.
    3. Report any defective gauges.
    4. If so equipped, change nitrogen bottles when pressure reaches 200 psig minimum.
    5. Check the mechanical relief device to see if it has operated.
    6. Clear fans and radiators of obstructions (birds nests, leaves, and so on).
    7. Operate fans and pumps. Report any that are not functional. Ensure all are set to "automatic."
    8. Check all covers and gaskets to ensure they are in place.
    9. Check for oil leaks and repair if possible. If a leak is new, excessive, or cannot be repaired, report to supervisor.

10. If a transformer uses a drying agent, check color and replace as necessary.
- Semi-Annually:
  1. Perform a thermal (infrared) inspection. Be sure to check cooling fins, connections, and other possible areas of high heat.
  2. Perform a total combustible gases check and record readings.
  3. Perform a Dissolved Gas Analysis for 345 kV and above.
- Annually:
  1. Perform same inspection items as for semi-annual inspections.
  2. On transformers equipped with heat exchangers, externally clean and flush.
  3. Perform oil tests on all transformers. Tests should include the following (or equivalent) tests:
     a. Visual examination
     b. Color
     c. Dielectric breakdown voltage
        D877 for under 230 kV
        D1816 for 230 kV and above
        D1816 can be used for all
        Interfacial tension
        Neutralization number
        Moisture
        Metal particulate (Units with pumps)
        Atomic metal particulate (Units with pumps)
  4. Dissolved Gas Analysis (DGA)
     a. Annually for 230 kV and below
     b. Semi-annually for 345 kV and above
  5. Nitrogen blanket O_2 analysis

Testing (By Voltage Class)

- Low voltage (less than 600 V)

Test	Interval
1. Megohmmeter tests	
a. Insulation Resistance (1 minute)	2–4 years

- 69 kV and below

Test	Interval
1. Megohmmeter tests	2–4 years
a. Polarization index (PI)	
b. Step voltage (if PI is low)	
2. Winding insulation power/dissipation factor tests	2–4 years
3. Bushing insulation power/dissipation factor tests	2–4 years
a. Overall pf/df (condenser bushings)	
b. Capacitance (condenser bushings)	

  c. Tap (reversed) pf/df (condenser bushings)
  d. Hot collar tests (solid or compound-filled)
 4. Core excitation current   2–4 years
 5. Transformer turns ratio   2–4 years
 6. DC winding resistance   2–4 years
- 115 kV to 230 kV

 Test   *Interval*
 1. Megohmmeter tests   2–3 years
  a. Polarization index
  b. Step voltage (if PI is low)
 2. Winding insulation power/dissipation factor tests 2–3 years
 3. Bushing power/dissipation factor tests   2–3 years
  a. Overall pf/df (condenser bushings)
  b. Capacitance (condenser bushings)
  c. Tap (reversed) pf/df (condenser bushings)
  d. Tap capacitance (condenser bushings)
 4. Core excitation current   2–3 years
 5. Transformer turns ratio   2–3 years
 6. DC winding resistance   2–3 years
- 345 kV to 500 kV

 Test   *Interval*
 1. Megohmmeter tests   1–2 years
  a. Polarization index
  b. Step voltage (if PI is low)
 2. Winding insulation power/dissipation factor tests 1–2 years
 3. Bushing power/dissipation factor tests   1–2 years
  a. Overall pf/df (condenser bushings)
  b. Capacitance (condenser bushings)
  c. Tap (reversed) pf/df (condenser bushings)
  d. Tap capacitance (condenser bushings)
 4. Core excitation current   1–2 years
 5. Transformer turns ratio   1–2 years
 6. DC winding resistance   1–2 years

Sampling and testing procedures should follow established standard procedures as outlined in the text. The interfacial tension, acidity, color, visual, and dielectric breakdown voltage tests are not being covered by the DGA test. The DGA only looks for certain gases, not sludge, acid, moisture, fibrous contamination or other conductive contamination. Moisture may be detected accurately by the moisture content test and carbon, metallic fibers and other conductive materials can be detected by the power factor and dielectric breakdown tests. Sludge and acidity must be measured by their respective tests. The significance of these tests is outlined in the text and by various sources. All the above tests can be performed in the shop by maintenance personnel.

INSULATING LIQUID SAMPLING

Purpose of Liquid Sampling

Liquid samples are usually taken for two main purposes: to determine the properties of the liquid (such as moisture and other contaminants) and to measure the types and quantities of gases found in the liquid which can be analyzed to detect internal problems occurring in the transformers.

A good, clean, representative sample is required to obtain good test results. If the sample container itself is contaminated, or if the equipment used for taking the sample is contaminated, the test results will be erroneous. For example, if the sample container is left open or is not tightly sealed, dirt and/or moisture can enter. Either of these contaminants can cause a sample to test in an unsatisfactory way.

Liquid Sampling Precautions

General precautions to observe when taking liquid (oil) samples are as follows:

1. Take mineral oil and silicone samples from the sampling valve located at the bottom of the tank.
2. Askarel or other heavy synthetics should have their samples taken from the top of the tank.
3. Do not attempt to sample a transformer while it is under a vacuum.
4. Do not use rubber stoppers on sample bottles. Use cork only if it is protected with foil.
5. Take samples on a dry day and when the temperature of the oil is higher than the ambient temperature (to prevent condensation).
6. Do not use rubber or plastic hoses to drain, fill, or sample askarel. (Askarel has a strong solvent-type action and breaks down most plastics and all rubber or rubber composition materials.)
7. Use polyvinyl chloride (Tygon®) tubing with mineral oil, and use metal or Tygon® tubing with askarel. Do not use the same hoses, sample bottles, or test cups for both askarel and mineral oil.
8. Check liquid-level indicators to ensure there will be enough fluid left after the sample is drawn. If liquid level is low, add sufficient liquid to bring it to the proper operating level. If it is not possible to add liquid immediately, note the level and postpone taking samples until arrangements can be made to add liquid.
9. Ensure that the temperature of the liquid being sampled is warmer than the ambient temperature. If the liquid sample is cooler than the

ambient temperature, there is a possibility of condensation being absorbed into the liquid.
10. Do not sample when the relative humidity is 75% of more, and do not sample in the rain or snow.
11. Record the relative humidity, ambient temperature, and the liquid temperature prior to sampling.

Sample Jars and Preparation

Insulating liquid samples are taken in 1-qt glass jars designed especially for that purpose.

Proper maintenance and storage of jars ensure that samples are representative of the liquid being tested. The quart sample jars are primarily used to collect samples for the following tests:

1. Visual inspection
2. Color
3. Dielectric breakdown voltage
4. Neutralization number (NN or acidity)
5. Interfacial tension (IFT)
6. Power factor
7. Moisture

Before taking liquid samples, make sure the jars are free from all contamination, such as dirt and moisture. The following procedure is recommended by ASTM and transformer manufacturers, to ensure that sample bottles are free of contamination, clean, and dry (for routine samples):

1. Flush sample jars with Stoddard solvent, trichlorotrifluorethane, precipitation naphtha, or other solvents as approved.
2. Wash sample bottles with a strong soap, such as trisodium phosphate, thoroughly rinse with distilled water, and drain for 10 minutes.
3. Dry jars in upright position at approximately 110°C for one hour or more.
4. Clean glass stoppers in the same manner. If the covers have vinyl or cork liners, do not reuse them.
5. New covers and vinyl or cork liners should be oven-dried for 30 minutes at 110°C immediately before placing them on bottles.
6. After drying has taken place, immediately place stopper on each sample bottle tightly, as it is removed from the oven. Take care not to touch the lip of the bottle or inside the cover. Use gloves while handling hot sample bottles.

7. Store jars in a dust-free, heated cabinet (approximately 38°C by manufacturer's recommendations) until they are ready to use.

Liquid Sampling Procedures

Liquid sampling procedure (ASTM D-923) is as follows (using sample jars):

1. Carefully clean all contamination from the oil sample valve.
2. Remove the sample valve plug/cap and flush valve by draining one to two quarts of liquid into a waste container.
3. If sampling askarel, rinse thief out with a sample fluid or flush sampling valve with askarel.
4. If sample bottle does not fit under valve, connect a length of polyvinyl chloride (Tygon®) tubing from valve to sample bottle. Do not use plastics or rubber for askarel.
5. Immediately seal the sample jar and place it into a carton to protect it from sunlight.

ASTM recommends the use of a bleed-type device to collect samples whenever the sample is to be used for referee tests, Figure 15–7.

Samples from similar equipment in the same location can be taken with the same sampling device if the sampling device is free from water or other contamination, and the sampling tube is thoroughly flushed out before the actual sample is taken.

Sampling for Power Factor Testing

Samples taken for power factor tests can be collected according to ASTM D-923 or can be taken directly into the test cell. The test cell or sample bottle (if used) must be clean if good results are to be obtained.

Before sampling (using a three-electrode test cell), dismantle the test cell completely and wash all components with trichlorotrifluoroethane, petroleum ether, or pentane. (Some two-electrode test cells may not come apart for cleaning. Test cells of this type should also be cleaned as well as possible with trichlorotrifluoroethane, petroleum ether, or pentane.)

Wash the cell out with a mild soap and rinse with hot tap water, then with cold tap water and rinse several times with distilled water. Prior to cleaning, place stopper on the opening to the inner electrode. This will prevent moisture from entering the thermocouple and the inner electrode. Place the test cell in a heated oven (110°C) for a minimum of one hour. If the test cell is made from Monel®, do not dry for more than 90 minutes. After drying, assemble the cell wearing cotton gloves. Store the test cell, cleaned and dried, in a storage cabinet that is dust-free, clean, and heated to about 38°C. The test cell is now ready to be used for liquid sampling.

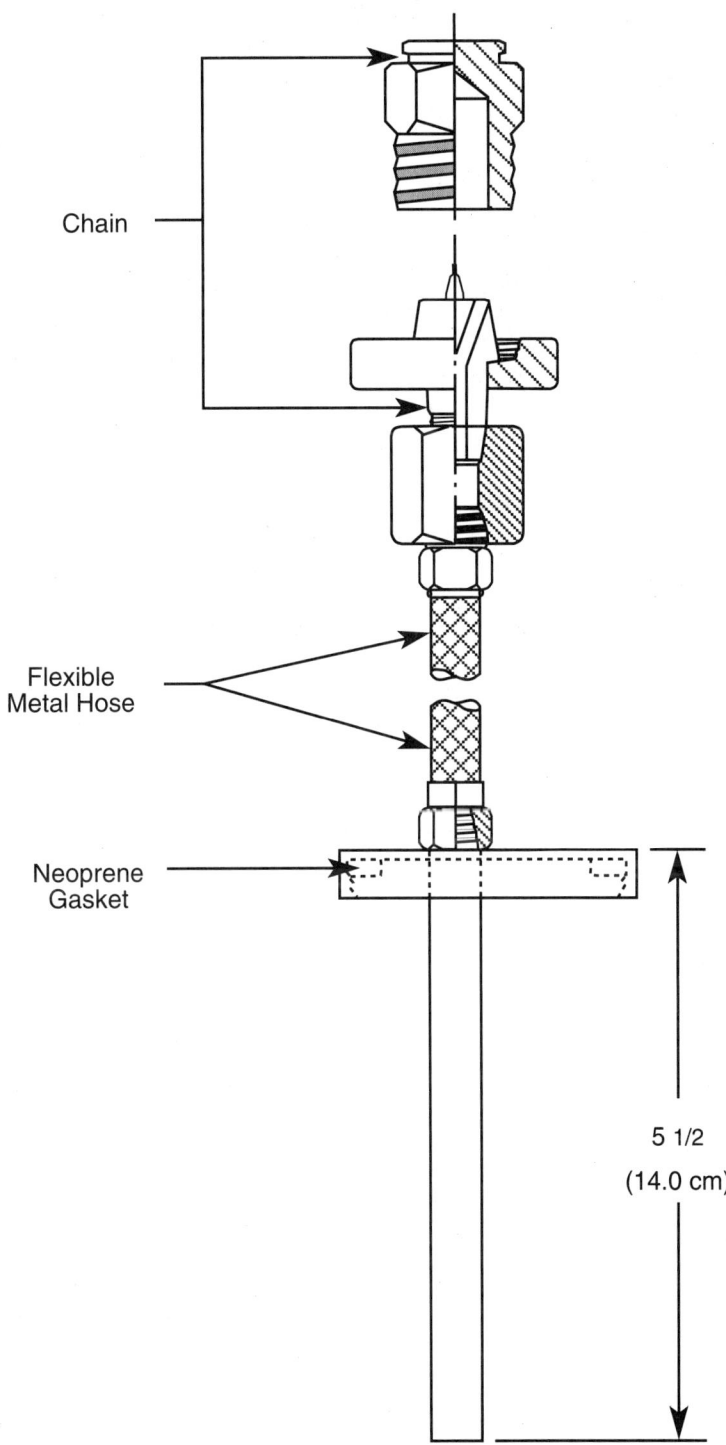

FIGURE 15–7
Bleed-type sampling device
(Copyright ASTM. Reprinted with permission)

To obtain a sample of liquid in a test cell for power factor tests, use the following procedure:

1. Carefully clean all contamination from the drain valve or sampling valve.
2. Remove the cap/plug from the drain valve or sampling valve.
3. Flush valve by draining one to two quarts of liquid into a waste container.
4. If the test cell cannot be held under the valve, connect a piece of polyvinyl chloride (Tygon®) tubing from the valve to sample cell. Do not use plastics or rubber for askarel.
5. Rinse the test cell with about one inch of oil after the drain valve has been flushed. Discard the oil rinse into a waste container.
6. Collect enough oil to fill the cell to about ¾ inch above the top of the cylinder inside the test cell.

    When more than one transformer is being sampled and tested, the test cell may be used without being cleaned if:
    a. The measured value of the previous sample was less than the specified value.
    b. The test cell is rinsed once more with one inch of fluid after flushing the sample valve.

If the previous sample tested higher than the specified value, the test cell needs to be recleaned before being used for more tests.

Sampling for Gas-in-Oil Analysis

ASTM D-923 - Rigid Syringe Oil samples for gas-in-oil analysis should be taken by the ASTM D-923 method with use of an approved sealed syringe, Figure 15–8. These syringes require no cleaning before use. To prepare the transformer for sampling, observe the following:

1. If a permanent sampling valve does not exist, install a temporary sampling valve on the drain valve.
2. If it is not possible to mount the sampling valve on the drain valve, replace the drain valve plug with a plug that has a predrilled hole in it (No. 22 bit).
3. Carefully clean the drain and the sampling valve to remove all contamination.
4. Remove the sampling valve plug and flush the valve by draining between one and two quarts of liquid into a waste container.
5. Ensure that the transformer is under positive pressure to use this sampling method (.5 psi or more).

Sampling Procedure for the Rigid Syringe—The ASTM-recommended sampling procedure for the rigid syringe is as follows:

1. Attach the adapter plug to the sample valve of the transformer and flush into a waste container until all air is removed, Figure 15–8.
2. Insert the auxiliary sampling stopcock and flush. Ensure that no bubbles are present, Figure 15–9.
3. Attach the syringe with the stopcock in the fill position to the auxiliary stopcock (still in the flush position).
4. Turn the auxiliary stopcock to the fill position and allow 5 to 10 ml (milliliters) of oil to enter the syringe, Figure 15–10.

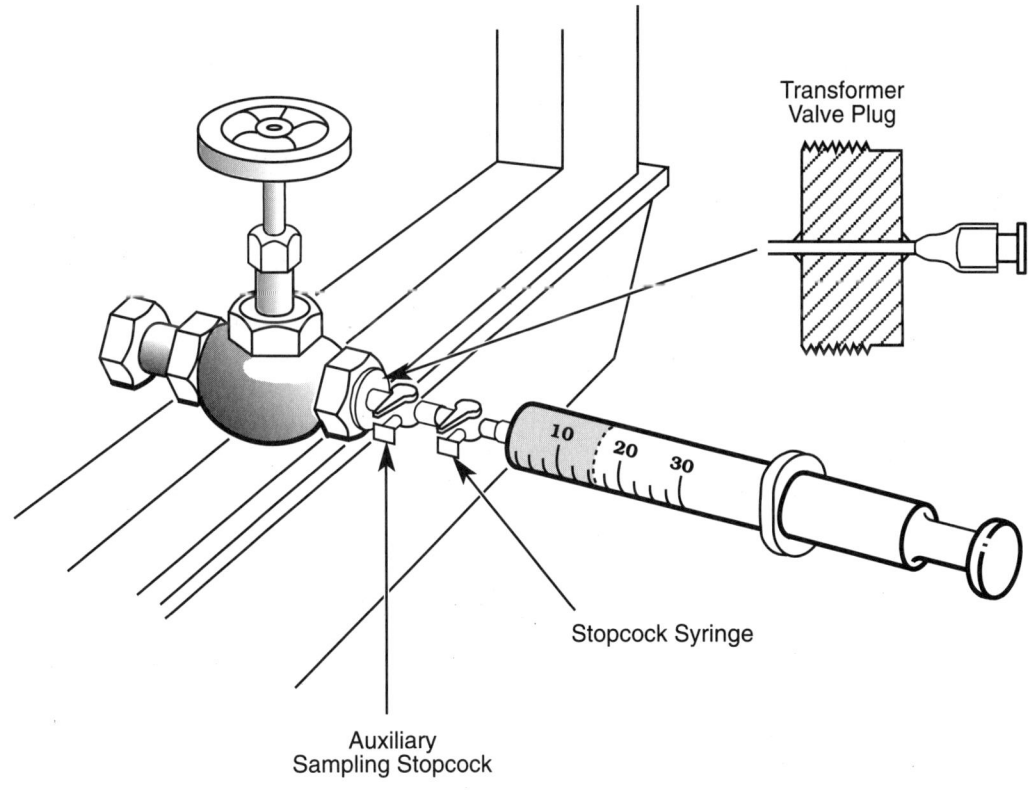

FIGURE 15–8
Syringe-type sampling device (rigid) (Copyright ASTM. Reprinted with permission.)

FIGURE 15–9
Stopcock position (flush position) (Copyright ASTM. Reprinted with permission.)

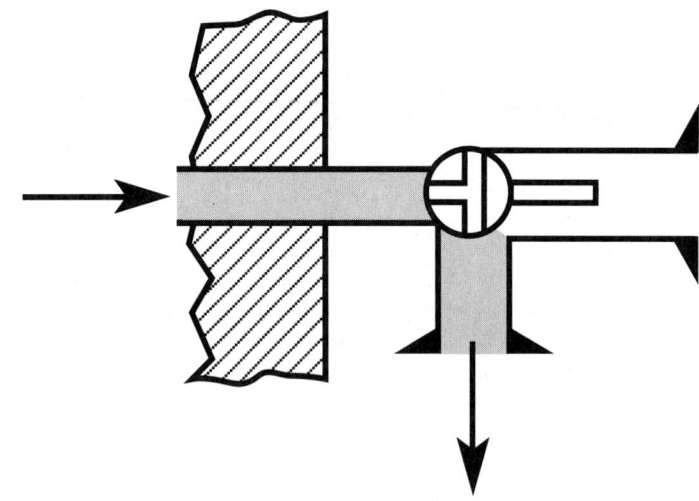

5. Turn the auxiliary stopcock to the flush position and remove the syringe from the auxiliary stopcock, Figure 15–11.
6. With the syringe held vertically, depress the plunger to the zero mark with the syringe in the fill position, Figure 15–12.
7. While holding the syringe to the zero mark, close the syringe stopcock, Figure 15–13.
8. Reattach the syringe to the auxiliary stopcock while still depressing plunger to zero mark.
9. Turn the auxiliary stopcock to the fill position while flushing the syringe stopcock, Figure 15–14.
10. Turn both the syringe and the auxiliary stopcocks to the fill position and allow the oil pressure to push the piston back, until the syringe is filled with oil. *Do not pull the piston back because air bubbles can be drawn into the syringe,* Figure 15–15.

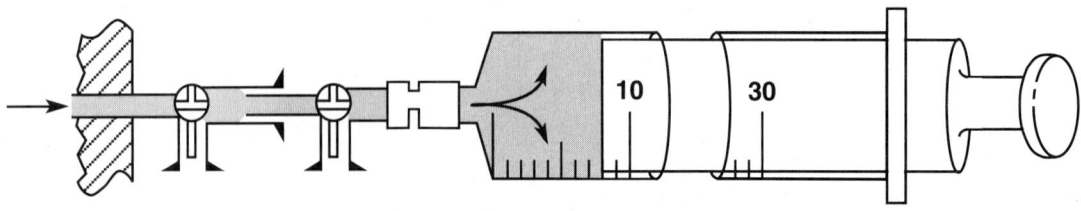

FIGURE 15–10
Stopcock position (fill position) (Copyright ASTM. Reprinted with permission.)

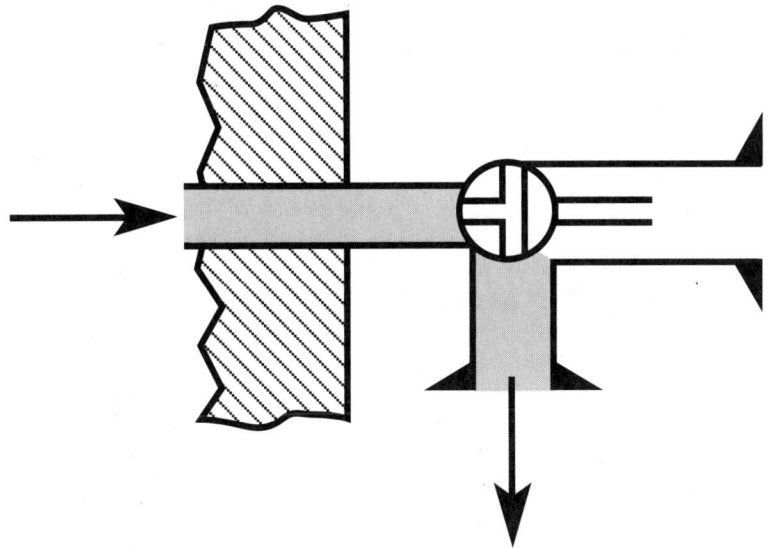

FIGURE 15–11
Stopcock position (flush position, syringe removed) (Copyright ASTM. Reprinted with permission.)

11. Return both the syringe and the auxiliary stopcocks to their flush position and separate the syringe from the auxiliary stopcock, Figure 15–16.

Once the sample has been taken, hold it with the needle of the syringe up and inspect for air. If the sample shows an air space, it must be flushed and the sample retaken. Squeezing out the air bubble could give a false high oxygen reading. At the same time, inspect the syringe around the plunger seal. If the seal is defective, air bubbles will be seen migrating past the seal towards the top of the syringe. If this condition is found, the syringe is defective and must not be used.

ANSI C57.104 - Metal Sample Container   Oil samples taken for gas chromatography can be taken in a metal sample container (ANSI C57.104). The sampling bottle is constructed of stainless steel with a shutoff valve at each end, Figure 15–17. Two lengths of polyvinyl chloride tubing are attached to the sample bottle, one at each end. Use the sample bottle as follows:

1. Before attaching the tubing to the drain/sample valve, carefully clean the valve of all contaminants.
2. Flush the valve by draining one to two quarts of liquid into a waste container.
3. Attach one end of the sample bottle to the oil drain/sample valve of the transformer.
4. Place the other end of the sample bottle into a waste container.

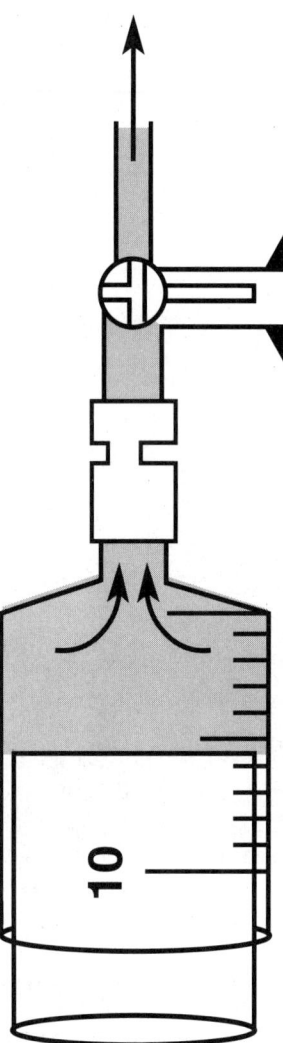

FIGURE 15–12
Stopcock position (fill position) (Copyright ASTM. Reprinted with permission.)

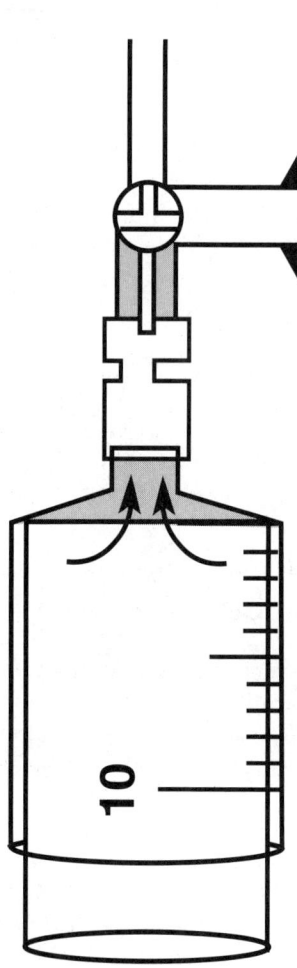

FIGURE 15–13
Stopcock position (closed position) (Copyright ASTM. Reprinted with permission.)

Transformer Protection

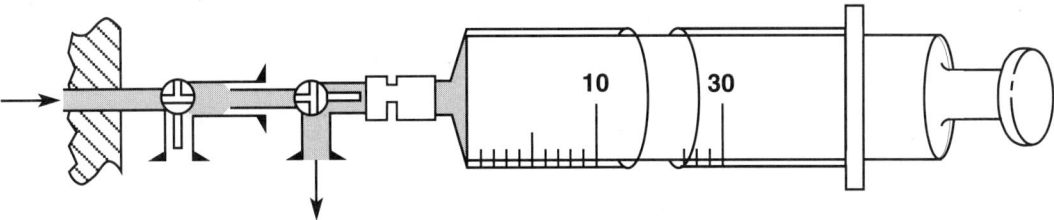

FIGURE 15–14
Stopcock position (auxiliary in fill position) (Copyright ASTM. Reprinted with permission.)

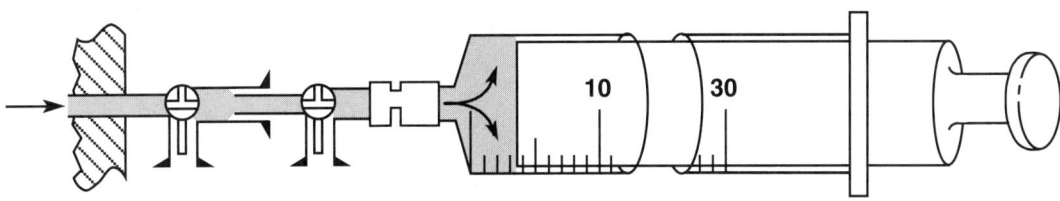

FIGURE 15–15
Stopcock position (auxiliary and syringe in fill position) (Copyright ASTM. Reprinted with permission.)

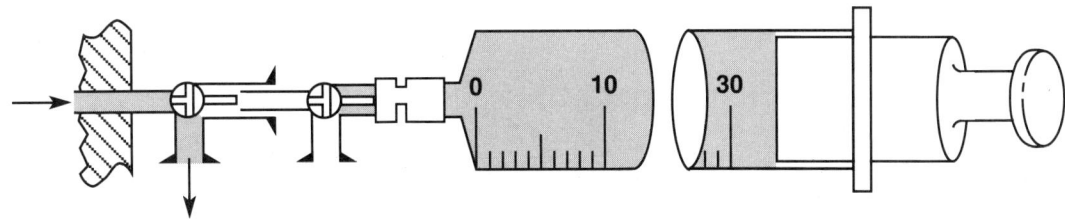

FIGURE 15–16
Stopcock position (auxiliary and syringe in flush position) (Copyright ASTM. Reprinted with permission.)

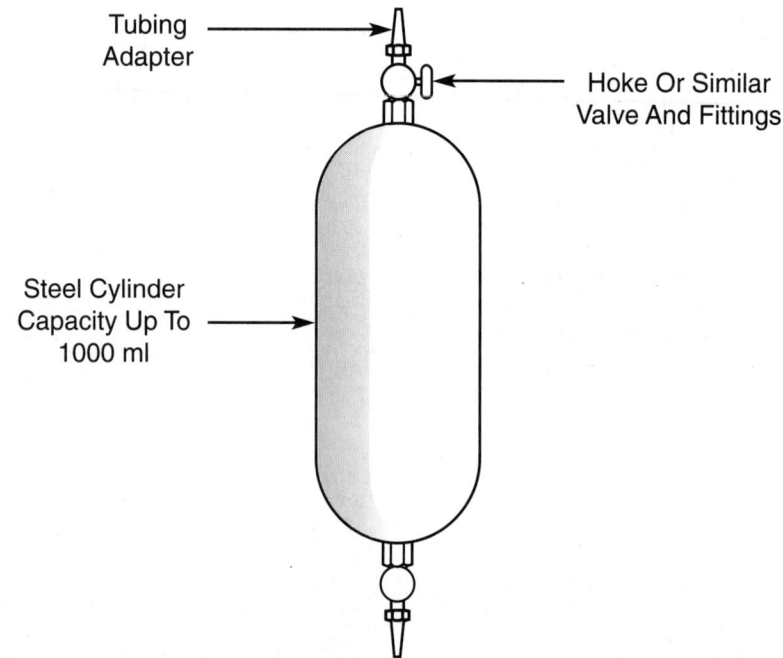

FIGURE 15-17
Metal sample bottle
(Courtesy of IEEE Std 242-1986, IEEE Recommended Practices for Protection and Coordination of Industrial and Commercial Power Systems)

5. Open both valves of the sample bottle.
6. Open the transformer valve.
7. Hold the sample bottle vertically. The oil will fill the bottle from the bottom and overflow into the waste container.
8. Allow the liquid to overflow until no air bubbles can be seen in the tubing and two quarts of liquid have been collected in the waste container.
9. Close the top valve of the sample bottle. Allow oil pressure to build up.
10. Close the bottom valve on the sample bottle.
11. Close the transformer valve.

TRANSFORMER INSULATING LIQUID TESTING

Insulating Liquids

Insulating liquid contained in equipment such as transformers, circuit breakers, regulators, and other substation equipment should exhibit certain characteristics in order to ensure satisfactory performance. Those characteristics must be maintained within certain limits prescribed for the equipment, in order for the equipment to give safe and reliable service.

The four basic functions of insulating liquid in substation equipment are (1) to provide electrical insulation, (2) to assist in arc-quenching, (3) to minimize chemical breakdown of interior components, and (4) to act as a heat transferring agent for cooling. To perform these roles, the following conditions of the liquid must be met:

1. The insulating liquid must have adequate dielectric strength to withstand the electrical stresses imposed in service.
2. The insulating liquid must retain a sufficiently low viscosity so that its ability to circulate and transfer heat is not impaired.
3. The insulating liquid must pour readily at low temperatures.
4. The insulating liquid must have a high flash point and fire point for safety.
5. Dielectric losses should not become excessive.
6. Oil should not be allowed to become so deteriorated or contaminated that it adversely affects other materials in the apparatus.
7. Deteriorated products should not be allowed to sludge the liquid sufficiently to impair its circulation through cooling ducts.

To ensure that proper conditions are maintained, it is necessary to perform certain "qualifying" tests on the liquid prior to placing equipment in service, and thereafter, to do periodic testing on the liquid so that any undesirable changes can be monitored and corrected before they can become detrimental to the equipment.

It is expected that insulating liquid will exhibit continual changes in its characteristics from the time it is refined. Once new liquid has been introduced into a piece of equipment, it will start to show different characteristics due to its contact with the apparatus' construction materials. When the equipment is finally energized, changes in the liquid's characteristics will be accelerated as the liquid begins to service-age. Affected by many variables, this process will continue for the life of the liquid. Only through an effective testing and maintenance program can the full potential life of insulating liquid be realized, and system integrity ensured.

Dielectric Breakdown Voltage Test (ASTM D-877/D1816)

The dielectric breakdown voltage test is the test most frequently performed on insulating liquids in the field. The test set is portable and easy to operate. There are two principal types of dielectric test sets. The most commonly used test set is the ASTM D-877. This test set utilizes two 1-inch flat disk electrodes spaced 1/10 inch apart. A more recent entry is the ASTM D-1816, also called the Verband Deutscher Elektritechniker (VDE) tester. The VDE uses two spherically capped electrodes that can be spaced either .040" or .080" apart.

Both types of test sets apply a 60-Hz voltage across two electrodes under the insulating liquid. The voltage at which the actual breakdown occurs is the amount of electrical stress the oil takes at that particular voltage and frequency.

The VDE's spherical electrodes more closely approximate the actual internal conditions of a transformer than the flat disk tester does. The VDE test set is also more sensitive to moisture and other contaminants. The flat disk tester will detect moisture from roughly 30 ppm and above. By comparison, the VDE will detect moisture from about 15 ppm and above. The dielectric breakdown test also indicates the presence of other contaminants such as dirt, carbon, metal, or other conducting particles. A low dielectric breakdown indicates that some insulating ability of the liquid has definitely been lost.

On the other hand, if an insulating liquid has a high dielectric strength, this does not mean it is free from contaminants or moisture. Up to 99% of a transformer's moisture may be contained in the cellulose insulation. One percent of the total moisture shows up in the liquid. Sludge may be present in a transformer, or acid may be in the oil, and not be indicated by the dielectric tests.

The dielectric breakdown voltage test, whether by the flat disk or VDE method, should not be the only test performed on insulating liquid. Other tests used in combination will yield a more accurate picture of the liquid's serviceability.

The two types of testers are operated in much the same manner. A representative sample of liquid is placed into a test cup, covering the electrodes. A voltage is applied across the electrodes until current arcs between the two electrodes. The actual test voltages, rate of voltage rise, and procedural details vary between the two test methods. Figure 15–18 shows a portable dielectric breakdown voltage test set that has interchangeable test cells for performing a variety of tests on insulating liquids and gases.

Before testing, whether using the VDE or flat disk tester, inspect the test set's electrodes for any evidence of pitting or accumulation of carbon. Check the electrode gap to ensure proper spacing. If any sample tests below the minimum test requirement, drain the test cell and flush with clean, dry oil of the same type being tested. It is also suggested that when not in use, the oil cup be filled with a good quality sample of the oil type normally tested.

ASTM D-877 This test method is generally considered adequate for acceptance testing of unprocessed insulating liquids and for routine maintenance testing in equipment that is rated 230 kV and below. Liquid that has been filtered, degassed, or dehydrated, or liquid that is in equipment rated 230 kV and above, should not be tested by this method.

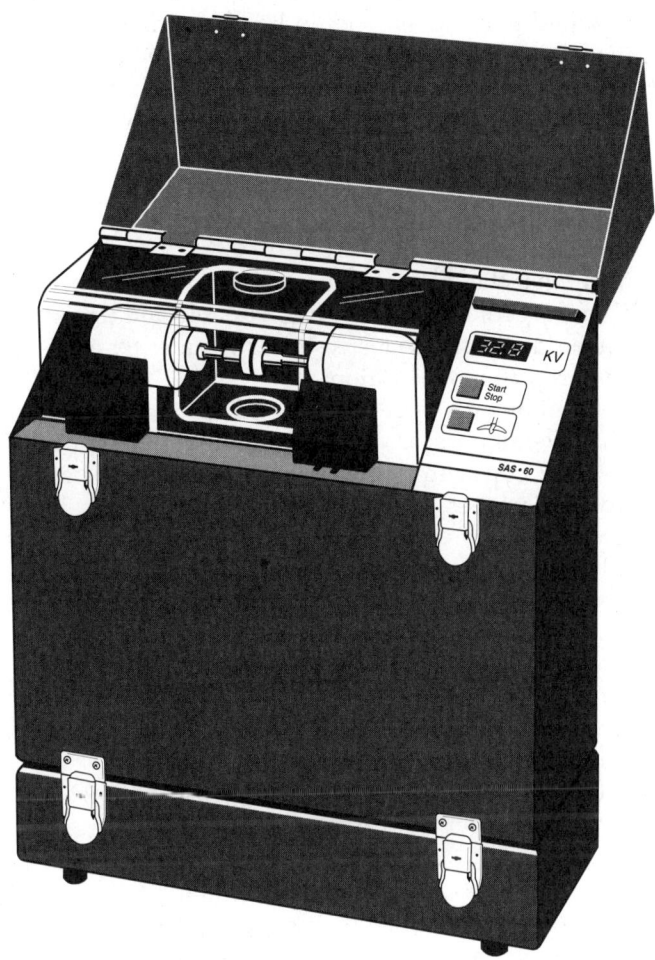

FIGURE 15–18
Dielectric breakdown voltage test set (Copyright ASTM. Reprinted with permission.)

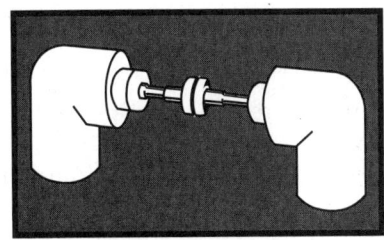

ASTM-877

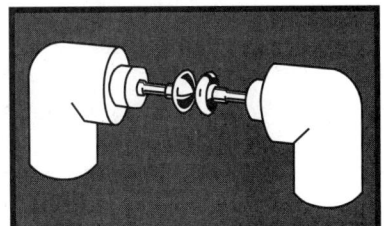

ASTM-1816

The liquid to be tested should be sampled according to the procedures outlined in ASTM D-923. The bottle should be tightly sealed and shielded from light until it is ready to be tested. Prior to the test, inspect the sample for moisture, sludge, metallic particles, or free water. The sample is unsatisfactory if these items are found.

The following guidelines are to be used for testing by the D-877 method (mineral oil):

1. Ensure that the oil comes to a temperature of not less than 68°F (20°C). Oil that is cooler than this gives variable and unsatisfactory results. The oil should also not be tested if it is warmer than 86°F (30°C), since more moisture will be in solution.
2. Prior to filling the test cup, gently invert the sample container and gently swirl several times. Do not shake to distribute impurities.
3. Rinse test cup with a small sample of liquid to be tested, and then discard oil in proper container.
4. Slowly fill the test cup to a level not less than 20 mm above tip of electrodes.
5. Place the test cup in the tester and allow the liquid to stand for not less than two minutes or more than three minutes.
6. Apply test voltage from zero in steps of 3000 V/sec, ± 20% until breakdown occurs and record the reading. (Occasional momentary discharges may occur. As long as these discharges do not cause the interrupting equipment to operate, they can be disregarded.)
7. Repeat steps 4 through 6 for five successive fillings of the test cup. Check for consistency in the readings.

NOTE: ASTM D-877 allows routine testing of all five tests on one cup of oil. A one-minute waiting period between tests is required. If the readings are inconsistent, five more tests must be done. These additional tests must be performed on five separate fillings of the cup. It is recommended that the test be done in this manner to begin with. This will ultimately reduce the amount of time to perform the tests.

Subtract the lowest reading from the highest reading and multiply by three:
  a. If the answer is less than or equal to the next-to-lowest reading, the readings are consistent. Average five readings and record this average as the breakdown voltage.
  b. If the answer is more than the next-to-lowest reading, the readings are inconsistent. Repeat steps 4 through 6 for five more successive fillings of the test cup. Average all 10 readings and record the average as the breakdown voltage.

8. Record the test method used next to the recorded breakdown voltage.

Example:

Readings:

    21 kV   30 kV
    29 kV   -21 kV
    26 kV   9 kV
    30 kV   x 3
    28 kV   27 kV

The value 27 kV is greater than the next-to-the-lowest reading (26 kV), therefore the readings are not consistent.

ASTM D-1816 This test method is more sensitive to moisture, especially when cellulosic fibers are present in the liquid. Insulating liquid that has been degassed, filtered, or dehydrated, or is to be used in equipment rated 230 kV and above, should be tested by this method rather than by the D-877 method. The VDE electrodes are not recommended for testing liquid in tank cars, tank trucks, or drums.

ASTM D-1816 can also be used to test the dielectric strength of oils rated less than 230 kV. The values will be lower because the D-1816 method is more sensitive to contaminants. Do not try to compare readings between the two methods.

As with the D-877 method, the sample tested must be representative of the equipment. ASTM D-923 outlines the required sampling procedure. The bottle should be tightly sealed and shielded from light until it is ready to be tested.

The following guidelines are to be followed for testing by the D-1816 method (mineral oil):

1. Allow the oil to come to a temperature of not less than 68°F (20°C). Oil that is cooler than this will give unreliable and unsatisfactory results. The oil should also not be warmer than 86°F (30°C) since more moisture will be in solution.
2. Prior to filling the test cup, gently invert the sample container and gently swirl several times. Do not shake to distribute impurities.
3. Rinse the test cup with a small sample of liquid to be tested and then discard oil in proper container.
4. Slowly fill the test cup with the remainder of the sample.
5. Place the test cup in the tester, turn on the stirrer, and allow the liquid to stand for at least three minutes.

6. Apply test voltage (500 V/sec rise) until breakdown occurs and record the reading. Allow the stirrer to operate during the test. This simulates the normal convection flow of the oil in a transformer during operation.
7. Repeat step 6 at one-minute intervals for five successive breakdowns on the same sample. (Disregard momentary discharges as long as they do not operate the circuit interrupting equipment.)
8. Check for consistency in readings. Subtract the lowest reading from the highest reading and multiply by three.
   a. If the answer is less than or equal to the next-to-lowest reading, the readings are consistent. Average the five readings and record the average as the breakdown voltage.
   b. If the answer is more than the next-to-lowest reading, the readings are inconsistent. Repeat steps 4 through 6 at one-minute intervals for five more successive breakdowns on the sample; then average all 10 readings, and record the average as the breakdown voltage. Refer to ASTM D-877 for an example.
9. Record the test method used next to the recorded breakdown voltage.

Color Testing (ASTM D-1500)

The color test is easily performed and is most useful as a screening test (separating those liquids that may require further testing). By comparing the color of insulating liquid to a color disk, it can be determined if deterioration or contamination has taken place. New mineral oils are generally "water white" and have a color value of less than 0.5.

By itself, the color of the liquid is of little importance. The primary value of this test is when a marked or radical color change occurs or when a high color number is obtained. This provides an indication that the liquid has been contaminated, but gives no indication of the specific contaminant. Radical color changes may be caused by electrical problems, such as arcing, but may also be caused by nonelectrical problems. Some of the nonelectrical problems may be pothead or bushing compounds, new liquid added to a dirty transformer, liquid that has been reclaimed, or uncured varnishes or polymers used as insulation.

The most common color test uses the Hellige comparator, which includes a comparator housing, a color disk, and a viewing depth cell. This method should be used in the following manner:

1. Fill viewing depth cell with sample of liquid to be tested and place it in the right-hand opening of the comparator.
   a. Fill the cell with at least 50 mm of liquid.
   b. If the color is cloudy, heat it to 10°F above its cloud point and observe its color.

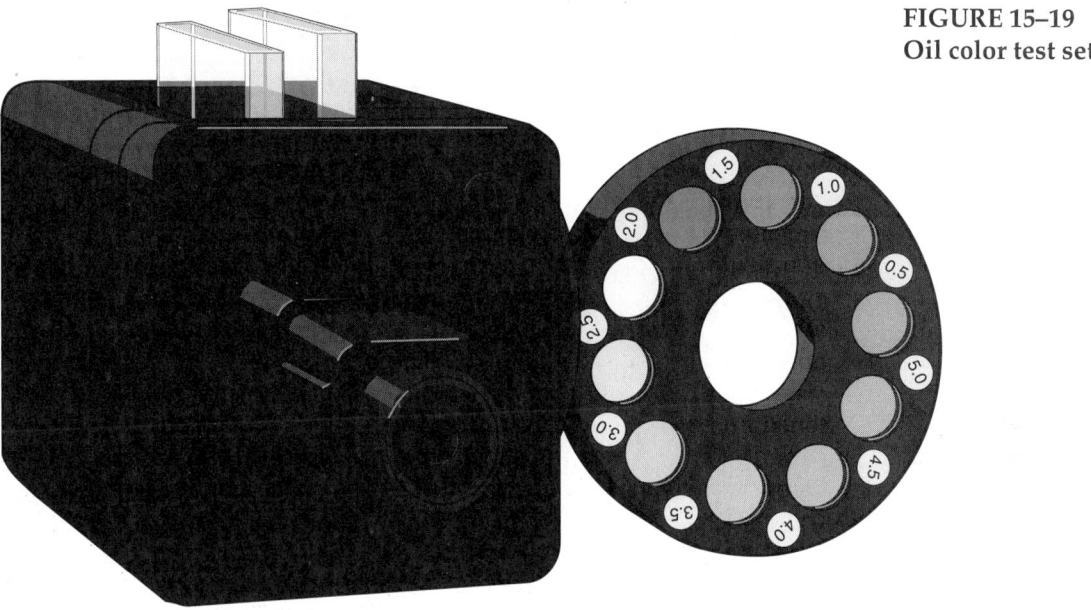

FIGURE 15-19
Oil color test set

2. While observing the side-by-side comparison zone (at a viewing distance of approximately 10 in) rotate the color disk until a color match is obtained between the sample being tested and one of the standards. Avoid direct sunlight as illumination for the comparator.
3. Observe the number in the upper right-hand corner of the comparator and record this value on the test sheet.
4. If a color match falls between two adjacent numbers, record the higher number as the test value.

Figure 15-19 shows a colorimeter used for color testing mineral oils.

Visual Examination (ASTM D-1524)

The visual condition of insulating liquid can be checked at the same time the color test is made. Good insulating liquid (such as mineral oil) is clear and water white. Liquid that is cloudy indicates the presence of moisture, carbons and/or sludge. The insulating liquid may also be inspected for free water, metallic contaminants, particles of insulation, or other suspended materials. An unusual or foul odor of the liquid may also indicate deterioration.

Acid Neutralization Number Test (ASTM D-1534)

The acid neutralization number (NN) is used to measure the approximate total acidity of an insulating liquid to determine what chemical changes are occurring. Values obtained from the NN are compared with values obtained when the liquid was new. These chemical changes are a result of deterioration by reacting with substances that the liquid has been in contact with, such as oxygen, copper, and iron. All insulating liquids are subject to chemical change as they age in service; however, the exact degree of change can be affected by many variables, ranging from the initial refining process, to the conditions the liquid is subjected to while in service. As these changes take place, decomposition products are formed, starting with peroxides, continuing with acids, and eventually ending in sludge.

Monitoring acidity offers an aid in determining when a liquid should be reclaimed or replaced, before decomposition becomes detrimental to the equipment. A low neutralization number represents a low level of acidic components and thus a minimum of chemical decomposition of the liquid. Neutralization number is expressed as milligrams of potassium hydroxide per gram of oil (mg KOH/g oil).

ASTM D-1534 is a field test designed to approximate the acidity of transformer oil. The following procedure (Gerin Ampule method) is used when field testing for acidity:

1. Pour the correct amount of oil sample in a glass cylinder according to the test sequence listed below.
2. Add neutral solution to approximately the 50 ml mark.
3. Add the contents of the specified ampule.
4. Insert stopper and shake vigorously for about 30 seconds.
5. When some clear liquid begins to separate out of the emulsified mix, note its color. If pink can be seen, record neutralization number, as listed below. If no pink is seen, continue using the next test listed until pink is seen or the oil is determined to be unserviceable (Note: only two ampules may be added for any one test).
6. Refer to Table 15–1 for guidelines in selecting ampule and neutral solution mixture.

NOTE: The Gerin Ampule method varies slightly from the ASTM standard in that the KOH is contained in pre-measured ampules. The performance of the test is essentially the same as in the standard.

Figure 15–20 shows the graduated cylinder, the 10 ml and 50 ml graduation marks, and the ampule. When performing the test, be certain to allow the two liquids to settle before determining the color. The color should be determined at the bottom of the graduated cylinder.

TABLE 15-1   Altering the Ampule Strength Ratings

Volume of Test Oil Sample	KOH Ampule Values When Testing Mineral Oil (Mg KOH Oil)				
	1A	1-1/2A	3A	6A	10A
5 ml	.200	.300	.600	1.200	2.00
7.5 ml	.133	.200	.400	.800	1.33
10 ml	.100	.150	.300	.600	1.00
12.5 ml	.080	.120	.240	.480	.800
15 ml	.066	.100	.200	.400	.666
17.5 ml	.057	.086	.171	.343	.571
20 ml	.050	.075	.150	.300	.500
25 ml	.040	.060	.120	.240	.400

Interfacial Tension Test (ASTM D-971/D-2285)

The interfacial tension (IFT) test is a measure of the attraction between oil and water. Soluble contaminants and oil decay produce sludge and cause a decrease in the IFT value. The IFT test is very good for detecting the presence of sludge in insulating liquids.

ASTM D-971 (DuNouy Tensiometer)   This IFT test is normally performed under laboratory-type conditions, although it is suitable for use in a mobile laboratory. The surface tension of distilled water is first measured (about 72 dynes/cm). Then, 20 mm of oil being tested is added while a platinum ring is submerged under the water. After a wait of 30 seconds the interface between the oil and the water is broken by the platinum ring. As the ring breaks through the interface, the scale reading is calculated to the nearest IFT reading.

ASTM D-2285 (Interfacial Tension Test)   The gravity-drop or point interfacial tension test is much more suitable for field testing than the DuNouy Tensiometer. Distilled water is injected from a calibrated orifice (needle into a sample of oil). The water is brought out to a bubble on the end of the orifice and allowed to "age" slightly. The bubble of water is made larger until it drops from the needle. The higher the amount of contaminants in the oil, the smaller the water bubble will be.

One of the advantages of this test method is that it is readily adaptable in field use. Another advantage is that the scale readings are in dynes/cm and require no conversion. This allows fewer mistakes in interpreting the readings and makes the test less subjective. The cost of this type of equipment is much less than a DuNouy unit due to the cost of the platinum ring for the DuNouy unit. The last advantage is that there are indications that the point IFT is a more reliable indicator of sludge and other soluble contaminants than the average IFT.

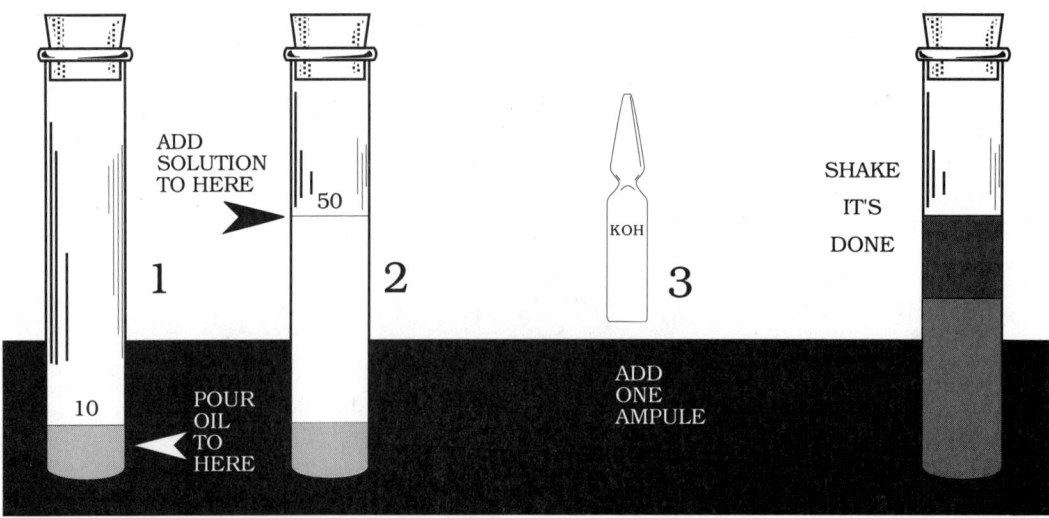

FIGURE 15–20
Gerin ampule method (Courtesy of the Gerin Corporation)

- Oil Interfacial Tensiometer
  This was designed for field testing of oil interfacial tension. It is recommended that deionized water be used instead of the ASTM required distilled water. This is because distilled water is hard to keep and some types of algae will grow in it. Ensure that the barrel is emptied after use of distilled water to prevent deposits of algae from forming.

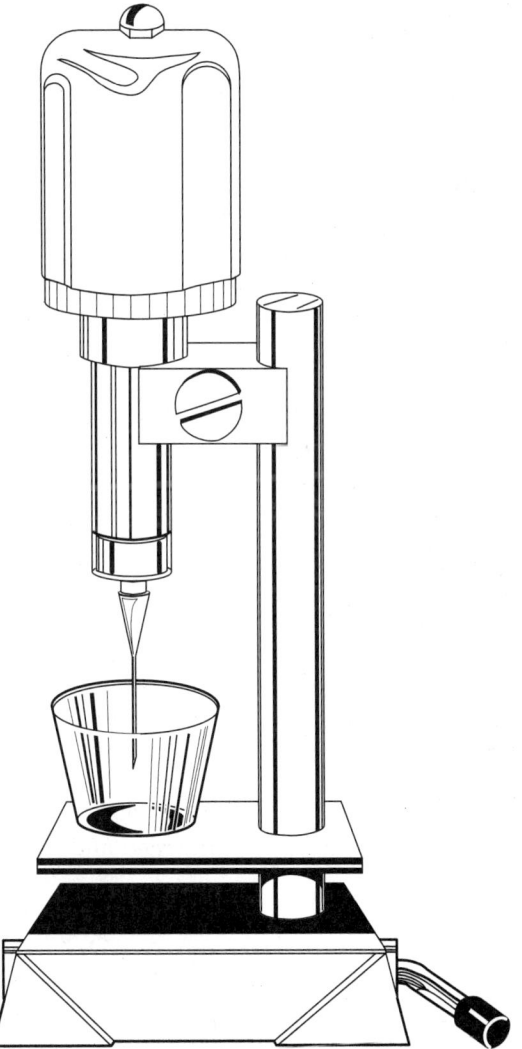

FIGURE 15–21
Oil Interfacial Tensiometer

- Drop-in-Air Calibration:
  Place the selected needle on the barrel. Make several trial drops to clean the needle. Then make 10 drops and record the dial reading for each drop. The 10 readings should be within 0.3 units of each other and plus or minus 0.2 units of the number indicated on the needle (i.e., 8.5 or 9.2). If a larger difference occurs, check for leaks or for dirt in the needle orifice.

- Drop-in-Oil Test:
  1. Place the needle ½ inch below oil surface.
  2. Make trial run by dropping one bubble of water within 30 seconds.
  3. Take ¾ of this reading and use as age reference number (i.e., 28 units drop lts ¾ = 21 units).
  4. Start first test by obtaining ¾ bubble (21 units) on dial within 15 seconds.
  5. Let water bubble age for 45 seconds.
  6. Continue to rotate dial until water bubble drops. Rotate dial smoothly and evenly. Record value.
  7. Repeat steps 2 through 6 four more times.

- Results:
  1. Readings should be within five percent of one another.
  2. Take average of five readings.

- Gerin Model IT-9 Series V:
  The Gerin Model IT-9 Series V is similar in operation to the oil interfacial tensiometer, Figure 15–21. The difference is that selection of the needle is determined by the specific gravity of the oil, Figure 15–22.

  To determine the specific gravity of the oil, fill the calibrated cylinder and insert one of the thermo-hydrometers into it. Allow the thermo-hydrometer to seek its own level. Note the scale reading at the miniscus level of the oil. If the reading is .878 or less, use the brown needle. If the reading is greater than .878, use the yellow needle. Mineral oil typically has a specific gravity of .886.

  The drop-in-air test is usually performed and the values should be:

  Brown Needle   7.7 dynes
  Yellow Needle 6.4 dynes

  If this standard is not met, then a correction factor must be calculated and applied to all readings. The correction factor formula is as follows:

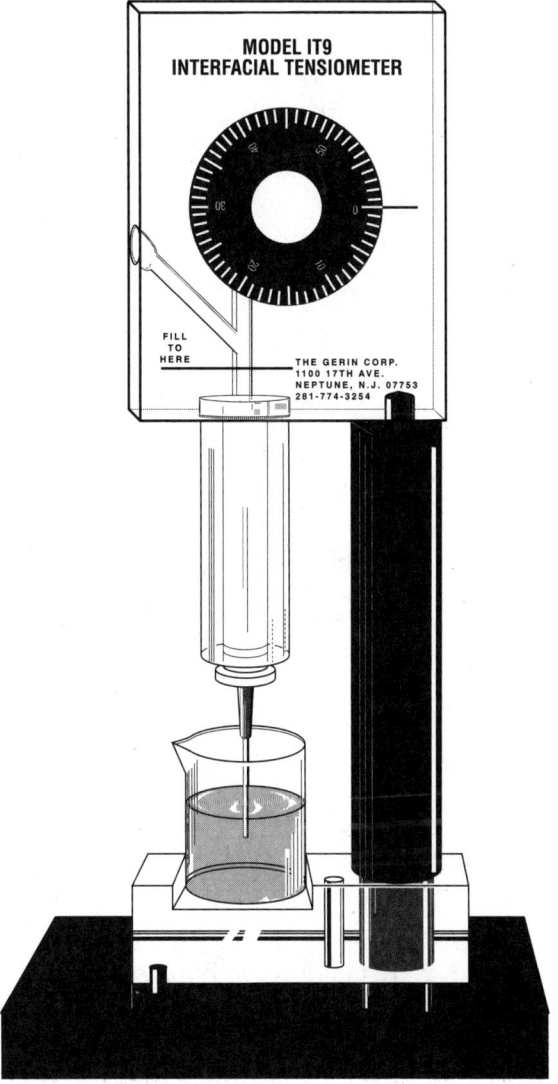

FIGURE 15–22
Gerin model IT-9 series V

1. Take 10 readings (drop high and low)
2. Average 8 remaining readings

3. Correction factor = $\dfrac{\text{Standard}}{\text{Average}}$
4. Repeat the drop-in-oil test.
5. Apply correction factors.

Moisture Content Test (ASTM D-1533)

The moisture content test measures the amount of moisture in oil. Water may be present in oil in many different forms. Moisture that is in the form of free water, 30 ppm or more, is detected by the dielectric breakdown strength test. Dissolved water may be present in oil and not be detected by the dielectric breakdown test. The actual amount of water, in parts per million, can be measured by the moisture content test.

The Karl Fischer method is one of the most commonly used methods of testing for moisture content. The test is performed by using Karl Fischer reagent to dehydrate a special solvent. When the solvent is dehydrated (neutralized), it will turn a chromate-yellow color. A galvanometer measures the depolarization of a platinum cathode. When the solvent is neutralized, a pre-measured sample of oil containing water is added, causing the solvent solution to become out of balance again. Another amount of Karl Fischer reagent is added until the solution neutralizes again, as indicated by the color and galvanometer. ASTM D-1533 gives the proper mixtures for the reagent and the solvent. Because of the need for pure chemicals and the careful mixing of the chemicals, the moisture content test is usually performed under laboratory conditions.

Evaluation of Test Data

Since judging the value of insulating oil with just one test is not possible, a series of tests must be performed on the oil. Based on the composite evaluation of the various test results, mineral oils that have been service-aged can be classified as to their continued serviceability. ANSI/IEEE C57.106 has four general classifications for service-aged oils (Table 15–2):

> Group I: Oils that are satisfactory for continued use.
> Group II: Oils that require minor reconditioning, such as the removal of moisture and insoluble contaminants by filtering or centrifuging.
> Group III: Oils in generally poor condition. These oils require reclaiming or disposal.
> Group IV: Oils that are in such poor condition that disposal is the best procedure.

These ANSI/IEEE classifications tell how poor an insulating oil is, but tell nothing about the effects the oil is having on the interior of the transformer. During the period of 1946 through 1957, ASTM ran a series of tests to determine the relationship of the neutralization number (NN), the interfacial tension (IFT), and the formation of sludge in a transformer. Five hundred in-service transformers were periodically tested and examined by ASTM for sludge formation during the 10-year period, as illustrated in Table 15–3.

Table 15-2 Classification of Service-Aged Transformer Oils (Values in Parentheses Show the Limits Established in the 1958 Survey)

(Courtesy of IEEE C 57.106, 1986, IEEE Guide for Acceptance and Maintenance of Insulating Oil in Equipment)

Test	Group I *Oils for Continued Service*	Group II *Oils to be Reconditioned*	Group III *Oils to be Reclaimed*	Group IV *Oils to be Scrapped*
Dielectric breakdown voltage, D877, kV, min	24 (23)	23.5 (22)	22	17
Neutralization number, mg KOH per g oil, max	0.36 (0.40)	0.40	0.40 (0.40)	0.75 (1.40)
Interfacial tension, dynes/cm, min	21 (19)	21	18 (18)	16 (14)
Power factor, 60 Hz 25°C, max	1.0 (1.4)	1.2 (1.8)	1.6 (1.4)	3.3
Moisture content, ppm, max	25	35	60	75

The study found that an increase in the neutralization number was followed by a drop in the interfacial tension. When these two tests are used together, mistakes caused by using only one or the other can be eliminated and the approximate life of an insulating oil can be determined. Based on the ASTM study, all 500 transformers had visible sludge formation by the time the neutralization number had risen to .60 mg KOH/g oil and the interfacial tension had dropped to 14 dynes/cm or below.

S. D. Myers Incorporated has combined the results of the ASTM study with the relationship between the neutralization number and the interfacial tension. Combining these factors has resulted in the oil quality index

Table 15-3 Relationship of NN and IFT to Sludging Based on ASTM Study of 500 In-Service Transformers

NN MG KOH/G Oil	IFT Dynes/Cm	% of 500	Units Sludged
.60 and above	Below 14	100	500
	13-16	85	425
.21-.60		72	360
	16-18	69	345
.11-.20		38	190
	18-20	35	175
	20-22	33	165
	22-24	30	150
.03-.10	Above 24	0	0

number (OQI), which is very useful in classifying insulating oils and their effects on the insulation, Figure 15–23.

To determine the oil quality index number, the interfacial tension is divided by the neutralization number (and a pure numerical value is arrived at much like the polarization index test performed on solid insulation). The insulating oils can then be classified into seven categories, according to their color and OQI number.

With these seven categories, oil can now be evaluated as to how well it is performing its various functions (such as cooling, insulating, and prevention of sludge buildup). Oils in the good range (pale yellow color and an OQI of 300–1,500) are performing all of their functions effectively. Proposition A oils (yellow coloring and an OQI of 276–600) are still performing all of their functions well, but should the interfacial tension drop below 27, sludge may begin appearing in solution. However, a transformer with oil in the marginal range (bright yellow color and an OQI of 160–318) may not be receiving proper cooling, and acids may be forming sludge deposits in insulation voids. Oils in the Proposition A and marginal categories can benefit from servicing, Table 15–4.

Transformer Protection

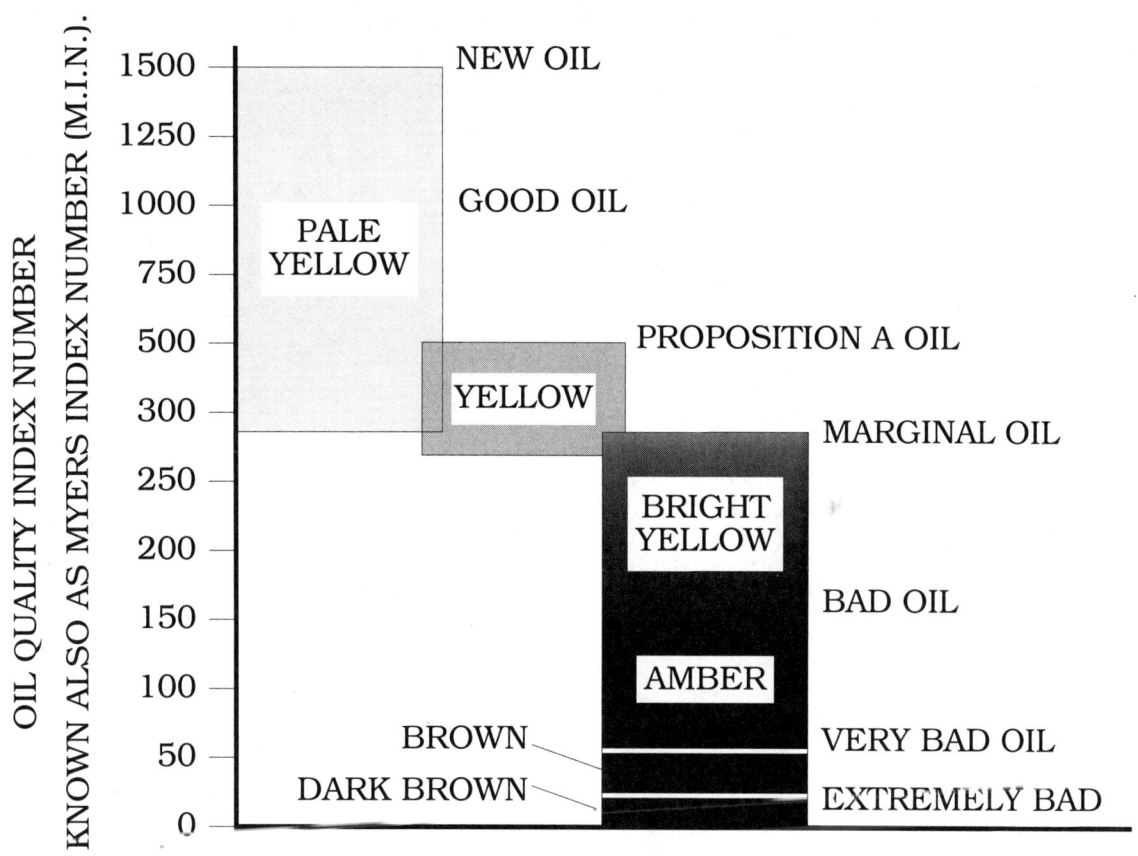

FIGURE 15–23
Oil Quality Index System (Reprinted with permission from Transformer Maintenance Institute Division of S.D. Myers, Inc., Akron, OH)

Table 15-4 Test Limits for Transformer Oil Before and After Reconditioning and Reclaiming

(Courtesy of IEEE C 57.106, 1986, IEEE Guide for Acceptance and Maintenance of Insulating Oil in Equipment)

Test	*Oil Requiring Reconditioning*	*Oil After Reconditioning*	*Oil Requiring Reclaiming*	*Oil After Reclaiming*
Dielectric breakdown voltage, D877, kV, min	23.5	27	22	28
Neutralization #, mg KOH oil, max	0.40	0.27	0.40	0.15
Power factor, 60 Hz 25°C, %, max	1.2	0.70	1.6	0.36
Interfacial tension, dynes/cm, min	21	26	18	30
Dielectric breakdown voltage, D1816, 0.04 in gap, kV, min	-	23	-	24
Moisture content, ppm, max	35	23	60	19
Visual examination	Test made	-	Test made	-

One of the most useful tables in the ANSI/IEEE Standard C57.106 is Table 15-5, *Acceptable Limits For Service-Aged Oil Versus Voltage Class.* As the various tables and recommendations are reviewed in the standard, it appears that they are conflicting. This table gives the recommendations for oil, based upon the voltage-class of the equipment it is in. As the voltage-class of the equipment increases, the minimum requirements are also increased. Therefore, oil that is suitable for continued service in equipment rated 34.5 kV may not be suitable for use in equipment rated 115 kV.

Transformer oils in the next three categories require immediate servicing. By the time insulating oil has deteriorated to the bad category (amber and OQI of 45–159), sludge is most probably beginning to form on trans-

Table 15-5 Acceptable Limits for Service-Aged Oil Versus Voltage Class
(Courtesy of IEEE C 57.106, 1986, IEEE Guide for Acceptance and Maintenance of Insulating Oil in Equipment)

Test	69 kV and below	Above 69 kV through 288 kV	325 kV and Above
Dielectric breakdown voltage, D877, kV, min	25	29	31
Neutralization number, mg KOH per g oil, max	0.39	0.33	0.28
Interfacial tension, dynes/cm, min	20.5	21.7	30.7
Power factor, 60 Hz 25°C, max	1.0	0.7	0.3
Moisture content, ppm, max	30	26	15

former parts. Oils that fall into very bad (brown coloring and an OQI of 22–44) and extremely bad categories (dark brown coloring and an OQI of 3–21) indicate the possibility of a transformer failure in the near future. Oil that is black with an OQI of 0 indicates that a failure has occurred.

If the interfacial tension and the neutralization number are poor and the coloring of the oil is good, the contamination may be from some cause other than oxidation. If the interfacial tension and the neutralization number are good and the coloring of the oil is bad, the cause may be overheating and/or arcing in the transformer.

Other Insulating Liquids

Although mineral oil is widely used as an insulating liquid, it is not ideal under all conditions or situations. The *National Electrical Code (NEC)*® requires that it be considered a flammable liquid in transformers; in addition, it is very susceptible to the formation of sludge. Various manufacturers have produced a variety of insulating liquids that meet or exceed the properties of oil in certain areas.

Askarel is the primary example of a synthetic alternative insulating liquid. Askarel is nonflammable, nonsludging, and extremely stable. However, the stability of askarel has caused much concern to environmentalists because it will not readily decompose on its own. Askarel is known by the abbreviation of its proper name, polychlorinated biphenyls (PCB(s). It was named as a hazardous material in the 1976 Toxic Substances Control Act and is the subject of much debate. Other insulating liquids that are not as potentially harmful to the environment have been developed to replace askarel as an insulating liquid. They are silicone, R-temp, tetrachloroethylene, and Freon®.

Askarel Askarels are still found in transformers and other equipment, and will be used for some time to come. Because askarels are still in use, they must be properly maintained.

The primary test performed on askarel is the dielectric breakdown voltage test. Askarel is nonsludging, so acidity is not usually a problem. The major enemy of askarel is water. Most recommended tests performed in the field are designed to locate water in the askarel.

Askarel samples should come from the top of the liquid, because askarel has a higher specific gravity than water. The dielectric breakdown test (D-877) should show a minimum strength of at least 26 kV. The askarel also should be visually inspected for particulates and free water; it should be free of clouding.

Askarel may be a yellow or even brownish color, samples that show other colors (reddishness, green, or cloudiness) may be contaminated with insulation or moisture. Askarel that is dark brown or black in color indicates severe arcing inside the transformer.

Other tests that are commonly performed on askarels are the moisture content and the power factor tests. The moisture content in askarel is performed with the Karl Fischer method (ASTM D-1533) and the maximum moisture content should be 70 ppm. Power factor tests are performed in a regular test cell; the liquid should be tested further if the power factor is 5% or higher.

A neutralization number test may be performed on askarel. Even though askarel is nonsludging, a sudden increase in acidity could be an indication of arcing. Arcing releases hydrogen chloride gas from the askarel and when it combines with any moisture present, hydrochloric acid is formed. When askarels (and all other insulating liquids) are being tested for dielectric breakdown strength, power factor, or other tests, the container, test cell, and other parts of the test set should be thoroughly cleaned prior to and after conducting tests. Also, a separate container should be maintained for each type of insulating liquid.

R-Temp® R-Temp® is a high-temperature hydrocarbon liquid that is often used as a retrofill for askarel transformers. R-Temp® and similar insulating liquids have properties similar to those of mineral oil, but have greater fire resistance. High-temperature hydrocarbons have a cost advantage over silicone liquids and, under heavily loaded conditions, heat transfer is improved, thus allowing a slightly higher kVA rating under some conditions.

The following precautions apply to high-temperature hydrocarbons:

1. Avoid energizing a transformer filled with R-Temp® or similar liquid if the temperature of the top oil is -15°C or less.
2. Do not mix R-Temp® and mineral oil because the mineral oil reduces the fire point of the R-Temp®.

When testing for dielectric breakdown strength (ASTM D-877), the minimum dielectric should be 26 kV. The neutralization number (D-974) should be a maximum of .10 mg KOH/g oil. Interfacial tension should be a minimum of 28 dynes/cm and moisture content should be a maximum of 50 ppm. The value given for the interfacial tension is tentative because ASTM does not specify this test for R-Temp®. Since the high-temperature hydrocarbons are still susceptible to sludging, the IFT test should be valid.

Tetrachloroethylene Tetrachloroethylene is a dry cleaning fluid in its unpurified state. In transformers it is not prone to sludging and is absolutely nonflammable. Westinghouse introduced a purified form of tetrachloroethylene for use as an insulating liquid, marketed under the name Wescosol®.

Extreme care is required in the handling of Wescosol® and similar insulating liquids because they evaporate into a heavy vapor. This vapor collects in low-lying areas and can cause severe reactions (headaches, confusion, lack of coordination, nausea, and even death), depending on the concentration of the vapor. Work should be performed in a well-ventilated area to prevent the vapor from collecting.

The dielectric breakdown voltage test is the primary test performed on tetrachloroethylene. New liquid should test to at least 30 kV (dielectric strength) and service-aged liquid should have a minimum dielectric strength of 26 kV. When sampling the liquid, sample under a positive pressure to prevent air from being drawn into the tank.

Freon® (R-113) Freon® is used in General Electric Vaportrans® transformers. Freon® is nonsludging and nonflammable, and because of its cooling efficiency, it operates at a much lower temperature than mineral oil transformers (45°C rise as compared to 55°C and 65°C).

Freon® (R-113) should be handled with the same precautions as refrigerant-grade Freon®. Since it is heavier than air, it will collect in low-lying areas and will displace oxygen. If it is inhaled there may be no immediate symptoms; however, unconsciousness and death can occur later. Work with Freon® only in well-ventilated areas to prevent it from collecting.

The dielectric strength (ASTM D-877) should be a minimum of 25 kV.

DC TESTING OF TRANSFORMERS

General Safety Precautions

Good safety practices must be followed during all tests, and during connection for testing, to ensure the protection of workers and of the transformer. Be sure to refer to the individual transformer manufacturer's literature and company procedures to ensure total compliance with all safety considerations and parameters. Some of the basic safety precautions are as follows:

1. Safety is the responsibility of the individual; therefore, review all safety considerations.
2. Never work on a transformer alone and be sure to restrict access to the transformer site.
3. Ground the transformer tank and leave it grounded at all times. The winding and bushings should be grounded except when electrical tests are performed on that winding.
4. Never perform electrical tests on a transformer when a vacuum is being applied to the transformer.
5. Always short circuit any current transformer secondaries to avoid dangerously high voltages that develop across the secondary terminals.
6. Ground the transformer's bushings to drain off any high-voltage charge remaining after the application of voltage to the bushing leads during tests. For example, the foil layers of condenser bushings, including the potential tap can hold a high voltage charge unless grounded after testing. Be sure to allow time for the residual voltage to decay.
7. Only qualified personnel should be involved in the transformer inspections, and under the full control of qualified supervision, to avoid possible misapplication of tests, that can result in personal injury.
8. The transformer tank of a sealed unit should be filled with oil, and a dry nitrogen blanket in place, when electrical tests are being conducted.
9. Obtain electrical clearances prior to performing routine or annual tests on in-service transformers. Be certain everyone who needs to know is informed.

10. Make sure that all possible sources of voltage are disconnected. A good rule of thumb is that there should be, whenever possible, two disconnecting devices in series, opened on both the primary and secondary of the transformer. One of these devices should be a type of disconnecting device in which its position can be visually verified.
11. Apply locks and security tags to all disconnecting devices' operating mechanisms to prevent the accidental reclosure of the device.
12. Extreme caution must be exercised while working on top of the transformer to avoid injury from slips and falls. The use of handrails or guards is highly recommended. When using ladders to obtain access to the top of a transformer, be sure to tie off the ladder to ensure its stability.
13. If required, wear rubber gloves when operating or connecting the test equipment.
14. Ensure that the test equipment is properly grounded to a good earth ground.
15. Observe caution with operators who wear heart pacemakers, since high-voltage discharges and strong electrical and magnetic fields may interfere with the proper operation of pacemakers.
16. Prevent contact with energized parts of the test equipment and the apparatus under test by using safety caution signs, barriers, warning lights, and by securing the test area.
17. Treat all terminals of high-voltage equipment as a potential for electric shock hazard.
18. Upon termination of a test, gradually reduce the test voltage to zero before the power supply is turned off.
19. At the end of the testing, account for all tools and equipment.

Insulation Resistance Testing   The use of DC to test solid insulation complements the AC tests that may be performed. DC stresses the solid insulation differently than AC does, and the information derived from the test is also different.

Insulation Resistance Testing Methods   One method of measuring the insulation resistance of a transformer is the time-resistance method using a megohmmeter. The megohmmeter is convenient to use and reads out directly in megohms. The instrument that is used should have a minimum output of 1,000 V or more, with a full-scale reading of 20,000 megohms or more. Since most megohmmeters are not accurate at full scale, the resistance should be capable of being read at midscale or less. The measuring lead should be supported from the ground and separated from all other leads and grounded objects. This separation will prevent inaccuracies caused by measuring conductor insulation resistance instead of the transformer insulation resistance.

To obtain uniform results in testing, insulation resistance testing should follow a regular procedure. The following guidelines will prove helpful:

1. Do not test unsealed, free-breathing or open transformers in high humidity; otherwise the readings may be erroneous.
2. Do not test a transformer while a vacuum is applied. A vacuum, unless it is perfect, reduces the insulating resistance.
3. Ensure that the tank and the core iron are grounded. Do not disconnect the core ground.
4. Disconnect all connections to the transformer. This is to include:
   a. High- and low-voltage bushing terminals
   b. Neutral bushing terminals
   c. Lightning arresters
   d. Meters
   e. Fans
5. Bushings can also be guarded if the megohmmeter is so equipped.
   In addition, ensure that the bushings are clean and free of contaminants, to avoid surface leakage.
6. Do not test oil-filled transformers that have been drained. Insulation resistance values taken in oil are ¼ to ½ of the values taken while the same windings are in air. This could lead to erroneous conclusions about the unit.
7. Short circuit each winding of the transformer at the bushing terminals.
8. Insulation resistance measurements are made between each winding and all other windings, with all other windings grounded. Windings that are not under test are not to be left ungrounded (floating) during insulation resistance measurements.
9. Windings that are solidly grounded must be ungrounded to measure the insulation resistance of that particular winding. In the case of internally grounded windings that cannot be ungrounded, the insulation resistance to ground cannot be measured.
10. Insulation resistance measurements should be taken from winding to winding, and winding to ground, for a duration of at least one minute.

Test Connections Two-winding transformers:

1. High to low and ground (H-LG), Figure 15–24
2. Low to high and ground (L-HG)
3. High and low to ground (HL-G)

Three-winding transformers:

1. High to low, and tertiary grounded (H-LTG), Figure 15–25
2. Low to high, and tertiary grounded (L-HTG)
3. Tertiary to high, and low grounded (T-HLG)

Transformer Protection

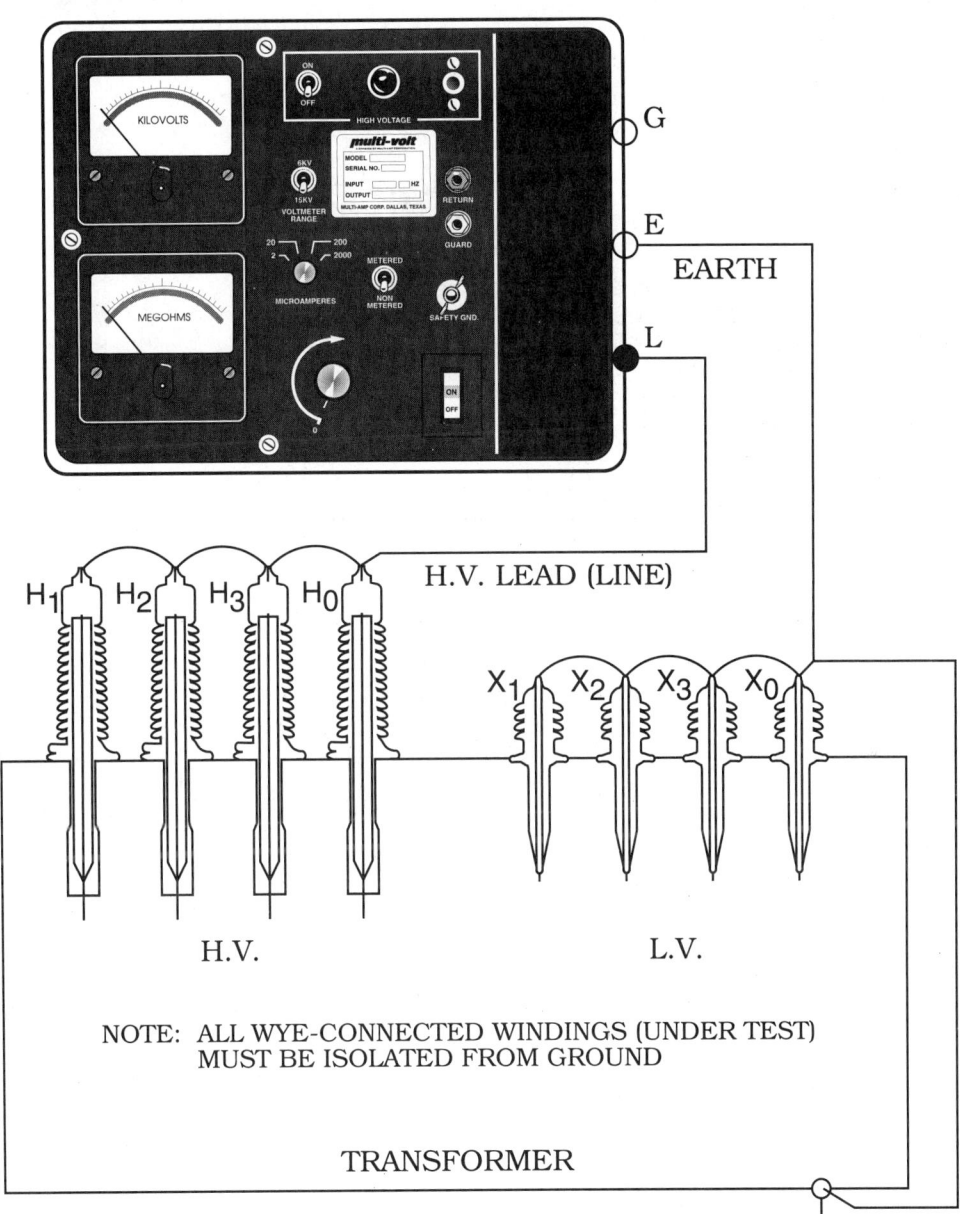

FIGURE 15–24
Insulation resistance testing: (2-winding transformer)
high to low and ground

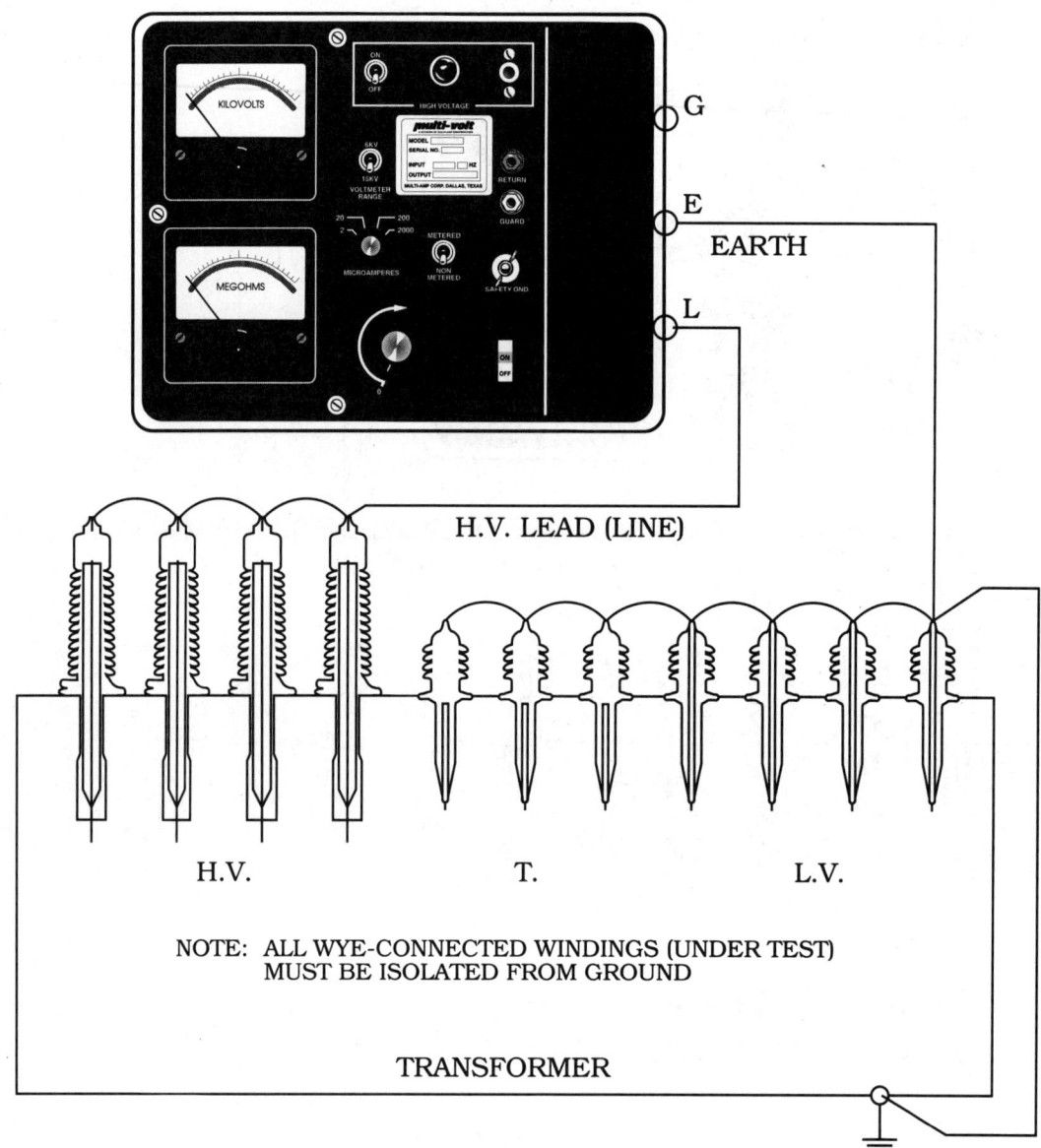

FIGURE 15-25
Insulation resistance testing: (3-winding transformer) high to low and tertiary grounded

4. High and low to tertiary grounded (HL-TG)
5. High and tertiary, to low grounded (HT-LG)
6. Low and tertiary to high grounded (LT-HG)
7. High and low, and tertiary to ground (HLT-G)

As an example, a three-winding transformer is measured as in the first step. The high-voltage terminal of the transformer is connected to the line terminal of the megohmmeter. The low-voltage and tertiary terminals of the transformer are short-circuited to each other and then to ground. The earth terminal of the megohmmeter is connected to transformer ground. The guard terminal of the megohmmeter is left unused (floating). If the results are minimum, the guard circuit should be utilized and the test repeated to check for surface leakage problems.

After the test is completed, ground all terminals for a sufficient length of time to bleed off any charges that may have built up during the test.

Insulation resistance tests are greatly affected by temperature, moisture, and contamination. Since moisture and contamination are two items that are looked for in the tests, the only item to be corrected for is temperature. Insulation resistance values rise as the temperature falls. A transformer will have a higher insulation resistance when it is cold than when it is hot. For this reason, all insulation resistance test results must be corrected to a common temperature base, usually 20°C. Use the values in Table 15–6 for temperature correction of insulation resistance tests. Note that these correction factors are different from the correction factors for power factor testing. Do not use these correction factors for anything but insulation resistance testing.

Minimum Insulation Resistance   Table 15–7 shows test voltages and minimum insulation resistance values for performing tests with a megohmmeter. These values are those recommended by the InterNational Electrical Testing Association (NETA) and are much more conservative than the values obtained by the above formulas. In actual practice, the measured insulation resistance should be many times greater than the calculated minimum, so the two are not greatly different.

Polarization Index   Polarization Index (PI) is a numerical value obtained by dividing the one-minute insulation resistance value into the ten-minute insulation resistance value. When moisture is present in insulation, the leakage current is greater, thus masking the effect of the decaying absorption current. This causes the PI to be lower for insulation in a more deteriorated condition.

The primary advantage of the PI is that both readings are taken under the same conditions. This makes the PI test a reliable indicator of insulation condition, even if the previous insulation resistance test results are not available.

Table 15-6 Insulation Resistance Conversion Factors for Conversion of Test Temperatures to 20°C

TEMPERATURE		TRANSFORMER	
°C.	°F	Oil	Dry
0	32	.25	.40
5	41	.36	.45
10	50	.50	.50
15	59	.75	.75
20	68	1.00	1.00
25	77	1.40	1.30
30	86	1.98	1.60
35	95	2.80	2.05
40	104	3.95	2.50
45	113	5.60	3.25
50	122	7.85	4.00
55	131	11.20	5.20
60	140	15.85	6.40
65	149	22.40	8.70
70	158	31.75	10.00
75	167	44.70	13.00
80	176	63.50	16.00

Typically, a transformer should have a PI of 2.0 or greater. PI values between 1.0 and 2.0 are marginal, with values closer to 1.0 (1.1 to 1.0) being poor. Insulation that falls below 1.0 PI is considered to be in dangerous condition. Once again, the results obtained with oil-filled transformers can be masked by the effect of the oil.

Table 15-7 Transformer Insulation Resistance Test Voltage

Transformer Coil Rating (Volts)	Minimum Test Voltage (Volts dc)	Recommended Minimum Insulation Resistance in Megohms	
		Liquid Filled	Dry Type
0 - 600	1,000	100	500
601 - 5,000	2,500	1,000	5,000
5,001 - 15,000	5,000	5,000	25,000

Example:

1-minute test = 30 megohms
10-minute test = 80 megohms

Polarization Index: $\dfrac{80}{30} = 2.67$ \hfill (1)

Dielectric Absorption Ratio The dielectric absorption ratio (DAR) is done at the same values, time, and manner as the dielectric absorption test. The same advantages that apply to the DAR also apply to the PI because both are ratios. Commonly, the DAR is used as a first check of the insulation being tested to determine if the test should be continued. The 30-second insulation resistance value is divided into the 60-second value. The same grading of insulation applies to the DAR and to the PI.

Example:

30-second test = 7 megohms
60-second test = 16 megohms

Dielectric absorption ratio: $\dfrac{16}{7} = 2.28$ \hfill (2)

Bushing Insulation Resistance An insulation resistance test may also be performed on bushings and arresters. With the transformer entirely disconnected from all lines, connect the line terminal of the megohmmeter to the conductor terminal of one bushing or arrester. Connect the earth terminal to the mounting flange of the device being tested or to the tank. Surface leakage can be eliminated from the results by connecting the guard terminal of the megohmmeter to a conductive collar placed around the arrester or bushing.

If the measured value is higher than the minimum calculated value (same value as for transformer windings), the bushing is satisfactory. If the measured value is lower, disconnect the bushing from the winding and retest.

Step-Voltage Test The step-voltage test is used to test insulation at two voltages. The step-voltage test does not stress the insulation, but will reveal moisture and contamination in the insulation.

The preferred method of performing the step-voltage test is to apply voltages at two different levels, usually at a 5 to 1 ratio. The lower voltage (as an example, 500 V) is applied first for one minute, and then the higher voltage (2,500 V) is applied for one minute. There should be less than a 25% difference between the two readings if the insulation is in good condition. A difference of more than 25% is an indication of excessive moisture. Good, dry insulation will have insulation resistance readings that are very close to each other. Once again, however, the results obtained with oil-filled transformers can be masked by the effect of the oil.

Winding Resistance Testing

The purposes for making winding resistance measurements can be placed into several general areas such as:

1. Calculation of I^2R component of conductor losses.
2. Calculation of winding temperatures at the end of a temperature test.
3. As a base for assessing possible damage.

This measurement is usually conducted, for field purposes, to evaluate and determine if any changes have occurred in the current-carrying path. Some examples of the changes are as follows:

1. Tap changer connections or contacts working loose (high resistance or open).
2. Bushing connections working loose.
3. Shorted or open windings.
4. Winding interconnections working loose.

These changes can occur during shipment, installation, or after severe fault interruptions.

Performing the Winding Resistance Testing The winding resistance measurements should be conducted using test equipment such as the Wheatstone or Kelvin bridge or modern digital low-resistance ohmmeters, which are capable of measuring fractional ohms accurately.

The instruments that employ the bridge method are highly accurate and are used for measurement of resistances up to 10,000 Ω, and as low as a few microhms.

- Wheatstone Bridge—Figure 15–26 illustrates the Wheatstone bridge principle of resistance measurement. The circuit consists of four resistance arms, a source of current, and a null detector. The measurement of the unknown R_x is made in terms of the three known resistances. The three resistors R_A, R_B, and R_S are adjusted for zero current in the null detector circuit. When the bridge is balanced, as indicated by a null reading of the detector, the unknown resistance is obtained. The formula for calculating the unknown R_x resistance is as follows:

$$R_X = \frac{R_A R_S}{R_B} \tag{3}$$

 where
 R_A and R_B = Values of the ratio resistors
 R_S = Value of the standard resistor
 R_x = Unknown resistance to be measured

- Kelvin Bridge—The Kelvin bridge is similar to the Wheatstone bridge; however, the Kelvin circuit contains an additional set of ratio arms (Figure 15–27). This arrangement permits a four-terminal measurement of resistance elements, thus eliminating the effects of lead and contact resistance errors often associated with low-resistance measurements.

 When the bridge is balanced, the unknown resistance is given in the following equation (if R_A/R_B is exactly equal to R_a/R_b):

$$R_X = R_S \frac{R_A}{R_B} \tag{4}$$

To further explain this, if the resistances of R_x and R_S are small, so that the resistance of the yoke (link) connecting them is comparable, then the value of the yoke is significant. However, if the ratio A/B is equal to the ratio a/b, the correction is negligible. This can be demonstrated after the bridge is balanced, by opening the yoke (link); if the inner and outer ratios are equal, the bridge will remain balanced.

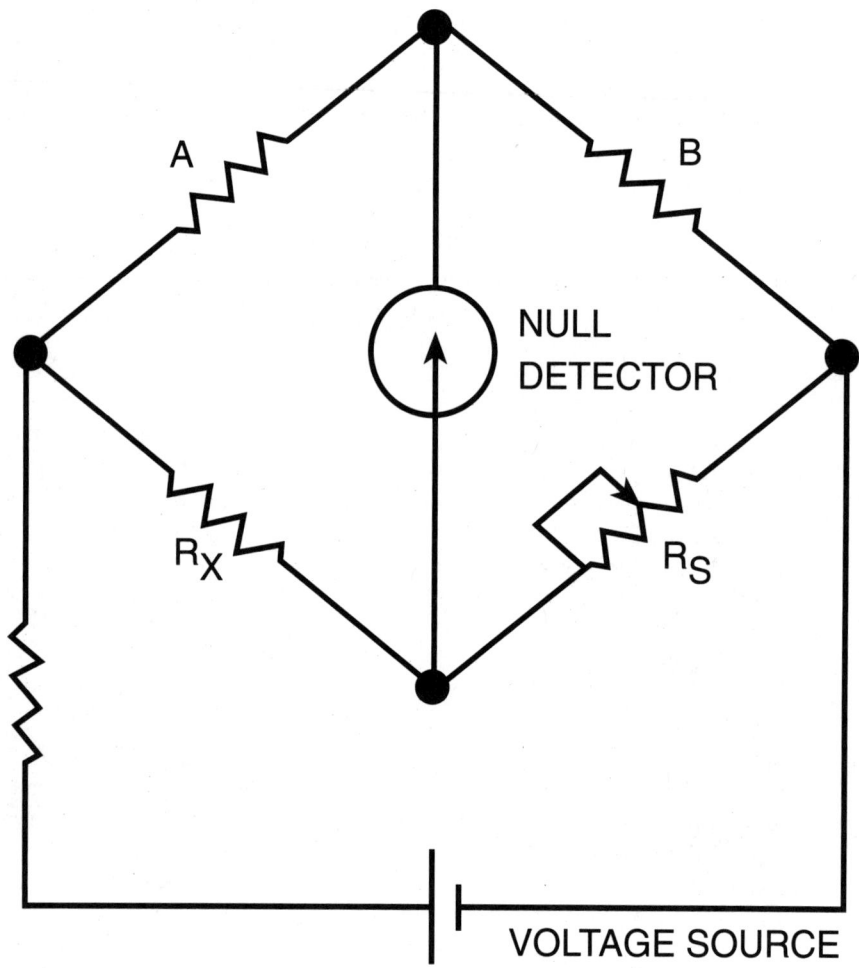

FIGURE 15–26
Wheatstone Bridge

The resistance of the test leads between the bridge terminals and the voltage terminals of the resistors can contribute to a ratio unbalance. One method employed in some Kelvin bridges to balance the lead effect is to provide small adjustable resistors. Another method is to provide a high-resistance shunt around the a or b arm until A/B = a/b with the yoke removed. At balance point, the yoke is replaced and the main bridge balance is readjusted.

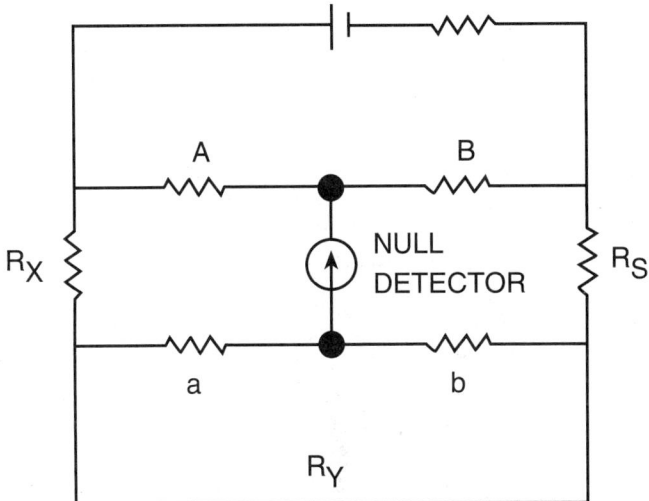

FIGURE 15–27
Kelvin bridge

R_X - UNKNOWN RESISTANCE
R_S - VALUE OF STANDARD RESISTOR
R_A - VALUE OF RATIO RESISTOR
R_B - VALUE OF RATIO RESISTOR
R_a - VALUE OF RATIO ARMS
R_b - VALUE OF RATIO ARMS
R_Y - VALUE OF YOKE

- Digital Ohmmeters—There are digital instruments available on today's market which are extremely accurate and usually employ the Kelvin bridge principle of low-resistance measurements. The voltage drop across the resistance under test is measured by an accurate digital voltmeter reading directly in ohms. An example of this type of instrument is the DLRO® manufactured by Biddle Instruments, Figure 15–28.

Testing Precautions Regardless of the actual instrument used in measuring the low resistance, several precautions based upon ANSI/IEEE C57.12.90 must be observed.

1. Cold resistance measurements should not be made on a transformer when the ambient temperature is varying rapidly or erratically.
2. The ambient temperature around a deenergized transformer must be known.

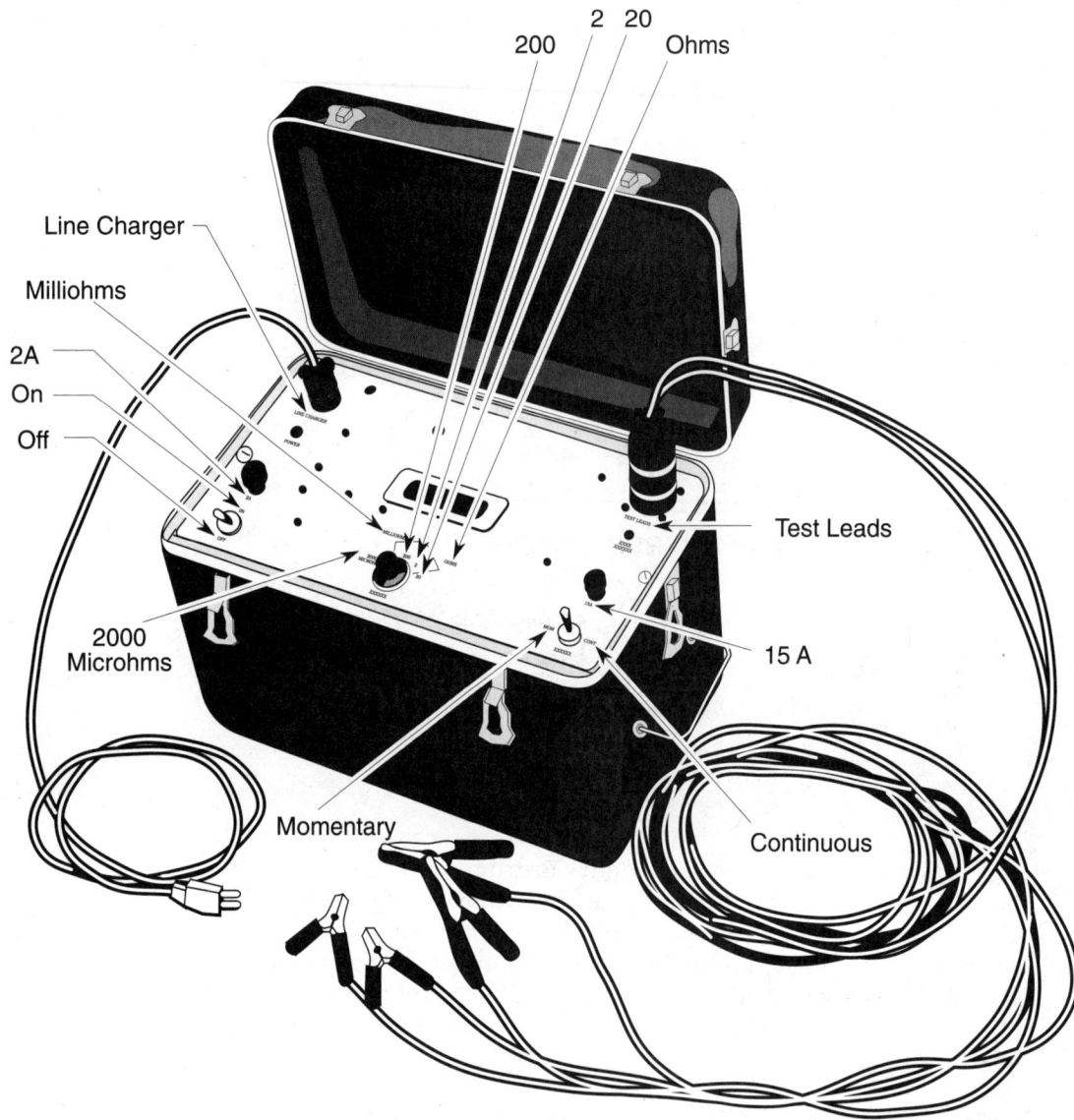

FIGURE 15–28
Digital low-resistance ohmmeter, showing connection cables and lead supplied (Biddle™ Digital Transformer Ohmmeter)

3. The ambient temperature should be fairly constant for several hours to ensure that the winding or windings are at ambient temperature (usually eight hours).
   a. For transformer windings immersed in insulating liquid, the winding can usually be assumed to be at the liquid temperature provided that:
      (1) The windings have been under the insulating liquid from 3 to 8 hours with no excitation (current) in the windings.
      (2) The temperature of the insulating liquid has stabilized, and the difference between top and bottom temperature does not exceed 5°C.
   b. For dry-type transformers or transformer windings out of insulating liquid, the temperature of the windings should be recorded as the average of several thermometers or thermocouples inserted between the coils. The measuring points should be as nearly as possible in actual contact with the winding conductors.
4. When R_X is highly inductive (as it is in a transformer winding), sufficient time must be allowed for the readings to stabilize. The inductive effect and low resistance of the winding may cause the time constant L/R to become quite large. The time required for stabilization can vary from several minutes to several hours if a unit that supplies a pulsed current is used.

Transformer ohmmeters are manufactured by various companies. These units have the advantage of applying a steady current through the winding. This will greatly reduce the amount of time required to charge the windings. Some units also are constructed to simplify testing delta-delta transformers, Figure 15–29.

Conversion of Resistance Measurements Since resistance varies as a function of temperature in most conducting materials, resistance readings are normalized to one temperature. The normal or standard reference temperature is equal to the rated average winding temperature rise plus 20°C. For a 55°C rise transformer, the standard temperature is 75°C and for a 65°C rise transformer, the standard temperature is 85°C. Transformer windings can be made of either copper or aluminum, requiring two separate constants for temperature conversion calculations. The constants are referred to as zero-resistance temperature constants, measured in °C. The zero-resistance temperature for copper is −234.5°C, and for aluminum −225°C. These constants are based upon a temperature in which resistance is theoretically zero, thus a base for calculation, and comparison of results.

Copper Winding

$$75°C \ \ R_{75} = R_m \times \frac{234.5 + 75}{234.5 + T_m} \tag{5}$$

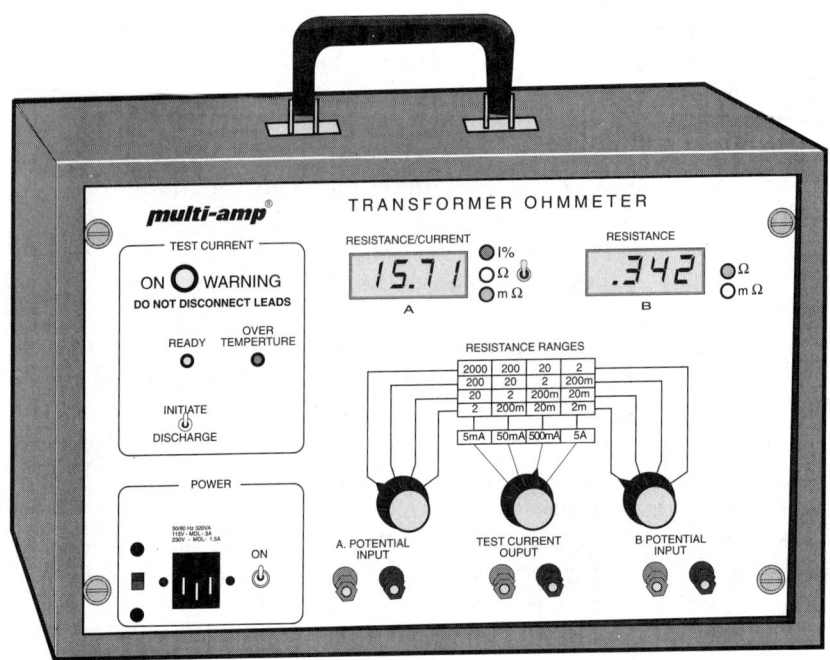

FIGURE 15–29 Transformer ohmmeter (Multi-Amp® Transformer Ohmmeter)

$$85°C \quad R_{85} = R_m \times \frac{234.5 + 85}{234.5 + T_m}$$

Aluminum Winding

$$75°C \quad R_{75} = R_m \times \frac{225 + 75}{225 + T_m} \tag{6}$$

$$85°C \quad R_{85} = R_m \times \frac{225 + 85}{225 + T_m}$$

where:

R_{75} or R_{85} = corrected resistance

R_m = measured resistance @ ambient temperature

T_m = ambient temperature in °C

Table 15–8 lists some simple conversion factors based upon copper windings corrected to 75°C.

When calculating the winding resistance, it is very important to note that most manufacturers report the winding resistance measurements as if all three windings are connected in series.

On transformers that are wye-connected, take the readings from phase-to-neutral (A-N; B-N; C-N) and sum the readings. For delta-connected

Table 15-8   Temperature Conversion Factors for Copper for Converting to 75°C

TEST TEMP °C	MULTIPLY BY	TEST TEMP °C	MULTIPLY BY	TEST TEMP °C	MULTIPLY BY
-20	1.443	21	1.211	61	1.047
-19	1.436	22	1.207	62	1.044
-18	1.430	23	1.202	63	1.040
-17	1.423	24	1.197	64	1.037
-16	1.416	25	1.193	65	1.033
-15	1.410	26	1.188	66	1.030
-14	1.404	27	1.184	67	1.027
-13	1.397	28	1.179	68	1.023
-12	1.391	29	1.175	69	1.020
-11	1.385	30	1.170	70	1.016
-10	1.379	31	1.166	71	1.013
-9	1.373	32	1.161	72	1.010
-8	1.366	33	1.157	73	1.007
-7	1.360	34	1.153	74	1.003
-6	1.354	35	1.148	75	1.000
-5	1.349	36	1.144	76	0.997
-4	1.343	37	1.140	77	0.994
-3	1.337	38	1.136	78	0.990
-2	1.331	39	1.132	79	0.987
-1	1.325	40	1.128	80	0.984
0	1.320	41	1.123	81	0.981
1	1.314	42	1.119	82	0.978
2	1.309	43	1.115	83	0.975
3	1.303	44	1.111	84	0.972
4	1.298	45	1.107	85	0.969
5	1.292	46	1.103	86	0.966
6	1.287	47	1.099	87	0.963
7	1.282	48	1.096	88	0.960
8	1.276	49	1.092	89	0.957
9	1.271	50	1.088	90	0.954
10	1.266	51	1.084	91	0.951
11	1.261	52	1.080	92	0.948
12	1.256	53	1.077	93	0.945
13	1.251	54	1.073	94	0.942
14	1.245	55	1.069	95	0.939
15	1.240	56	1.065	96	0.936
16	1.236	57	1.062	97	0.934
17	1.231	58	1.058	98	0.931
18	1.226	59	1.055	99	0.928
19	1.221	60	1.051	100	0.925
20	1.216				

* Use for dc winding resistance test

calculated from correction factor = $\dfrac{234.5 + 75}{234.5 + t_m}$

transformers, each reading taken will actually measure two windings in series which are in parallel with the winding under test. Thus, the readings are obtained by measuring from phase-to-phase (A-B; B-C; C-A); then the readings are summed and multiplied by 1.5. In either case the separate recorded readings should be saved for future comparisons. The results should indicate no more than a + 2% tolerance from factory results.

AC TESTING OF TRANSFORMERS

Transformer Turns Ratio

Turns ratio testing of transformers provides a direct means to detect open connections, shorted or open turns in the transformer windings, incorrect polarity connections, and to verify precisely the no-load voltage ratio of the transformer at all tap positions.

The turns ratio of a transformer is defined as the number of turns in one winding in relation to the number of turns in the other winding of the same phase, which is also equal to the no-load voltage ratio. Defined mathematically:

$$\frac{E_p}{E_s} = \frac{N_p}{N_s} = TR \qquad (7)$$

Where

E_p = voltage primary
E_s = voltage secondary
N_p = number of turns primary
N_s = number of turns secondary
TR = turns ratio (primary to secondary)

For reliable operation, and parallel connections of power and distribution transformers, it is extremely important that the turns ratio be determined very precisely. The turns ratio is required to be within + ½ percent (.005) of the indicated nameplate voltage ratio.

The nameplate voltage ratio (NP) of a transformer is defined as the line-line voltage of the primary winding to the line-line voltage of the secondary winding at no-load.

$$\frac{\text{H.V. (Primary) Line-line Voltage}}{\text{L.V. (Secondary) Line-line Voltage}} = NP \text{ or } \frac{E_p}{E_s} = NP \qquad (8)$$

Since the no-load voltage ratio is directly proportional to the turns ratio, an accurate measurement of the turns ratio will provide the no-load voltage ratio, without anyone having to work with dangerously high voltages.

When a transformer is excited by its low-voltage winding, the no-load voltage ratio is nearly equal to the turns ratio. The difference between the two ratios is caused by a voltage drop in the winding that results from the magnetizing current flowing through the winding. Normally this difference is less than 0.1% or .001. (This same condition exists if the high-voltage winding is excited, and a voltage drop is developed in the winding.)

By exciting a transformer winding with a known voltage, and applying this same voltage to a reference transformer (in which the exact turns ratio is known), a balance circuit can be used.

When both the transformer under test and the reference transformer voltage ratios are equal, the turns ratio is obtained.

Transformer turns ratio test sets (TTR®) are designed to operate on the above principle. In addition, the test sets are designed so that during the ratio tests, polarity, open or short-circuited turns, and the vectorial relationships of the various transformer windings are checked. Several different types of turns ratio test sets are manufactured; however, they operate generally by the same principle. A typical turns ratio test set is shown in Figure 15–30.

A single-phase turns ratio test set can be operated by a hand crank generator, with an electric driven motor or by 120 VAC.

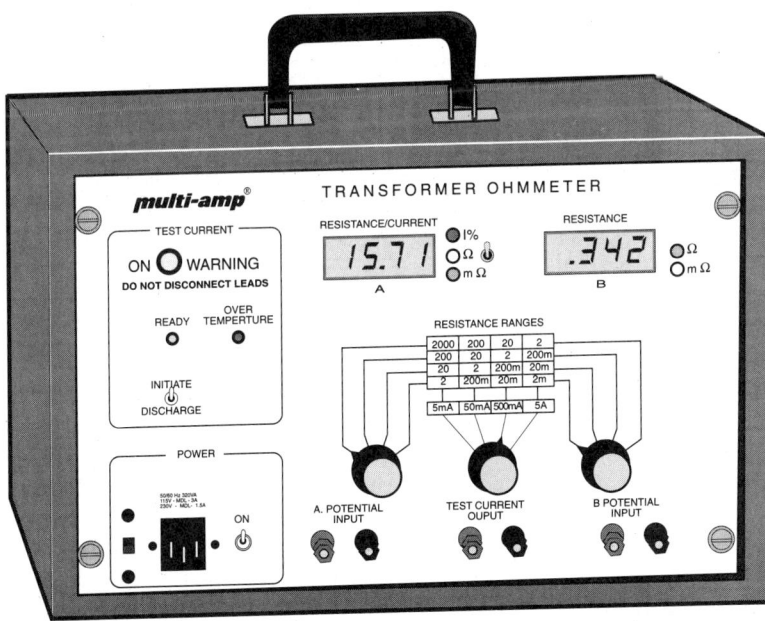

FIGURE 15–30
Transformer turns ratio test set (Biddle™ TTR®)

Theory of Operation Figure 15–31 shows a simplified schematic diagram of a Biddle TTR, which will be used to explain the theory of operation. The test set is arranged so that the transformer to be tested, and the adjustable ratio reference transformer in the test set, are excited from the same source voltage. The secondary windings are connected in a series opposing configuration through a null detector. When the ratio of the reference transformer is adjusted so that no current flows in the secondary circuit (null point), the voltage ratios of the two transformers are equal. Since no load exists on either secondary, and the no-load voltage ratio of the reference transformer is known, the voltage ratio of the transformer under test is obtained; thus, its turns ratio is also known.

Test Connections As with any test or test procedure, the connections are critical. Both single-phase and three-phase connections will be discussed in detail.

- *Connections for Single-Phase Transformers*
  Along with the theory of operation and correct reading of the results, the connections for a test are extremely important. Wrong connections

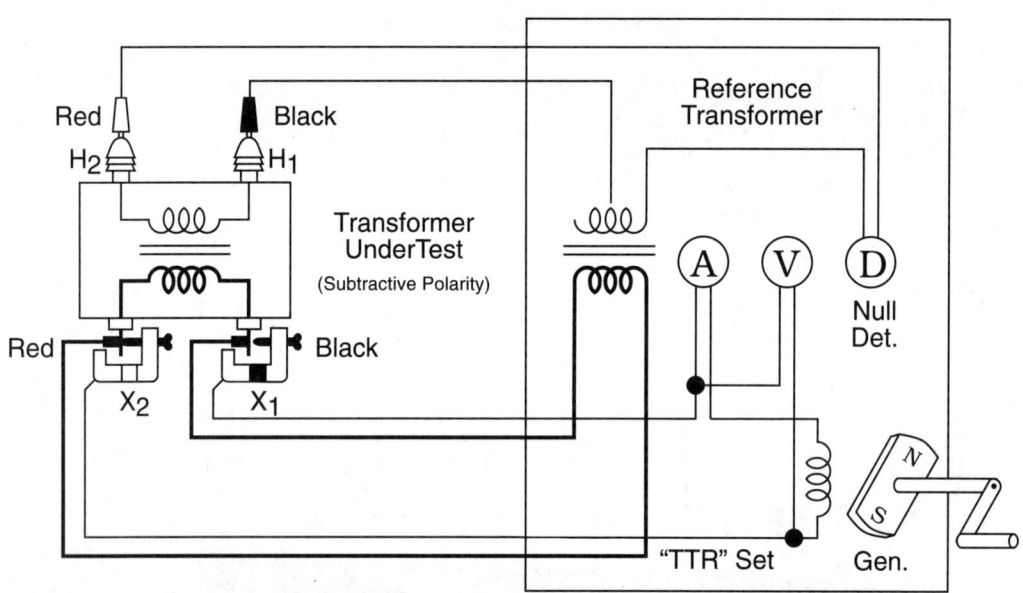

FIGURE 15–31
Simplified schematic diagram

result in improper operation and readings. The Biddle TTR has four leads marked X_1, X_2, H_1, and H_2.

1. The exciting lead (X_1) black is used to connect the tested transformer's low voltage winding to the primary of the reference transformer in the TTR test set. The black color code indicates the polarity and must be observed.
2. The exciting lead (X_2) red is used for the non-polarity connection of the tested transformer's low voltage winding to the non-polarity side of the primary of the reference transformer.
3. Secondary lead (H_1) black is used to connect the polarity side of the tested transformer's high voltage winding to the secondary polarity side of the reference transformer.
4. Secondary lead (H_2) red is used to connect the non-polarity side of the tested transformer's high voltage winding to the secondary non-polarity side of the reference transformer.

The first step in determining the proper connections is to examine the transformer's nameplate polarity and/or schematic diagrams. The proper winding and polarity must be observed. The single-phase distribution transformer, illustrated in Figure 15–32, shows that H_1 and X_1 are the instantaneous polarities of this transformer; thus the ratio of

$$\frac{H_1 - H_2}{X_1 - X_2} \text{ will be determined} \qquad (9)$$

The H_1 and X_1 polarities must be observed in the following procedures:

1. Black (X_1) of the test set is connected to the X_1 terminal of the transformer to be tested.
2. Red (X_2) of the test set is connected to the X_2 terminal of the transformer to be tested. (This connection determines the instantaneous current flow in the transformer low voltage winding from X_1 to X_2.)
3. Black (H_1) of the test set is connected to the H_1 terminal of the transformer to be tested.
4. Red (H_2) of the test set is connected to the H_2 terminal of the transformer to be tested. (This connection determines the instantaneous current flow in the reference transformer secondary winding.)

Figure 15–32 shows a subtractive polarity transformer connection; however, not all transformers are subtractive. When testing an additive transformer, it is necessary to interchange the secondary leads H_1 and H_2 to properly connect the test set (Figure 15–33).

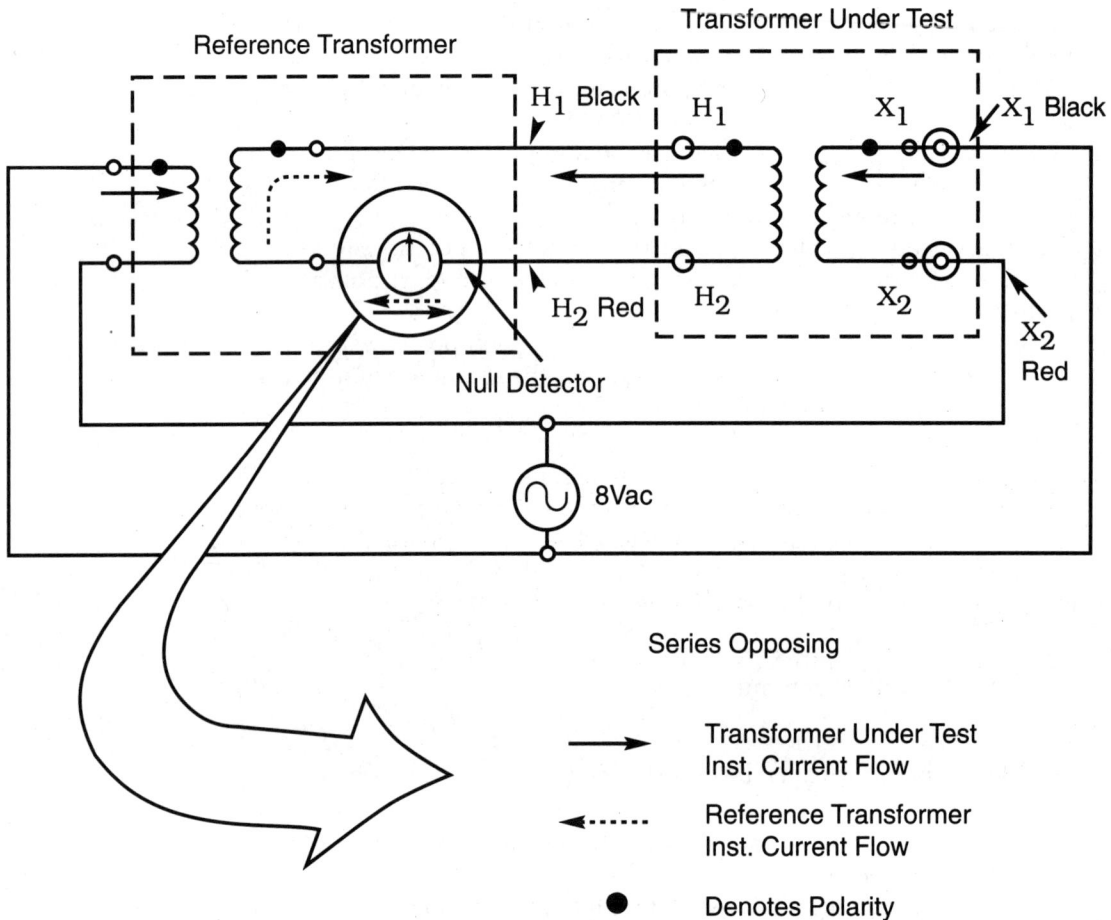

FIGURE 15–32
Correct connection for turns ratio test

- *Connections for Three-Phase Transformers*
  Three-phase transformers can be tested on a single-phase basis. Special attention is required to ensure proper connections so that the high voltage winding to low voltage winding (under test) is of the same phase. In addition, with three-phase transformers, the nameplate voltage ratio is not always equal to the measured turns ratio, such as with transformers connected wye-delta, or delta-wye.

Transformer Protection

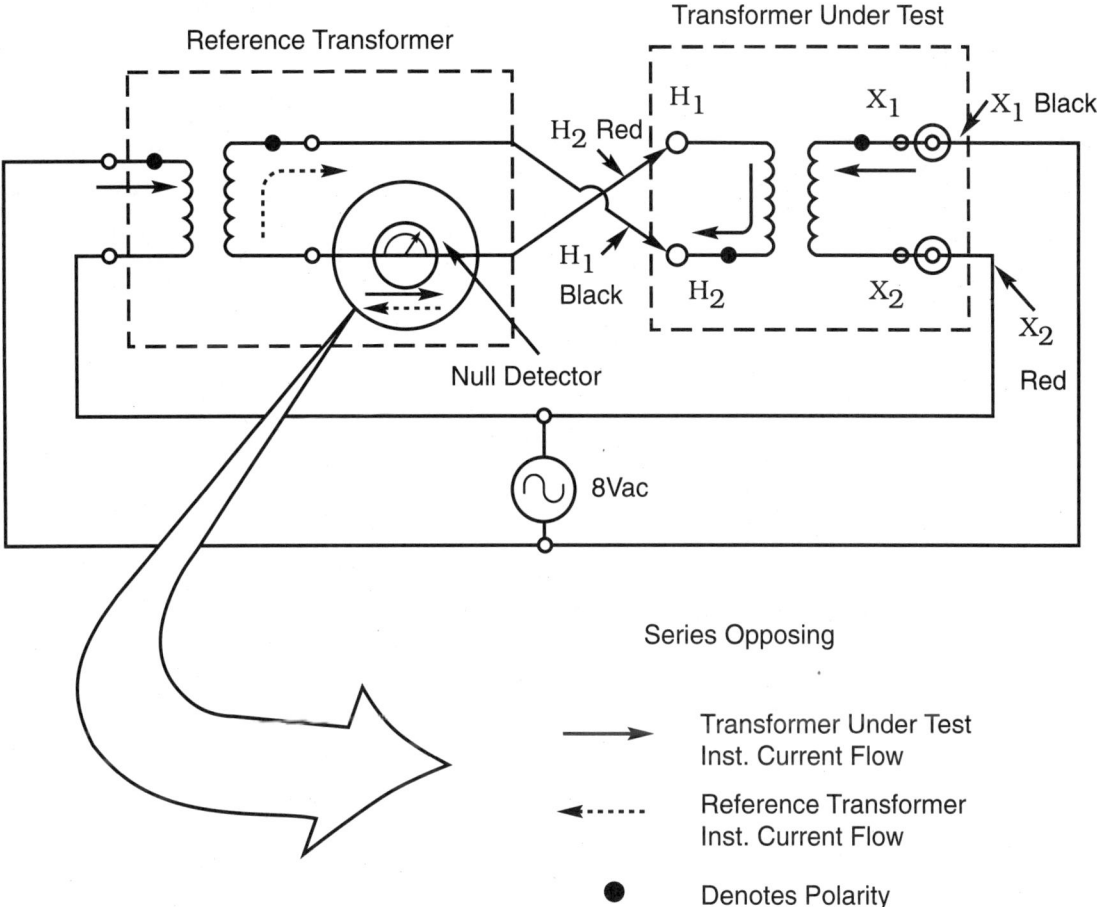

FIGURE 15–33
Correct connection for turns ratio test of an additive transformer

The first step in determining the connections for testing three-phase transformers is to analyze the polarity and schematic diagrams of the transformer. These can usually be found on the transformer nameplate. Figure 15–34 shows a three phase deluxe symbol transformer connected wye-wye. In order to test this transformer for its turns ratio, the polarity diagram is referenced (Figure 15–35).

TURNS RATIO TEST CONNECTION FOR WYE-WYE

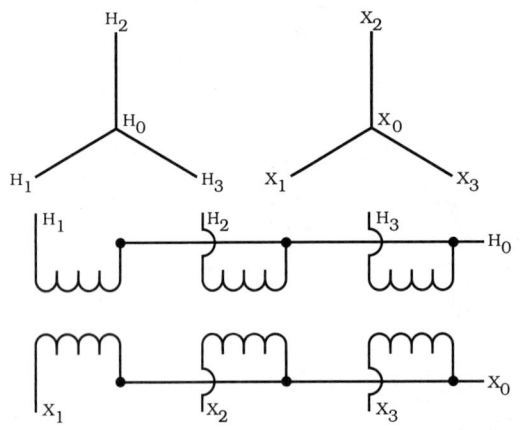

TYPICAL CONNECTION TABLE

TAP POSITION	•H_1 - H_0 •X_1 - X_0	•H_2 - H_0 •X_2 - X_0	•H_3 - H_0 •X_3 - X_0
1			
2			
3			
4			
5			

•DENOTES BLACK (POLARITY) LEAD FOR H AND X TERMINALS OF THE TEST SET.

0° ANGULAR DISPLACEMENT

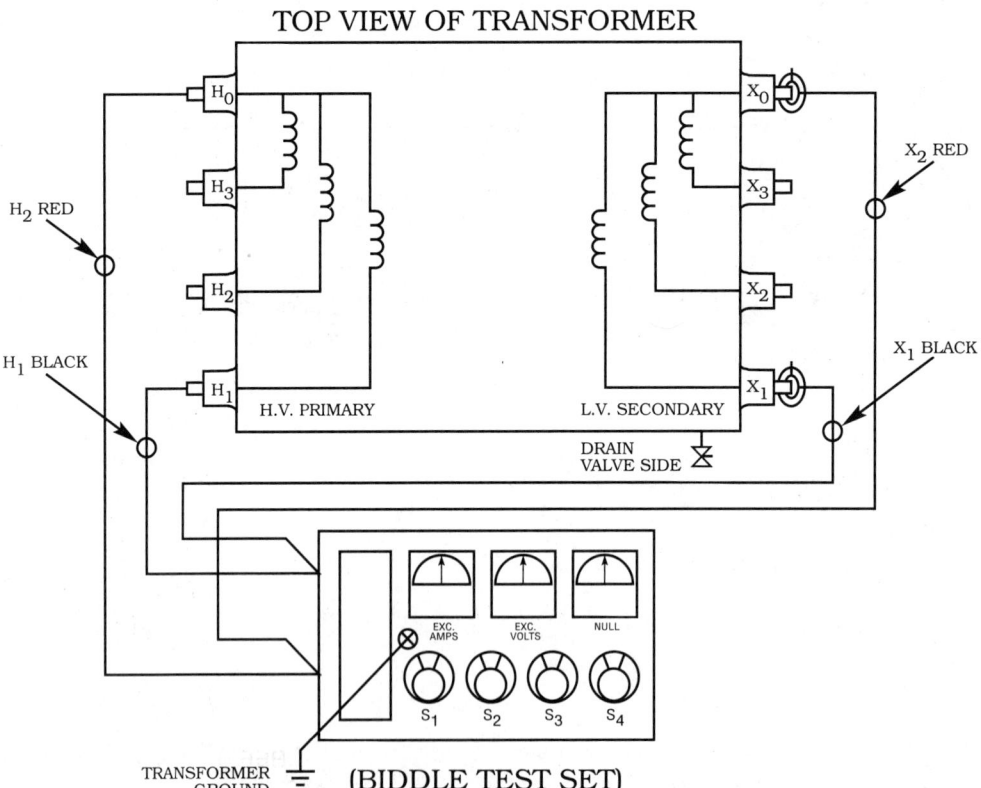

FIGURE 15–34
Connection diagram (shown connected to test H_1-H_0; X_1-X_0 ratio)

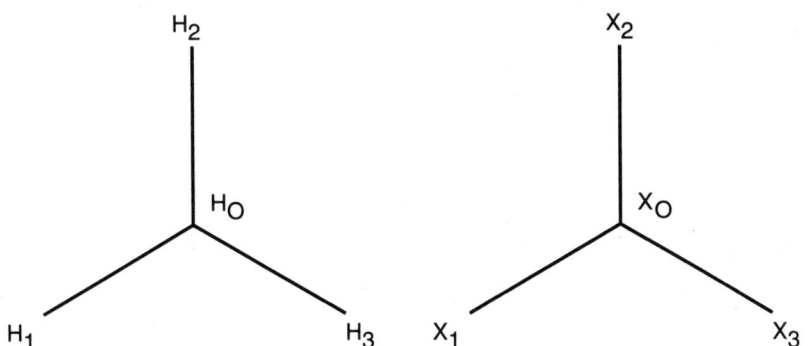

FIGURE 15–35
Polarity diagram

Testing A-Ø turns ratio: the test set must be connected so as to measure the turns ratio between windings H_1-H_0, and X_1-X_0. By examining the polarity diagram, it can be seen that the windings have the same relationship to the reference shown (Figure 15–36).

Using the reference, H_1-H_0 and X_1-X_0 have the same angular relationship. In addition, the schematic diagram should also verify this relationship (Figure 15–34c). Tracing out the windings H_1-H_0 and X_1-X_0 (indicated by arrows) will show that X_1-X_0 and H_1-H_0 are opposite each other and on the same phase of the transformer.

Connection procedure for testing A-Ø is as follows:

1. Connect black (X_1) lead of the test set to X_1 terminal of the transformer under test.
2. Connect red (X_2) lead of the test set to X_0 terminal of the transformer under test.
3. Connect black (H_1) lead of the test set to H_1 terminal of the transformer under test.
4. Connect red (H_2) lead of the test set to H_0 terminal of the transformer under test.

Quite often power transformer bushing terminals are not marked; however, most manufacturers provide a reference to indicate which terminal is H_1 and X_1. Notice that in Figure 15–34 the transformer drain valve is indi-

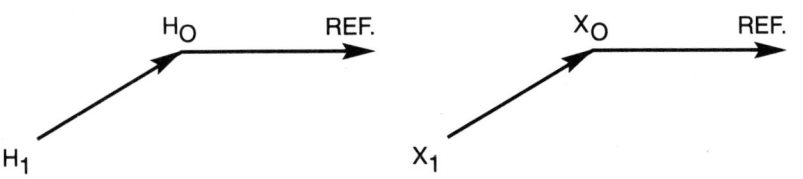

FIGURE 15–36
Schematic diagram

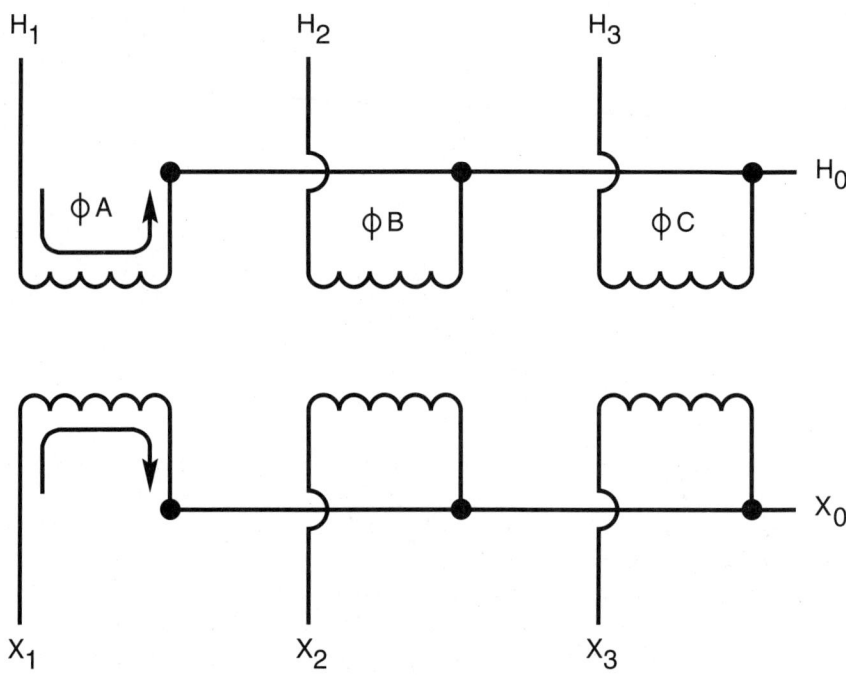

FIGURE 15–37 Schematic diagram

cated. This determines that H_1 and X_1 are the bushings located closest to that side. (This information is usually found on the transformer's nameplate.)

When testing a wye-wye-connected transformer, the nameplate voltage ratio can be determined by either line-line voltages or line-neutral voltages.

Example:

Nameplate Voltage: 12470/7200–480/277Y

$$\frac{12470 = \text{Line-Line Voltage Primary}}{480 = \text{Line-Line Voltage Secondary}} = 25.9{:}1 \qquad (10)$$

$$\frac{7200 = \text{Line-Neutral Voltage Primary}}{277 = \text{Line-Neutral Voltage Secondary}} = 25.9{:}1$$

The test set will measure the actual ratio of H_1-H_0 to X_1-X_0 = 25.9:1.

Figure 15–38 shows a three phase transformer connected delta-delta. Once again, the polarity diagrams are referenced in Figure 15–39.

Testing A-Ø turns ratio: the test set must be connected so as to measure the turns ratio between windings H_1-H_3, and X_1-X_3. By examining the polarity

TURNS RATIO TEST CONNECTION FOR DELTA-DELTA

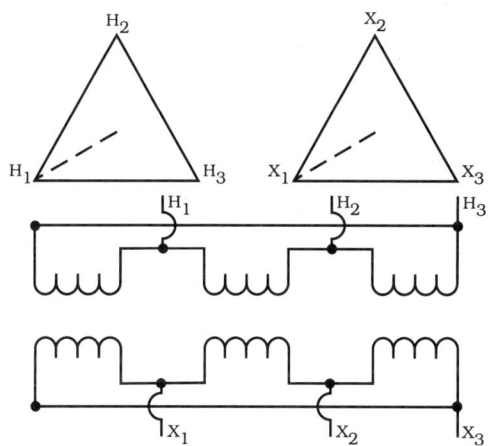

TYPICAL CONNECTION TABLE

TAP POSITION	•H_1 - H_3 •X_1 - X_3	•H_2 - H_1 •X_2 - X_1	•H_3 - H_2 •X_3 - X_2
1			
2			
3			
4			
5			

•DENOTES BLACK (POLARITY) LEAD FOR H AND X TERMINALS OF THE TEST SET.

0° ANGULAR DISPLACEMENT

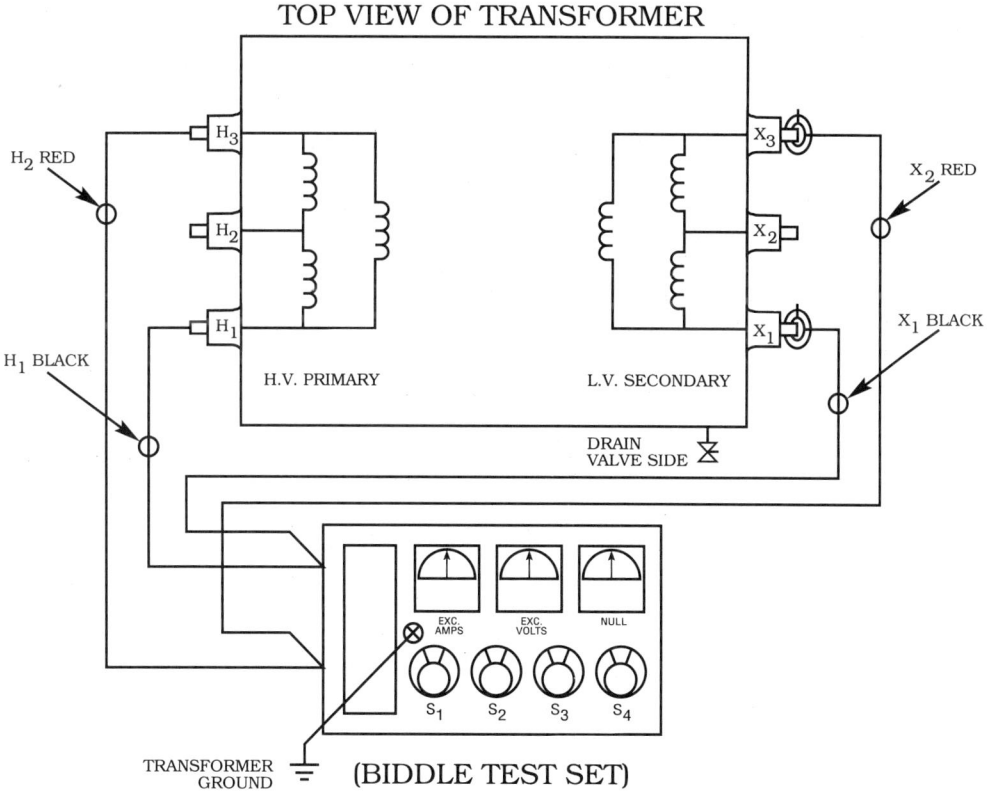

FIGURE 15–38
Connection diagram (shown connected to test H_1-H_3; X_1-X_3 ratio)

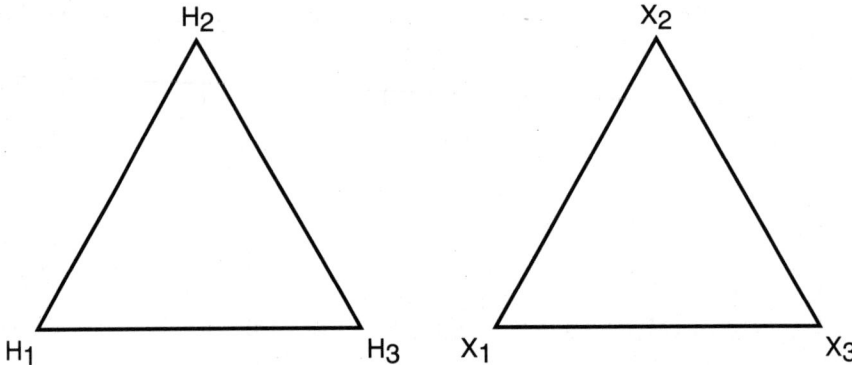

FIGURE 15–39 Polarity diagram

diagram, it can be seen that the windings have the same relationship to the reference shown (Figure 15–40). In addition, both windings have the same angular relationship to this reference. The schematic diagram will also verify this relationship (Figure 15–41). Tracing out the windings H_1-H_3 and X_1-X_3 (indicated by arrows) will show that H_1-H_3 and X_1-X_3 are opposite each other and on the same phase of the transformer.

Connection procedure for testing A-Ø is as follows:

1. Connect black (X_1) lead of the test set to X_1 terminal of the transformer under test.
2. Connect red (X_2) lead of the test set to X_3 terminal of the transformer under test.
3. Connect black (H_1) lead of the test set to H_1 terminal of the transformer under test.
4. Connect red (H_2) lead of the test set to H_3 terminal of the transformer under test.

When testing a delta-delta connected transformer, the nameplate voltage ratio is equal to the measured ratio:

$$\frac{\text{Line-Line Voltage Primary}}{\text{Line-Line Voltage Secondary}} = \text{Turns Ratio} \qquad (11)$$

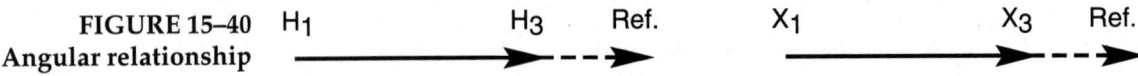

FIGURE 15–40 Angular relationship

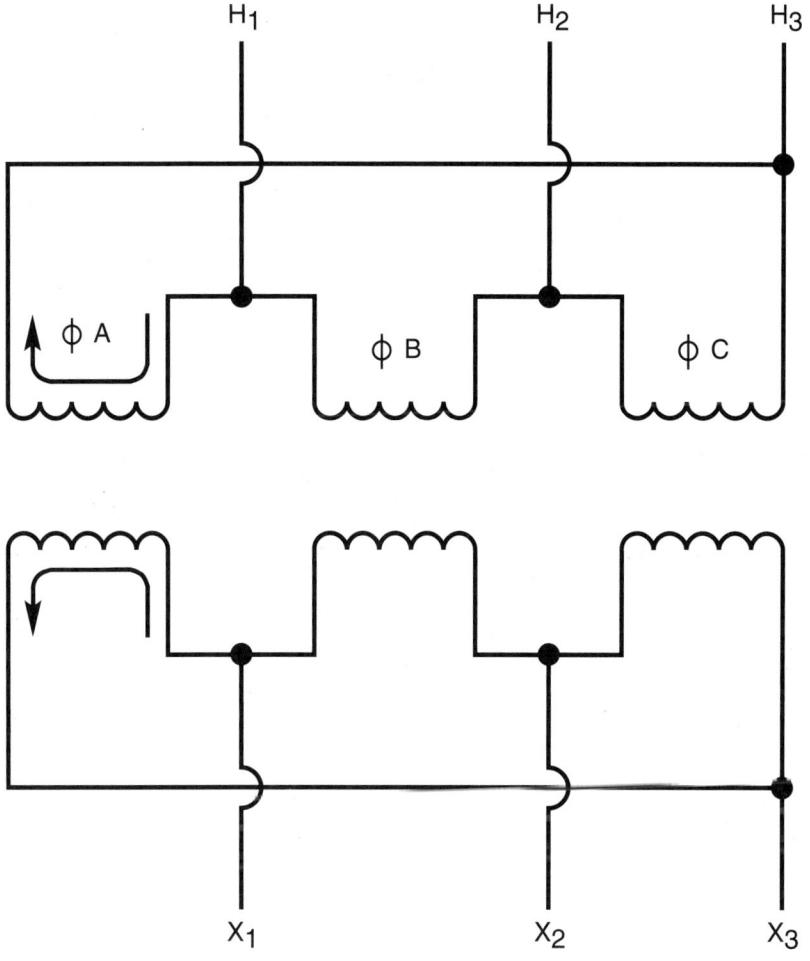

FIGURE 15–41
Schematic diagram

Example:

Nameplate Voltage: 12470/480

$$\frac{12470}{480} = 25.9:1 = \text{Turns Ratio} \qquad (12)$$

Figure 15–42 shows a 3 phase transformer connected delta-wye. Examine the polarity diagram (Figure 15–43).

TURNS RATIO TEST CONNECTION FOR DELTA-WYE

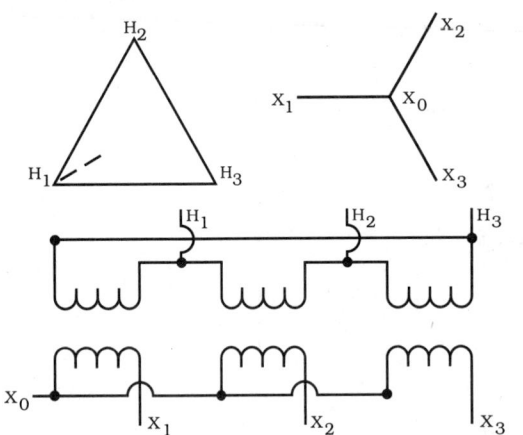

TYPICAL CONNECTION TABLE

TAP POSITION	• $H_1 - H_3$ • $X_1 - X_0$	• $H_2 - H_1$ • $X_2 - X_0$	• $H_3 - H_2$ • $X_3 - X_0$
1			
2			
3			
4			
5			

•DENOTES BLACK (POLARITY) LEAD FOR H AND X TERMINALS OF THE TEST SET.

30° ANGULAR DISPLACEMENT

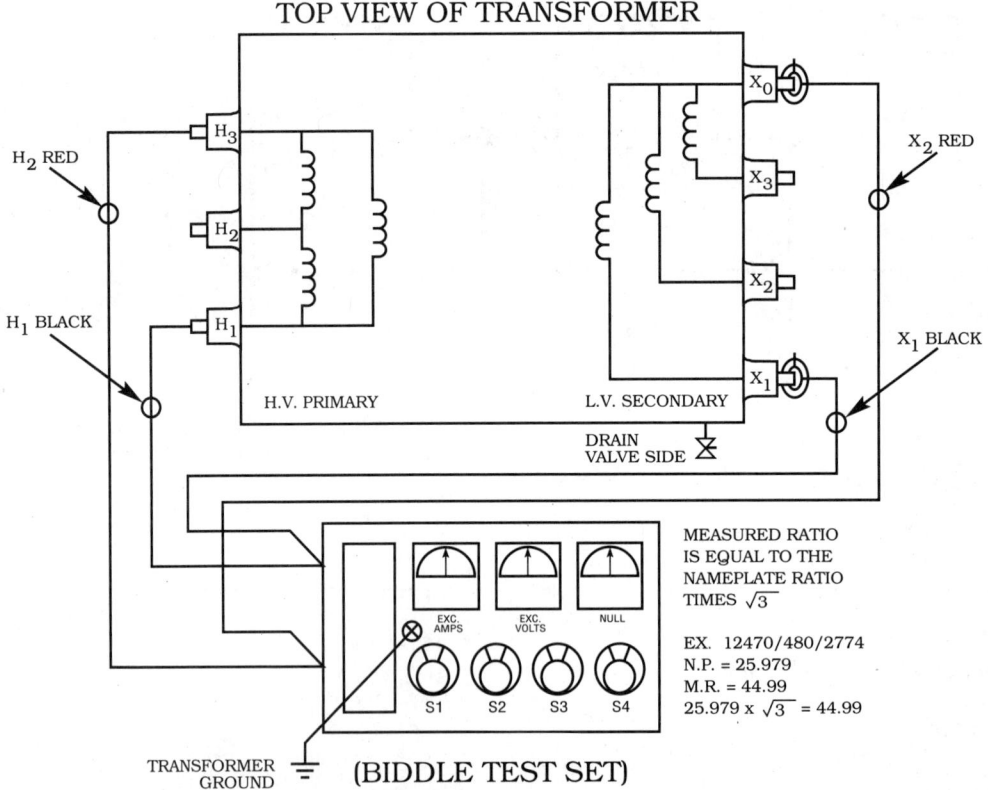

MEASURED RATIO IS EQUAL TO THE NAMEPLATE RATIO TIMES $\sqrt{3}$

EX. 12470/480/2774
N.P. = 25.979
M.R. = 44.99
25.979 x $\sqrt{3}$ = 44.99

FIGURE 15–42
Connection diagram

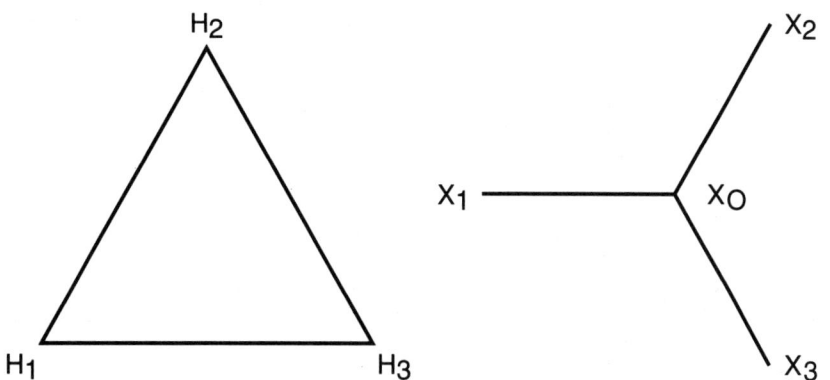

FIGURE 15–43
Polarity diagram

Testing A-Ø turns ratio: the test set must be connected so as to measure the turns ratio between windings H_1-H_3, and X_1-X_0. By examining the polarity diagram, it can be seen that the windings have the same relationship to the reference shown (Figure 15–44). In addition, both windings have the same angular relationship to this reference. Once again, the schematic diagram should verify this relationship (Figure 15–45).

Tracing out the windings H_1-H_3 and X_1-X_0 (indicated by arrows) will show that H_1-H_3 and X_1-X_0 are opposite each other and on the same phase of the transformer.

Connection procedure for testing:

1. Connect black (X_1) lead of the test set to X_1 terminal of the transformer under test.
2. Connect red (X_2) lead of the test set to X_0 terminal of the transformer under test.
3. Connect black (H_1) lead of the test set to H_1 terminal of the transformer under test.
4. Connect red (H_2) lead of the test set to H_3 terminal of the transformer under test.

When testing a delta-wye connected transformer, the nameplate voltage ratio must be multiplied by the $\sqrt{3}$ to obtain the measured ratio.

FIGURE 15–44
Angular relationship

FIGURE 15–45
Schematic diagram

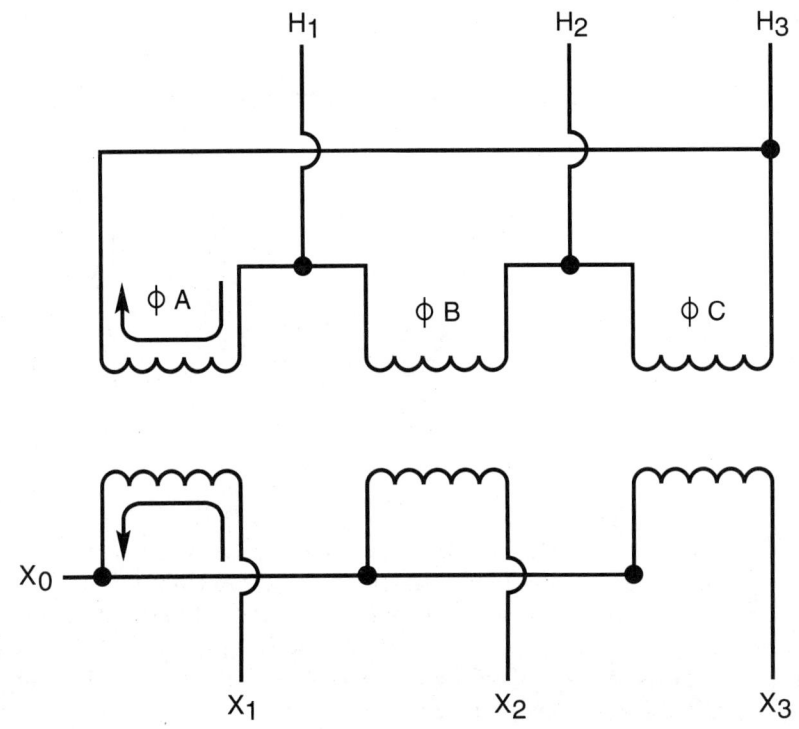

Example:

Nameplate Voltage: 12470/480/277Y

$$\frac{12470 \text{ (L–L)}}{480 \text{ (L–L)}} = 25.9 = \text{Nameplate Ratio} \tag{13}$$

Actually the test set is measuring the ratio of the delta primary (line-to-line) to the wye (line-to-neutral) secondary.

$$\frac{12470 \text{ (L–L)}}{277 \text{ (L–L)}} = 44.99:1 \tag{14}$$

In order to check the measured ratio against the nameplate voltage ratio, the nameplate ratio must be multiplied by $\sqrt{3}$.

$$25.979 \times \sqrt{3} = 44.99 \tag{15}$$

Transformer Protection

Figure 15–46 shows a 3Ø transformer connected wye-delta. Examine the polarity diagram (Figure 15–47).

Testing A-Ø turns ratio: the test set must be connected so as to measure the turns ratio between windings H_1-H_0, and X_1-X_2. By examining the polarity diagrams, it can be seen that the windings have the same relationship to the reference shown (Figure 15–48). Both windings are in phase, and they do have the same angular relationship. The schematic diagram can also be used to verify this (Figure 15–49). Tracing out the windings H_1-H_0 and X_1-X_2 (indicated by arrows) will show that these windings are opposite each other and on the same phase of the transformer.

Connection procedure for testing A-Ø is as follows:

1. Connect black (X_1) lead of the test set to X_1 terminal of the transformer under test.
2. Connect red (X_2) lead of the test set to X_2 terminal of the transformer under test.
3. Connect black (H_1) lead of the test set to H_1 terminal of the transformer under test.
4. Connect red (H_2) lead of the test set to H_0 terminal of the transformer under test.

When testing a wye-delta connected transformer, the nameplate voltage ratio must be divided by $\sqrt{3}$ to obtain the measured ratio.

Example:

Nameplate Voltage: 12470/7200/480

$$\frac{12470 \text{ (L–L)}}{480 \text{ (L–L)}} = 25.9 \text{ Nameplate Ratio} \tag{16}$$

Actually the test set is measuring the ratio of the wye primary (line-neutral) to the delta secondary (line-line).

$$\frac{7200}{480} = 15{:}1 \tag{17}$$

In order to check this measured ratio against the nameplate voltage ratio, the nameplate ratio must be divided by $\sqrt{3}$

$$\frac{25.9}{\sqrt{3}} = 15{:}1 \tag{18}$$

TURNS RATIO TEST CONNECTION FOR WYE-DELTA

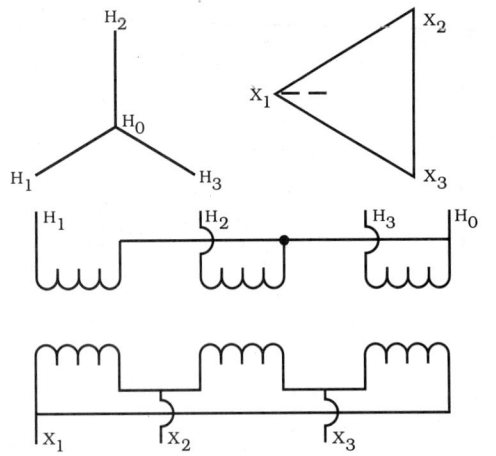

TYPICAL CONNECTION TABLE

TAP POSITION	•H_1 - H_0 •X_1 - X_2	•H_2 - H_0 •X_2 - X_3	•H_3 - H_0 •X_3 - X_1
1			
2			
3			
4			
5			

*DENOTES BLACK (POLARITY) LEAD FOR H AND X TERMINALS OF THE TEST SET.

30° ANGULAR DISPLACEMENT

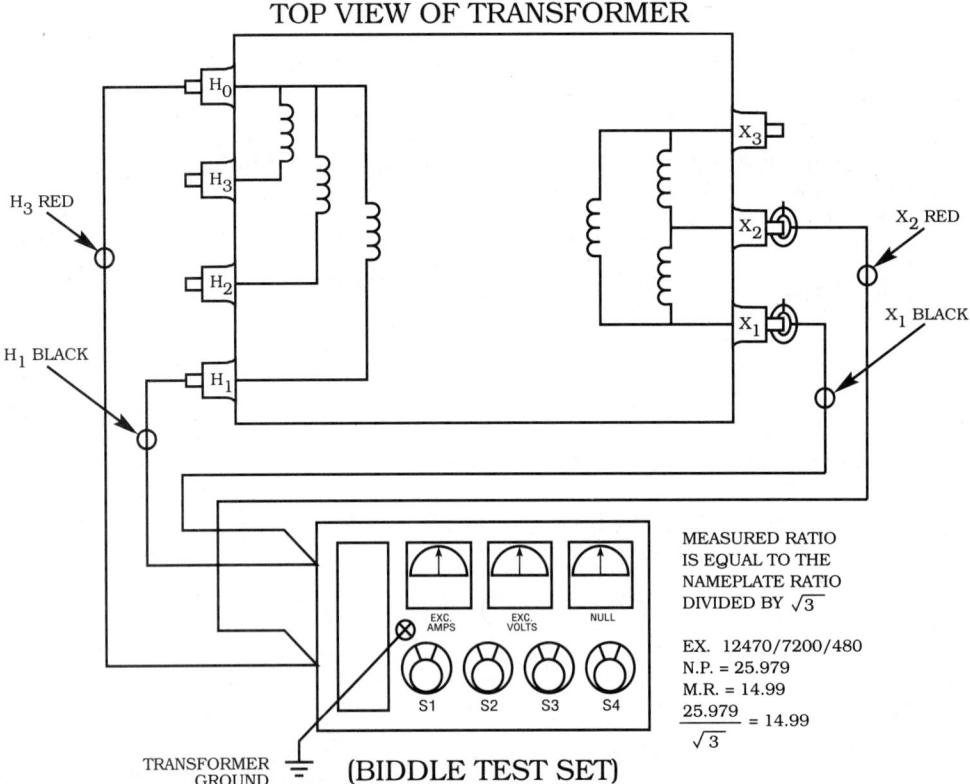

MEASURED RATIO IS EQUAL TO THE NAMEPLATE RATIO DIVIDED BY $\sqrt{3}$

EX. 12470/7200/480
N.P. = 25.979
M.R. = 14.99
$$\frac{25.979}{\sqrt{3}} = 14.99$$

FIGURE 15–46
Connection diagram (shown connected to test H_1-H_0; X_1-X_0 ratio)

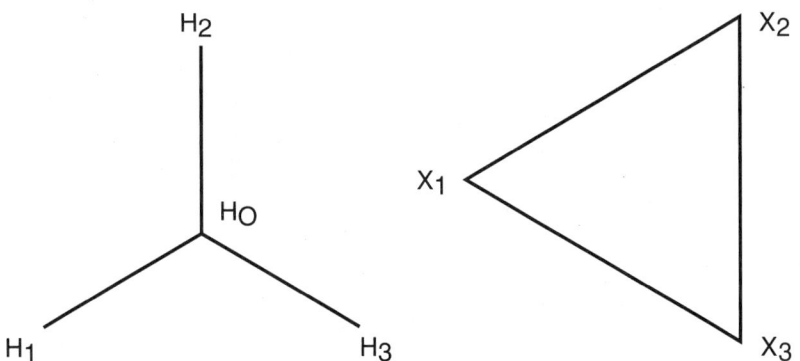

FIGURE 15–47
Polarity diagram

After each connection is made, a simple test can be performed to verify the connections on any transformer type or connection configuration as follows: With the decade switches at zero and the transformer under test connected correctly, a small turn of the test set's generator will cause the null meter to deflect to the left. If the null meter indicator deflects to the right, check for improper polarity connection.

Once again, there are several different manufacturers of turns ratio test sets, and numerous ways of connecting transformers. The Biddle single phase test set has been illustrated in all of the above examples; however, regardless of the test set used, always consult the operations manual for that test set, and follow the instructions contained therein. It should also be pointed out that on transformers equipped with adjustable tap positions, a turns ratio measurement should be performed on each winding, at each available tap position.

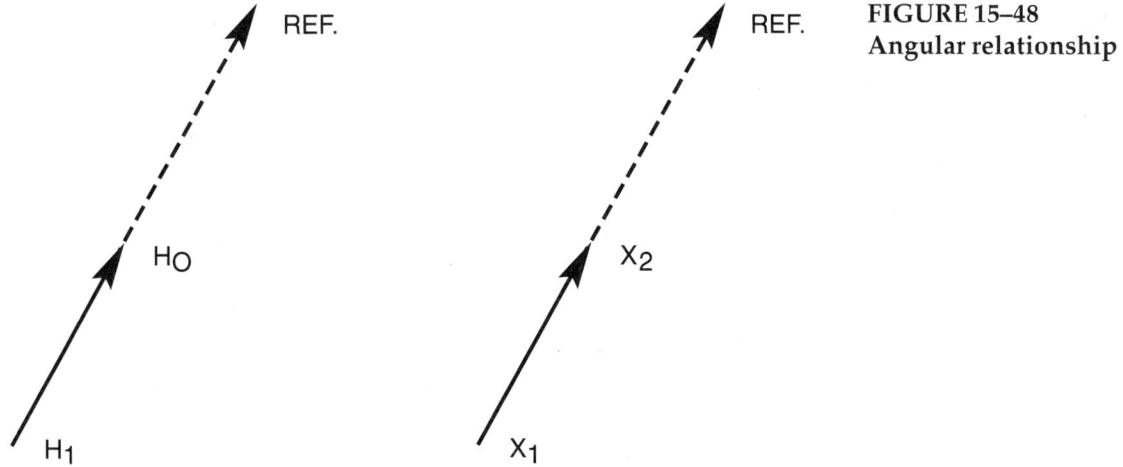

FIGURE 15–48
Angular relationship

FIGURE 15–49 Schematic diagram

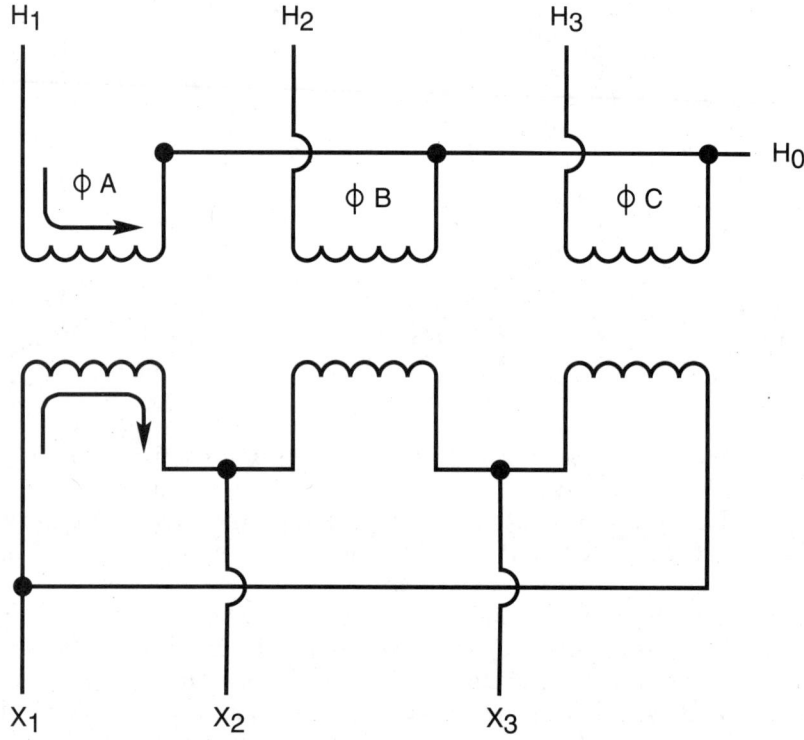

Power Factor Testing Fundamentals

Power factor, in power systems, refers to the proportion of power or work equivalent used, to the total product of the applied voltage and current.

$$\% \text{ PF} = \frac{\text{Watts}}{\text{Volts} \times \text{Amps (VA)}} \times 100 \tag{19}$$

Power factor is best described by using trigonometric functions, in that when a sinusoidal voltage is impressed on insulating materials (dielectric), the current flow through the insulation leads the applied voltage by an angle of nearly 90°. The angle at which this current leads the voltage is referred to as angle theta (θ). The cosine of this angle is the power factor, and when multiplied by 100, it is the percent power factor (Figure 15–50).

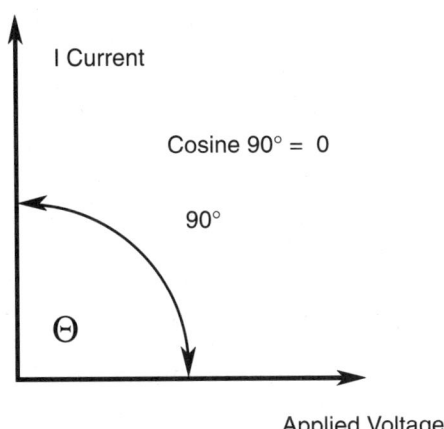

FIGURE 15–50
Phasor diagram

The power factor of insulation is expressed in terms of:

$$PF = \frac{\text{Watts absorbed in insulation}}{\text{Applied voltage} \times \text{charging current}} \qquad (20)$$

Thus, if the insulation is perfect, and has no losses (watts) when a voltage is applied, then the ratio above is zero. Perfect insulation has a power factor of 0 or 0%. When the insulation begins to deteriorate, the internal losses (watts) increase for the same identical applied voltage and current. This ratio then starts to increase from 0% to 5%, 10%, 20%, or greater.

There is no perfect insulation or insulator; thus all insulation materials have minor leakages. The greater the leakage path, the greater the power loss through that insulating system.

Acceptable power factor values are derived by comparing the results of the transformer (or other device) under test to previous test results for similar type devices. Acceptable maximum power factors of up to 2% have been found on older oil-filled transformers that have been in service; however, new or recently dried transformers should have a power factor of about .5%. On HV and EHV transformers, the power factors of properly dried and/or new transformers should be about .3% and .2% respectively. Askarel-filled power transformers frequently have power factor readings of 3% to 5% at 20°C.

For a specific transformer, the power factor values recommended by the manufacturer should be followed, rather than general values. *All power factor test results must be corrected to base temperature of 20°C for the purpose of evaluating the results.* For testing purposes, the temperature of the transformer is to be taken from the winding temperature indicator. If no winding temperature indicator is installed, then the oil temperature can be used.

The power factor values are affected by temperature, and it is recommended that transformers not be tested at or near freezing temperatures. A crack in an insulator is easily detected if it contains a conducting film of water. However, when the water freezes, it becomes nonconducting and the defect may not be revealed by the measurement.

It is for this reason that tests for the presence of moisture in solids intended to be dry should not be made at freezing temperatures. Moisture in oil, or in oil-impregnated solids, however, has been found to be detectable in power factor measurements at temperatures far below freezing, with no discontinuity in the measurements at the freezing point. If a transformer is tested when it is at a relatively high temperature, the power factor values may be high even after correcting to 20°C. If this should happen, allow the transformer to cool down to a temperature near 20°C and retest.

Most insulation measurements have to be interpreted based on the temperature of the specimen. The dielectric losses of most insulating materials increase with a rise in temperature.

It is important, therefore, to determine the power factor temperature characteristics of the insulation under test, at least in a typical unit of each design of apparatus.

Typical correction factors (from Doble Engineering) for mineral oil-filled transformers are listed in Table 15–9.

The interpretation of power factor measurements is usually based on past experience of the organization or individual conducting the tests, recommendations by the manufacturer of the equipment being tested and, most important, comparisons made as follows:

1. Between measurements on the same component after successive intervals of time.
2. Between measurements on duplicate components, or a similar part of one component, tested under the same conditions around the same time.
3. Between measurements made at different test voltages on one part of a unit; an increase in slope of a power factor voltage curve at a given voltage is an indication of ionization commencing at that voltage.

An increase of power factor above a typical value may indicate conditions such as those already discussed, any of which may be general or localized in character. If the power factor varies significantly with voltage, down to some voltage below which power factor is substantially constant, then ionization is indicated. If the "extinction voltage" is below the operating level, then ionization may progress in operation with consequent deterioration. Some increase of capacitance (increase in charging current) may also be

Table 15-9 Tables of Multipliers for Use in Converting Power-Transformers Power Factors at Test Temperatures to Power Factors at 20°C

(Courtesy Doble Engineering)

TEST TEMPERATURE		(1) DOBLE OIL AND OIL-FILLED POWER TRANSFORMERS (FREE-BREATHING AND OLDER CONSERVATOR TYPES)	(2) OIL-FILLED POWER TRANSFORMERS (SEALED, GAS-BLANKETED AND MODERN CONSERVATOR TYPES) THROUGH 161 kV (750 kV BIL)	(3) TRANSFORMER COMMITTEE OIL-FILLED POWER TRANSFORMERS (SEALED, GAS-BLANKETED AND MODERN CONSERVATOR TYPES) 230 kV, UP (ABOVE 750 kV BIL)
°C	°F			
0	32.0	1.56	1.57	.95
1	33.8	1.54	1.54	.96
2	35.6	1.52	1.50	.96
3	37.4	1.50	1.47	.97
4	39.2	1.48	1.44	.98
5	41.0	1.46	1.41	.98
6	42.8	1.45	1.37	.98
7	44.6	1.44	1.34	.99
8	46.4	1.43	1.31	.99
9	48.2	1.41	1.28	.99
10	50.0	1.38	1.25	.99
11	51.8	1.35	1.22	1.00
12	53.6	1.31	1.19	1.00
13	55.4	1.27	1.16	1.00
14	57.2	1.24	1.14	1.01
15	59.0	1.20	1.11	1.01
16	60.8	1.16	1.09	1.01
17	62.6	1.12	1.07	1.01
18	64.4	1.08	1.05	1.00
19	66.2	1.04	1.02	1.00
20	68.0	1.00	1.00	1.00
21	69.8	.96	.98	1.00
22	71.6	.91	.96	.99
23	73.4	.87	.94	.99
24	75.2	.83	.92	.98
25	77.0	.79	.90	.98
26	78.8	.76	.88	.97
27	80.6	.73	.86	.97
28	82.4	.70	.84	.96
29	84.2	.67	.82	.95
30	86.0	.63	.80	.95
31	87.8	.60	.78	.94
32	89.6	.58	.76	.94
33	91.4	.56	.75	.93
34	93.2	.53	.73	.93
35	95.0	.51	.71	.92
36	96.8	.49	.70	.91
37	98.6	.47	.69	.91
38	100.4	.45	.67	.90
39	102.2	.44	.66	.89
40	104.0	.42	.65	.89
41	105.8	.40	.63	.88
42	107.6	.38	.62	.87
43	109.4	.37	.60	.86
44	111.2	.36	.59	.86
45	113.0	.34	.57	.85
46	114.8	.33	.56	.84
47	116.6	.31	.55	.83
48	118.4	.30	.54	.83
49	120.2	.29	.52	.82
50	122.0	.28	.51	.81
52	125.6	.26	.49	.79
54	129.2	.23	.47	.77
56	132.8	.21	.45	.75
58	136.4	.19	.43	.72
60	140.0	.17	.41	.70
62	143.6	.16	.40	.67
64	147.2	.15	.38	.65
66	150.8	.14	.36	.62
68	154.4	.13	.35	.59
70	158.0	.12	.33	.55
72	161.6	.12	.32	
74	165.2	.11	.31	
76	168.8	.10	.30	
78	172.4	.09	.28	
80	176.0	.09	.27	

observed, above the extinction voltage, because of the short circuiting of numerous voids by the ionization process.

Power factor temperature characteristics, as well as power factor measurements at a given temperature, may change with deterioration or damage of insulation, suggesting that any such change in temperature characteristics may be helpful in assessing deteriorated conditions.

Test Sets Commercially-manufactured power factor test sets are available in voltages ranging up to 10,000 V. The selection of the test voltage to be used depends on many variables and must be made by individual companies for their specific requirements. The following are a few general points that should be considered:

1. The power factor test is not dependent on the use of over-potential as is the hipot test. The test voltage never exceeds the rating of the insulation.
2. Generally, the test voltage should be as high as efficiently possible and should meet the requirements of surface contact and unit stress on the insulation.
3. In energized substations above 13.8 kV, higher voltages (10 kV) may be helpful in swamping the effects of ambient 60 Hz noise.
4. At higher test voltages, the difference between corona losses and insulation losses becomes difficult to determine; therefore, the test voltage should not be any higher than necessary.

The Doble MEU 2500 test set is used for demonstration purposes in the following examples. Figure 15–51 shows a picture of the control panel for MEU 2500 and the three configurations that any power factor set utilizes. The final selection of which test set is used for a particular apparatus must be determined from either test specifications, manufacturer's recommendations, and/or company engineering standards.

Winding Power Factor Tests

When performing a power factor test on a transformer, the voltage rating of the winding under test must be considered, and the test voltage level selected accordingly. If a neutral is involved, the voltage rating of line-to-neutral must also be considered in the selection of voltage test level. In addition, the transformer to be tested must be isolated from the live bus, all external leads disconnected from the bushing terminals, all neutrals of each winding must be isolated from ground, and each winding short-circuited at its bushing terminals. During these tests a voltage will be induced in all windings. Therefore, treat all terminals as being energized.

Transformer Preparation for Test

1. Disconnect all external leads from the transformer bushing terminals (including neutral bushings). The bushings can be left connected, although the attached bus or cable will affect the readings.
2. Short circuit each winding (high, low, and tertiary) at its bushing terminals with bonding wires and connect a ground from a bushing in each winding to the station ground.
3. Clean all transformer bushings.
4. Record transformer NLTC position on the excitation current test form.
5. Move LTC to the lowest tap position and record tap number on the excitation current test form.
6. Connect power factor/dissipation factor test leads to the windings specified in the test procedure below.
7. Disconnect grounds from the transformer bushing terminals (leave windings shorted; see Item 2).

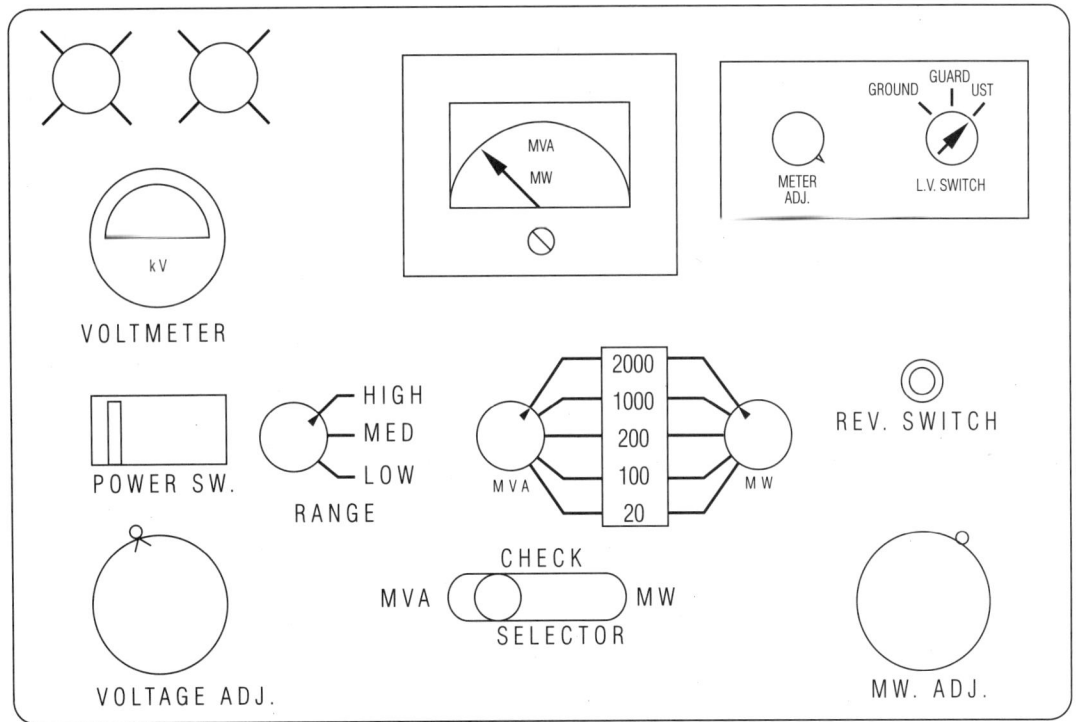

FIGURE 15–51
2.5 kV Doble test set (Copyright Doble Engineering Company)

Measurements

1. For a two-winding transformer, a measurement should be made from the high-side windings to ground and the low-side windings, using the grounded-specimen test grounded (GST-L-GND) configuration. This connection configuration will allow for testing the power factor of the winding under test with respect to ground, along with the power factor of the winding under test with respect to all other windings ($C_H + C_{HL}$), Figure 15–52.

   Connections:
   a. Short all high-voltage transformer bushing terminals together.
   b. Connect the H.V. lead of the test set to one high-voltage bushing terminal of the transformer.
   c. Connect the L.V. lead of the test set to one low-voltage bushing terminal of the transformer.
   d. Short all low-voltage transformer bushing terminals together.
   e. Connect the ground lead of the test set to the transformer's tank ground. Attach a safety ground.

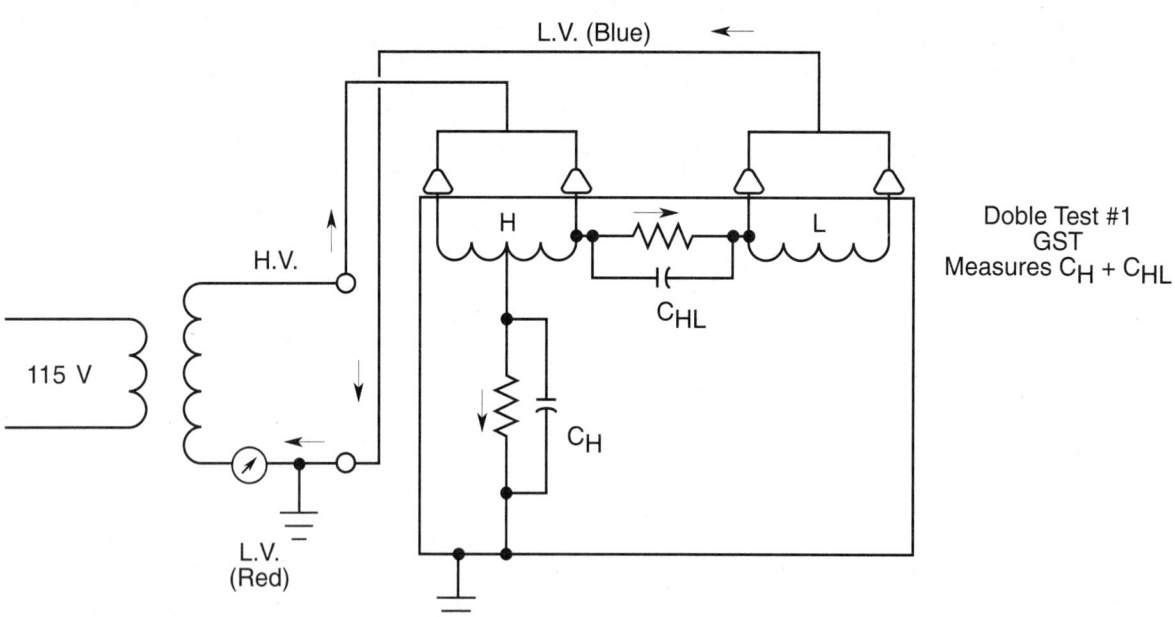

FIGURE 15–52
Three-phase, two-winding transformer (ground mode)

f. Place the low-voltage selector switch of the test set to the grounded position. In this position all leakage current between the H.V. winding to ground and the H.V. winding to L.V. winding is measured ($C_H + C_{HL}$).

2. Measurements should be made between the high-side windings and ground, with the low-side windings guarded. This connection configuration is known as grounded-specimen test, guarded (GST-L-GUARD) or, more simply, the guarded-specimen test. Using the GST-L-GUARD test configuration, the power factor of the winding under test with respect to ground can be measured (C_H) (Figure 15–53).

Connections (two-winding transformer):
a. Short all high-voltage transformer bushing terminals together.
b. Connect H.V. lead of the test set to a high-voltage bushing terminal of the transformer.
c. Connect L.V. lead of the test set to a low-voltage bushing terminal of the transformer.
d. Short all low-voltage transformer bushing terminals together.
e. Connect the ground lead of the test set to the transformer's tank ground. Attach a safety ground.

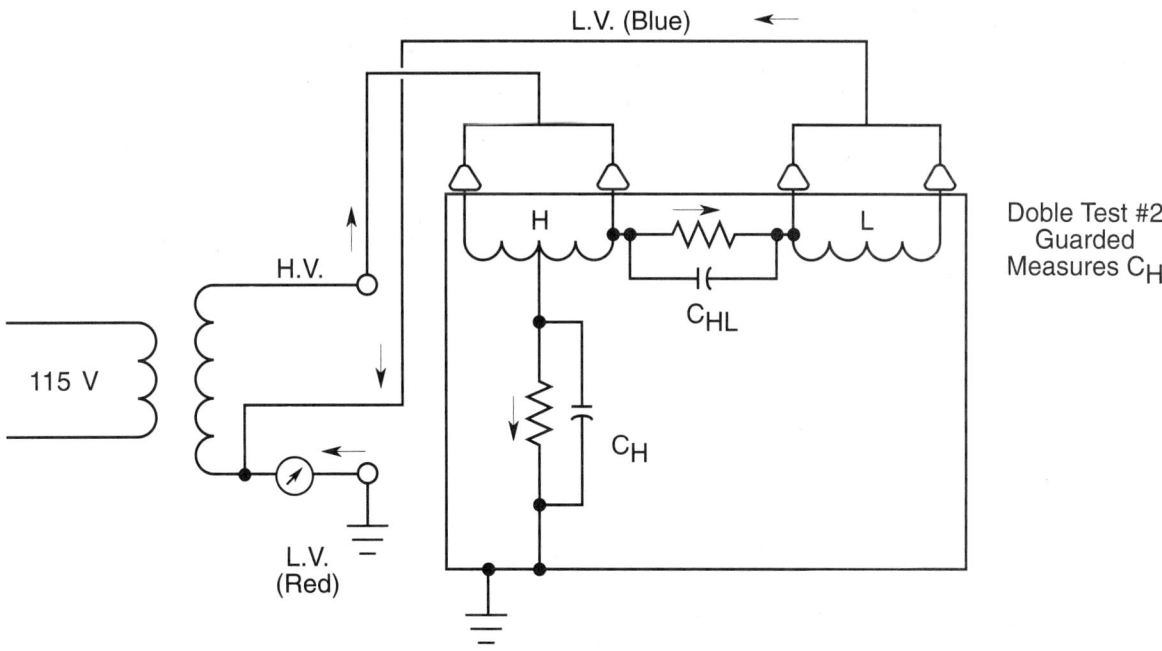

FIGURE 15–53
Three-phase, two-winding transformer (guard mode)

f. Place the low-voltage selector switch of the test set to the guard position. In this position, any leakage current between the H.V. winding and the L.V. winding is not measured (guarded); therefore, only the high-voltage to ground (C_H) is measured.

3. A test measurement should be made from the low-side windings to ground and the high-side windings using the grounded-specimen test grounded (GST-L-GND) configuration. This connection configuration will allow for testing the power factor of the winding under test with respect to ground, along with the power factor of the winding under test with respect to all other windings ($C_L + C_{HL}$), Figure 15–54.

Connections:
a. Short all low-voltage transformer bushing terminals together.
b. Connect H.V. lead of the test set to one low-voltage bushing terminal of the transformer.
c. Connect L.V. lead of the test set to one high-voltage bushing terminal of the transformer.
d. Short all high-voltage transformer bushing terminals together.
e. Connect the ground lead of the test set to the transformer's tank ground. Attach a safety ground.

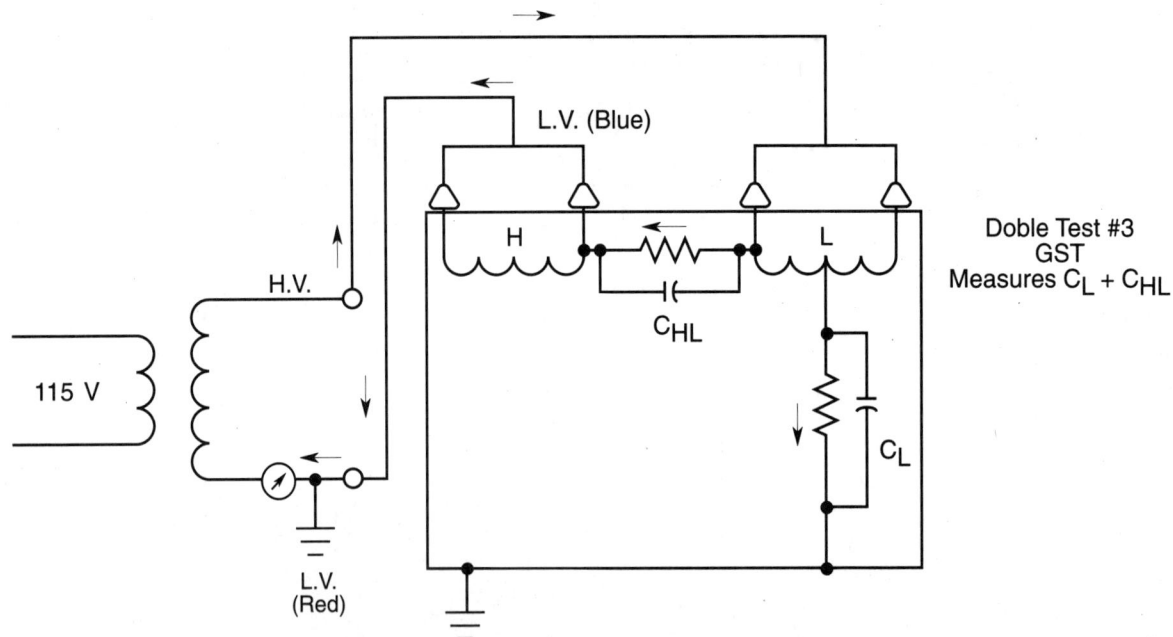

FIGURE 15–54
Three-phase, two-winding transformer (ground)

f. Place the low-voltage selector switch of the test set to the grounded position. In this position, all leakage current between the L.V. winding to ground and the L.V. winding to H.V. is measured ($C_L + C_{HL}$).
4. One final measurement should be made between all the low-side windings and the high-side windings (tank grounded). The GST-L-GUARD configuration can be used for this test (Figure 15–55).
Connections:
   a. Short all low-voltage transformer bushing terminals together.
   b. Connect H.V. lead of the test set to one low-voltage bushing terminal of the transformer.
   c. Connect L.V. lead of the test set to one high-voltage bushing terminal of the transformer.
   d. Short all high-voltage transformer bushing terminals together.
   e. Connect the ground lead of the test set to the transformer's tank ground. Attach a safety ground.
   f. Place the low-voltage selector switch to guard position. Leakage to ground from low-side windings is measured (C_L).

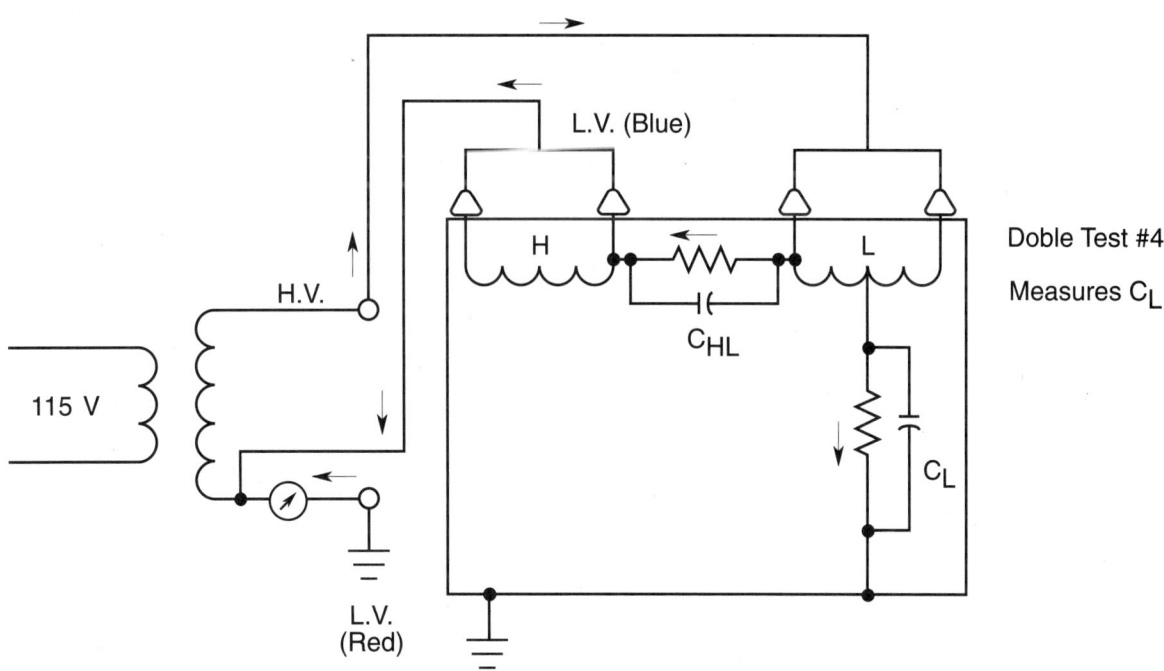

FIGURE 15–55
Three-phase, two-winding transformer (guard)

- *Three-Winding Transformer Power Factor Testing*

On three-winding transformers a measurement should also be made between each winding and ground, with the remaining winding guarded and the second remaining winding grounded, using the GST-L-GUARD configuration. This configuration will allow for testing the power factor of the winding under test with respect to the grounded winding and to ground, while isolating (guarding) the interwinding ($C_H + C_{HL}$).

Connections (H.V. to Gnd. to L.V., isolate T windings):

a. Short all high-voltage transformer bushing terminals together.
b. Connect H.V. lead of the test set to one high-voltage bushing terminal of the transformer.
c. Connect L.V. lead of the test set to one interwinding bushing terminal of the transformer.
d. Short all interwinding transformer bushing terminals together.
e. Connect the ground lead of the test set to one low-voltage bushing terminal of the transformer.
f. Short all low-voltage transformer bushing terminals together.
g. Tie the low-voltage bushing terminals to the transformer's tank ground.
h. Place the low-voltage selector switch to position guard. This will isolate the interwindings; thus, only the $C_H + C_{HL}$ will be measured.

Table 15–10 lists the test connections for two-winding transformers. Three-winding transformer and autotransformer test connections are shown in Tables 15–11 and 15–12, respectively.

Evaluation Since the PF of good insulation is extremely low, the mathematical formulas are at times difficult, and are not easy to interpret. In power factor test sets that measure the capacitance of the insulation system along with the power factor (by using the measured capacitance readings), the correlation between the different readings is easier. In fact, the PF from one reading to another should not vary significantly on good insulation systems. However, the capacitance will vary, because capacitance is dependent upon the amount of surface area and the dielectric material.

NOTE: Remember, all power factor test results must be corrected to 20°C for evaluation purposes. Table 15–9 shows correction factors for oil-filled transformers.

A test report from Doble Engineering is illustrated in Figure 15–56, p. 458. This form is typical of what is required for field use. Note that test No. 2 is subtracted from test No. 1. The results are then calculated to provide the loss C_{HL}. This same loss is measured when the UST test is performed on a transformer.

Table 15-10 Two-Winding Transformer Tests Doble M2H Test Set

Test No.	Test Mode	Winding Energized	Winding Grounded	Winding Guarded	Measure
1	GST	High	Low	None	High to low and ground
2 ****	GST	High	None	Low	High to ground
3	GST	Low	High	None	Low to high and ground
4	GST	Low	None	High	Low to ground

DISSIPATION FACTOR TEST SET

Test No.	Test Mode	Sw. Pos.	Winding Energized (Black Lead)	Low Voltage Red Test Lead on Winding	Measure
1	UST	3	High	Low	High to low
2	GST	4	High	Low (ground)	High to low & ground
3 ***	GST	5	High	Low (guard)	High to ground
4	UST	3	Low	High	Low to high
5	GST	4	Low	High (ground)	Low to high & to ground
6	GST	5	Low	High (guard)	Low to ground
7	GST	5	High & low	Off	High & low to ground

*** Reapply grounds which were removed in Step 7 of "Transformer Preparation for Test", relocate test leads and remove grounds.

Perform bushing tests as outlined in the procedure for "Bushing Tests". Perform "Transformer Exciting-Current Test" as outlined in the test procedure.

Table 15-11 Three-Winding Transformers Tests Doble Test Set

Test No.	Test Mode	Winding Energized	Winding Grounded	Winding Guarded	Measure
1	GST	High	Low	Tertiary	High to low and ground
2	GST	High	None	Low and tertiary	High to ground
***					
3	GST	Low	Tertiary	High	Low to tertiary and ground
4	GST	Low	None	High and tertiary	Low to ground
***					
5	GST	Tertiary	High	Low	Tertiary to high and ground
6	GST	Tertiary	None	High and low	Tertiary to ground
7	GST	All None	None	None	All windings to ground

Test No.	Test Mode	Sw. Pos.	Winding Energized (Black Lead)	Low Voltage Red Test Lead on Winding	Low Voltage Blue Test Lead on Winding	Measure
1	UST	3	High	Low	Tertiary (ground)	High to low
2	GST	6	High	Low (ground)	Tertiary (guard)	High to low and ground
3	GST	5	High	Low (guard)	Tertiary (guard)	High to ground
4	UST	3	Low	Tertiary	High (ground)	Low to tertiary
5	GST	6	Low	Tertiary (ground)	High (guard)	Low to tertiary and ground
6	GST	5	Low	Tertiary (guard)	High (guard)	Low to ground
***						
7	UST	3	Tertiary	High	Low (ground)	Tertiary to high
8	GST	6	Tertiary	High (ground)	Low (guard)	Tertiary to high and ground
9	GST	5	Tertiary	High (guard)	Low (guard)	Tertiary to ground
10	GST	5	All	Off	Off	All windings to ground

*** Reapply grounds which were removed in Step 7 of "Transformer Preparation for Test," relocate test leads and remove grounds.
Perform bushing tests as outlined in the procedure for "Bushing Tests."
Perform "Transformer Exciting-Current Test" as outlined in the test procedure.

Transformer Protection

Table 15-12   Two-Winding Transformer Tests Doble M2H Test Set

Test No.	Test Mode	Winding Energized	Winding Grounded	Winding Guarded	Measure
1	GST	High & low	Tertiary	None	High & low to tertiary & ground
2	GST	High & low	None	Tertiary	High & low to ground
****					
3	GST	Tertiary	High & low	None	Tertiary high & low & to ground
4	GST	Tertiary	None	High & low	Tertiary to ground

DISSIPATION FACTOR TEST SET

Test No.	Test Mode	Sw. Pos.	Winding Energized (Black Lead)	Low Voltage Red Test Lead on Winding	Measure
1	UST	3	High & low	Low	High and low to tertiary
2	GST	4	High & low	Low (ground)	High and low to tertiary and to ground
3	GST	5	High & low	Low (guard)	High and low to ground
***					
4	UST	3	Tertiary	High	Tertiary to high and low
5	GST	4	Tertiary	High (ground)	Tertiary to high and low and to ground
6	GST	5	Tertiary	High (guard)	Tertiary to ground
7	GST	5	All	Off	All windings to ground

***   Reapply grounds which were removed in Step 7 of "Transformer Preparation for Test," relocate test leads and remove grounds.
Perform bushing tests as outlined in the procedure for "Bushing Tests."

OIL ☐
ASKAREL ☐
AIR ☐
GAS ☐

TWO-WINDING TRANSFORMERS

FORM MM - ZW170

COMPANY	DIVISION	DATE	
LOCATION OF TESTS		AIR TEMP.	TOP OIL TEMP.
TRANSFORMER		WEATHER	% HUMIDITY
MFR. SERIAL NO.	AGE	TYPE/CLASS	KVA
FREE BREATHING ☐ SEALED ☐	GAS BLANKETED ☐	CONSERVATOR ☐	GALLONS OF OIL

	BUSHINGS	MFR.	TYPE	CLASS	DWG. NO.	CAT. NO.	KV	YEAR
HIGH SIDE KV Y☐ △☐								
LOW SIDE KV Y☐ △☐								
NEUTRAL						DATE LAST TEST		

COPIES TO — LAST SHEET NO.

OVER-ALL TESTS

	TEST CONNECTIONS				EQUIVALENT 10 KV READINGS						% POWER FACTOR		KEY TO INSULATION RATING: G - GOOD; D - DETERIORATED; I - INVESTIGATE; B - BAD (REMOVE OR RECONDITION)	INSULATION RATING
					MILLIAMPERES			WATTS						
TEST	WINDINGS ENERGIZED	WINDINGS GROUNDED	WINDINGS GUARDED	TEST KV	METER READING	MULTIPLIER	MULTI-AMPERES	METER READING	MULTIPLIER	WATTS	MEASURED	COR. 20°		
1	HIGH	LOW									— —	— —		
2	HIGH		LOW										C_H	
3	LOW	HIGH									— —	— —		
4	LOW		HIGH										C_L	
CALCULATED RESULTS					— —	— —		— —	— —				C_{HL} (TEST 1 MINUS TEST 2)	
					— —	— —		— —	— —		— —	— —	(TEST 3 MINUS TEST 4)*	— —

* CURRENT AND WATTS SHOULD COMPARE WITH THOSE OF C_{HL}

BUSHING TESTS

					EQUIVALENT 10 KV READINGS						% POWER FACTOR		COLLAR TESTS (WATTS/CURRENT)		INSULATION RATING
					MILLIAMPERES			WATTS							
LINE NO.	BUSH NO.	PHASE	BUSHING SERIAL NO.	TEST KV	METER READING	MULTIPLIER	MULTI-AMPERES	METER READING	MULTIPLIER	WATTS	MEASURED	COR. 20°	TOP		
HIGH SIDE 1															
2															
3															
4	N														
LOW SIDE 5															
6															
7															
8	N														
9															
10															
11															
12															
13															
14															
15															
16															
17															
18															
19	OIL SAMPLE												OIL TEMP. °C		

N - NEUTRAL

DIAGRAM

REMARKS

TEST SET NO. ______ TEST BY ______ CHECKED BY ______ SHEET NO. ______

FIGURE 15–56
Power factor test report (Copyright Doble Engineering Company)

Power Factor Testing Other Transformer Equipment

Other equipment can also be power factor tested, including bushings and insulating liquid. You should refer to manufacturer's instructions for connections and evaluation methods.

Core Excitation Current Testing

Core excitation is often referred to by other names, such as no-load loss, iron loss, excitation current, core loss and others. Regardless of the name applied, core excitation is the amount of energy required to excite the transformer with no load connected to the transformer's secondary.

Failures that originate in the transformer core are basically due to deterioration of the insulation layers between the laminations of the core. Failures around the bolts, loosening of the core clamps, slippage of the core clamps which cause short circuits of the laminations, disarrangement of the iron at the joints, a high-resistance ground between the core and the tank, and multiple core grounds all contribute to potential operating problems.

Measuring the excitation current of a transformer during field acceptance and routine preventive maintenance functions are effective in detecting winding and core faults, even though normal turns ratio and winding resistance test results were obtained. This test should not be confused with core loss testing which is more complicated and sensitive to the wave shape of the test source.

Core loss testing will detect the operating hazards as mentioned above; however, as a field test, under field conditions, performing the core loss test accurately is extremely difficult. Thus, a field test called *transformer excitation current* is used, and the results are analyzed on a comparison basis only. This test is usually conducted in the field, upon acceptance of the transformer to establish a baseline record, and may be compared to the factory results. The factory excitation current measurements are usually recorded at full-rated primary voltage, whereas field test voltage levels are a percentage of full-rated voltage.

Since the results of the core excitation current test are used to detect changes in the core, the same level of input voltage should be used from test to test to ensure compatible results. In addition, the same test method should be used.

Core excitation tests can be performed using the same test equipment as for power factor testing. The only limit to the use of power factor test sets for core excitation tests is that the required transformer excitation current may in some cases exceed the output capability of the test set. When this happens, the circuit breaker on the test set will trip. This also happens if there is a core fault.

To reduce the required source current, the primary winding of the transformer is excited. Shorted turns in the secondary winding will still be detected by energizing the primary winding. Consider the following guidelines when setting up the test:

1. Caution should be exercised in the vicinity of all transformer terminals. High voltage will be induced in all windings during the test.
2. Proceed only after fully understanding appropriate safety precautions.
3. Ensure that the transformer is deenergized and secured.
4. Disconnect all loads from the transformer.
5. Secondary winding terminals that are normally grounded during service should remain grounded during the test.
6. Always apply the exact same test voltage to each phase of a three-phase transformer winding. This will minimize errors due to any nonlinearity between voltage and current.
7. On three-phase transformers, the excitation current is generally similar for two phases and noticeably lower for the third phase, which is wound on the center leg of the core.
8. On single-phase transformers, the winding is normally energized alternately from opposite ends. This should be done on the individual phases of three-phase transformers if either the unit is suspect, or the exciting current measurements are abnormal.
9. The residual magnetism in the core will sometimes affect the measurements; the probability of residual magnetism interference should be considered if the current measurements are abnormally high.
10. Do not exceed the voltage rating of the winding under test: for delta-connected windings, the line-to-line voltage rating should not be exceeded; for wye-connected windings, the line-to-neutral voltage rating should not be exceeded.
11. At a minimum, excitation current test readings should be recorded with the transformer load tap changer (LTC) in each of the following positions:
    a. Neutral
    b. One step above neutral (raise)
    c. One step below neutral (lower)
    d. Fully raised
    e. Fully lowered

Test Connections for Single Phase Transformers For a single-phase transformer, Figure 15–57 illustrates the test connections required to perform exciting current measurements using the UST (ungrounded specimen test) configuration.

In Figure 15–57, the transformer is energized with the high-voltage lead connected to the H_1 terminal bushing and the low-voltage lead connected

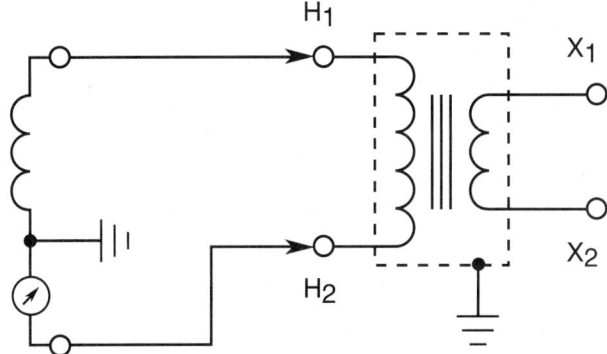

FIGURE 15–57
Core excitation insert on a test single-phase transformer

to H_2. In this configuration the excitation current of $H_1 - H_2$ is measured. In order to cancel out small interference currents (which may cause minor differences in results), or when the recorded exciting current measurement is questionable, the transformer should be rechecked by energizing the high-voltage winding with H_2 terminal bushing connected to the high-voltage lead. This configuration will allow for the measurement of the excitation current $H_2 - H_1$, which is compared to the original $H_1 - H_2$ value. This should also be done on each individual phase of a three-phase transformer if either the unit is suspect, or if the initial exciting current measurements are not within limits. Values obtained should be within +10% of each other.

Test Connections for Three Phase Transformers Figures 15–58 through 15–60 illustrate routine test connections for three-phase transformer excitation current measurements.

Figure 15–58 illustrates testing a transformer with a wye-connected winding. This test is similar to testing a single-phase transformer. In this case H_1 to H_0 is energized and the X winding is floating. If the X winding is normally grounded, it should remain grounded during these tests. Test voltage is applied to the winding (up to the rated voltage of the winding) and the exciting current at that voltage is recorded. This process is repeated for the other two windings (H_2 to H_0 and H_3 to H_0).

Figure 15–59 illustrates a routine test conducted on a three-phase, delta-connected transformer, in which the unenergized or static winding $H_2 - H_3$ is bypassed or shunted during the test. In most cases, the shunting of the static winding has very little effect on the measured currents. The normal, expected results on three-phase transformers (especially wye-connected) are typically two high and similar currents, and one lower current (normally on the center core leg). An example of this pattern would be:

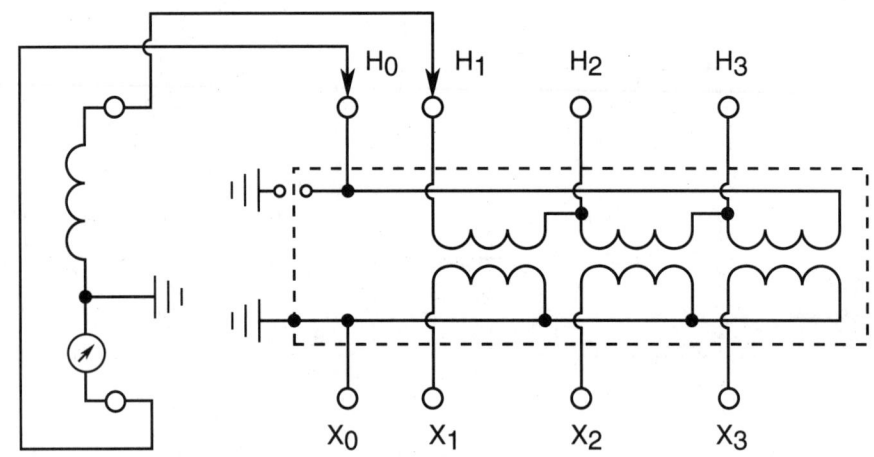

FIGURE 15–58
Core excitation current test three-phase wye-connected transformer (UST mode)

Note: Open H₀ to ground link.
If the low voltage winding is normally grounded, it is to remain grounded during tests.

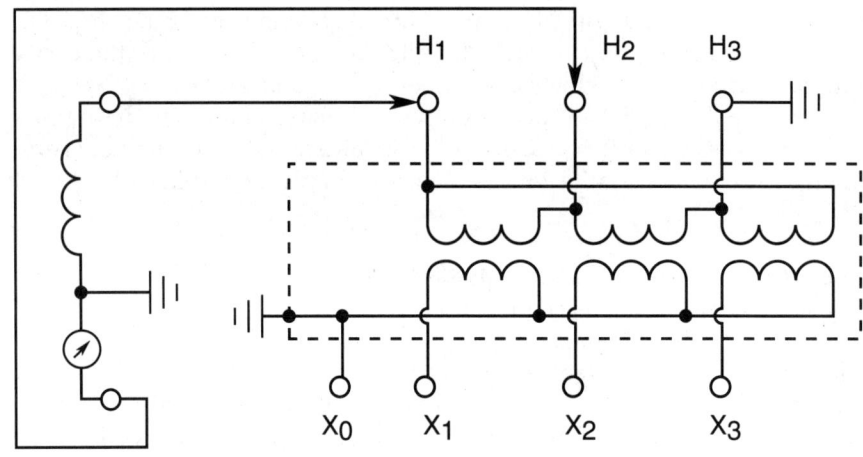

FIGURE 15–59
Core excitation current test three-phase delta-connected transformer (UST mode)

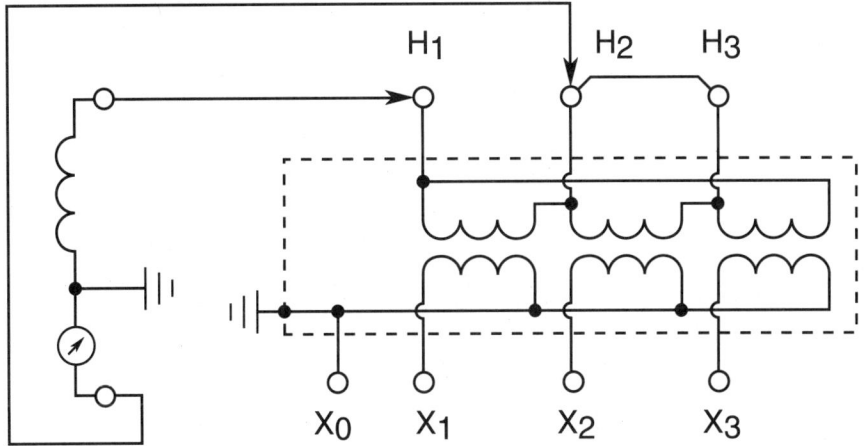

FIGURE 15–60
Alternate Delta Connection (UST Mode)

1. Phase A and C within 5% of each other.
2. Phase B within 65–90% of phases A and C.

In some instances this pattern cannot be obtained due to the shunting effect of the static winding. In order to eliminate the shunting effect, an alternate method can be used. Figure 15–60 illustrates this connection.

When the delta-connected transformer is connected as in Figure 15–60, the expected results will vary from the routine pattern.

Two phases of a delta-connected winding will produce two similar currents, with the third current being higher. An example of this pattern would be:

1. Phase A and C within 5% of each other.
2. Phase B within 5% greater than phases A and C.

Record all test data on a form similar to Figure 15–61.

EXCITATION-CURRENT TESTS

SINGLE PHASE

ENERGIZE	UST
H_1	H_2 (or H_0)
H_2 (or H_0)	H_1

THREE-PHASE WYE [1]

ENERGIZE	UST	PHASE
H_1	H_0	A
H_2	H_0	B
H_3	H_0	C

THREE-PHASE DELTA [1]

ENERGIZE	UST	GROUND	PHASE
H_1	H_2	H_3	B
H_2	H_3	H_1	C
H_3	H_1	H_2	A

THREE-PHASE AUTO

ENERGIZE	UST	PHASE
H_1	$H_0 X_0$	A
H_2	$H_0 X_0$	B
H_3	$H_0 X_0$	C

MFGR. _____ SERIAL NO. _____

NLTC POSITION (CHECK): 1(A) ___ 2(B) ___ 3(C) ___ 4(D) ___ 5(E) ___

TAP CHANGER FOUND/LEFT ON POSITION _____

TEST VOLTAGE: _____ kV[2]

LINE NO.	(3) ULTC POSITION	MILLIVOLTAMPERES									REMARKS
		PHASE A			PHASE B			PHASE C			
		METER READING	MULTI-PLIER	MVA	METER READING	MULTI-PLIER	MVA	METER READING	MULTI-PLIER	MVA	
1											
2											
3											
4											
5											
6											
7											
8											
9											
10											
11											
12											
13											
14											
15											
16											
17											
18											
19											
20											
21											
22											
23											
24											
25											
26											
27											
28											

NOTES:
1. IF THE LOW-VOLTAGE WINDING IS WYE CONNECTED. THEN X_0 IS CONNECTED AS IN SERVICE (USUALLY, THIS WOULD MEAN GROUNDING X_0).
2. ALL TESTS SHOULD BE PERFORMED ROUTINELY AT THE SAME VOLTAGE
3. INDICATE TAP POSITION RAISED, LOWERED OR NEUTRAL

SHEET NO ____

FIGURE 15–61
Excitation current test sheet (Copyright Doble Engineering)

SUMMARY

The recommended maintenance and testing procedures for a transformer depend on its size and voltage level. At a minimum, transformers should be periodically inspected for load current, voltage, temperature, and liquid level. Other tests, such as insulation resistance or power factor tests, will be performed periodically for large, high voltage transformers.

Insulating liquid tests are among the most cost effective of all tests. Transformer oil can be sampled annually, and tested to evaluate the continuing serviceability of the unit. Tests that should be performed include the dielectric breakdown voltage, color, acid neutralization number, and interfacial tension test.

Electrical testing can be performed to evaluate condition and/or diagnose incipient failures. The most common of these tests are the insulation resistance tests which use high voltage, low energy equipment to measure the resistance of the transformer's insulation. These tests can identify insulation fatigue and/or contamination. Winding resistance tests, also performed using direct voltage test equipment, can be used to identify high resistance connections inside the transformer which would not otherwise be found.

Alternating voltage can also be used to test the transformer. AC insulation power factor tests are more commonly used on large, medium or high voltage, liquid-filled transformers. These tests are extremely sensitive to deterioration of the transformer insulation.

The magnetic structure of the transformer can also be checked using AC equipment. Since even a small problem in the magnetic structure will change the amount of current the transformer draws when it is not loaded, the excitation test results will show such problems.

REVIEW QUESTIONS

1. Look at Figure 15–6. What is the relative humidity when the wet bulb temperature is 70°, and the dry bulb temperature is 95°? When the wet bulb is 70°, and the dry bulb is also 70°?

2. Oil samples for the dielectric strength test should be drawn into a _____ bottle.

3. What is the difference between a D-877 test and a D-1816 test?

4. What is the color value that is assigned to new mineral oil that is *water white?*

5. You are testing the insulation resistance of the 4160-V coil of an oil-filled transformer. The oil temperature is 35°C. The one-minute reading is 365 megohms. Is this acceptable? (Hint: See Tables 15–6 and 15–7)

6. You have performed a step voltage test on a dry type transformer. The 500-V insulation resistance is 560 megohms. The 2500-V reading is 336 megohms. Is this acceptable? If not, what may be wrong?

7. You are testing the turns ratio of a 13,800 to 480-V transformer. The result is 29.00. Is this acceptable? Why or why not?

8. A certain transformer is being tested for insulation power factor. The applied mVA is 1000. The mW reading is 2. What is the percent power factor of this transformer?

9. You are performing an excitation test on a three-phase power transformer. The results are A phase – 25 mA, B phase – 20 mA, C phase – 24.4 mA. Do these readings seem acceptable? Why or why not?

10. You are evaluating test results on some service-aged transformer oil. The color is a pale yellow, the NN is 0.09, and the IFT is 28. What is your evaluation based on the Oil Quotient Index approach?

Glossary

acceleration time Period of time that the initial inrush of current lasts.

alternating current The current flow when voltage is an alternating source. Abbreviated AC.

alternating voltage Pressure that pushes electrons in one direction and then the other.

ampere Unit used to measure current.

amortisseur windings Starting windings that prevent damage to the motor by locking the rotor into synchronous speed when the field current is applied. Also help to regulate speed when sudden step loads are applied or removed from the synchronous motor shaft. (Also called damper windings.)

Ampere's Law Any current-carrying conductor located in a magnetic field at right angles to the lines of force will be pushed by a force that is directly proportional to the flux density, the current, and the length of the conductor.

apparent power The sum of the real power and the reactive power.

armature The loop of wire that is free to rotate in a magnetic field of a generator or motor.

bars Segments of a commutator that are connected to the armature windings when the commutator is mounted on the armature shaft.

brush An electrical conducting component of a slip ring that slips over the surface of the ring as the armature rotates.

charges The positive and negative electric forces exhibited by protons and electrons.

commutator The high speed rotary switch of a motor or generator that is used to switch from one armature winding to the next, to keep the output at the brush at the same polarity.

counter emf The rotation-induced electromotive force in a motor that opposes the applied voltage. Abbreviated cemf.

current The flow of electrons through a conductor.

current limiting The fast clearing time that limits the amount of current flow in fuses.

cycle A complete alternating current flow.

damper windings See amortisseur windings.

D-dimension The distance, in inches, from the shaft center line to the bottom of the mounting feet of a motor.

delta voltage The voltage between any two terminals. (Also called phase-to-phase voltage.)

diamagnetic Substances that are slightly repelled by a magnet.

dielectric absorption ratio Ratio obtained when dividing the 60-second reading by the 30-second reading of a megohmmeter. Abbreviated DAR.

direct current When the source of voltage is a direct source. Abbreviated DC.

direct voltage The constant pressure that pushes electrons in one direction only.

dry type Term used when no liquid is used for insulating or cooling the transformer.

eddy current loss The result of voltages generated in the core iron that result in a flow of current through the core's laminations.

electron The negative charged electrical particle of an atom.

encapsulated or sealed windings machines Windings that are covered with or impregnated with a heavy coating of resin to protect them from moisture, dirt, and so on.

Faraday's Law The magnitude of the generated voltage is proportional to the number of lines of flux that are being cut each second.

ferromagnetic Materials that have weak magnetic properties.

flashing The term used to describe the monetary application of a voltage to the field to get the generator output started.

flux density The number of lines of force per unit area.

form factor The basis of rating applied to the type of DC supply for the motor.

form wound The armatures in which the windings are made to proper shape, insulated, and then installed on the rotor.

frequency The number of cycles per second of an alternating current flow.

Hertz The unit of frequency equal to one cycle per second. Abbreviated Hz.

hysteresis loss The term used to describe the constant reorientation of the polarity of the iron molecules in the core when the generator rotates and the magnetic flux shifts rapidly.

impedance The resistance and the opposition created by the formation of magnetic and electrostatic fields.

induction generator A generator that does not have a separate excitation source.

Lenz's Law A current set up by a voltage induced due to the motion of a closed circuit conductor will create a magnetic field which opposes the field which caused the current in the first place.

lines of force Imaginary lines used to illustrate the pattern of the magnetic field.

locked rotor current The term used to describe the initial inrush of current.

lodestones Naturally occurring rocks that attract iron and similar metals.

magnetic field The area around a magnet in which the magnetic force acts.

magnetic flux The magnetic lines of force that surround a magnet.

magnetism The force exerted by lodestones and similar substances.

magnets Objects that exhibit magnetic characteristics.

megohmmeter A type of insulation tester commonly called a Megger.

motoring A condition that appears when a generator becomes a motor, due to the failure of its prime mover, and starts to draw power from the system. (Also called reverse power.)

negative sequence The type of current that appears when the power system currents are unbalanced.

neutral point The point at which the voltage on the commutator segments are equal as they make contact with the brushes, therefore preventing arcing at the brushes.

neutrons Elements of an atom found in its nucleus.

north pole The north pointing pole of a magnet.

nucleus The center of an atom where protons and neutrons are found.

Ohm's Law   The fundamental relationship between voltage, current, and resistance.

open machines   Machines that have ventilated openings that permit the passage of external cooling air over and around the windings.

paramagnetic   Machines that have a very weak magnetic effect

phase-to-neutral voltage   The voltage between any terminal and the common point of the wye connection.

phase-to-phase voltage   See delta voltage.

polarization index   The ratio obtained when dividing the 10-minute reading by the 60-second reading of a megohmmeter. Abbreviated PI.

proton   A component of atom found in its nucleus and has a positive electric force.

random windings   Windings that are formed directly on the armature by winding insulated wire into the slots.

relays percent slope   The minimum ratio that is calculated by dividing the operating current by the smaller restraining current when the relay is just balanced.

residual magnetism   The portion of magnetism retained by a ferromagnetic material after magnetic force has been induced into it.

resistance   The opposition created by the friction of the movement of electrons or by the performance of real work.

retentivity   The ability of a material to retain magnetism.

reverse power   See motoring.

rigging   An assembly that attaches the brush holders to the frame of a motor or generator.

RMS   Abbreviation for root-mean-square values. Used to give an alternating voltage or current their equivalent DC values.

rotating exciters   Generators that are used to supply the DC excitation current to the main generator.

rotating magnetic field   The term used to describe the field created by allowing two or more magnetic fields to add in such a way that their net magnetic north and south poles seem to rotate.

rotor   The rotating shaft of a generator or motor.

salient-pole rotor   The type of rotor used when the prime mover is relatively slow.

sealed windings machines See encapsulated windings machines.

self-excited motor or generator The term used when the DC supply for the main poles are connected to the armature.

separately-excited motor or generator The term used when the DC supply for the main poles comes from an external source.

single-phase When only one voltage and one current are developed.

slip The percentage of synchronous speed at which the induction motor operates.

slip rings Components that are connected to the armature loop.

south pole The south pointing pole of a magnet.

static exciter Has no moving parts. Causes the generator output to drop very quickly resulting in a reduced current output when a short circuit occurs on the generator output terminals.

stator The stationary coils of a motor or generator.

synchronous generators Generators that have their own source of excitation and are driven at synchronous speed.

synchronous speed The speed at which an induction motor field rotates.

three-phase Where three voltages are equal in magnitude and are displaced by 120 electrical degrees.

torque The overall turning effect on the armature that causes the coil to turn on its axis.

totally-enclosed machines Machines that are constructed to prevent free exchange of air between the inside and outside of the case.

undercut The process wherein the mica is cut down to a reasonable depth after the commutator has been resurfaced.

VAR Acronym for volt-ampere-reactive.

volt The unit used to measure voltage when moving electrons.

voltage The force that is used to move electrons.

voltage drop The amount of voltage drop that naturally occurs on a generator.

voltage regulation The use of external circuits to compensate for droop.

wire-wound rotor One type of induction rotor whose key value is to control the speed of operation by placing a rheostat or stepped resistors in series with the windings.

Index

Note: Page numbers in *italics* reference non-text material.

A

Absorption
   current, insulation testing, 82
   tests, dielectric, 88–89
AC generator
   armature, 34–36
   construction of
      large, 42–48
      small, 37–41
   controls, 48–50
   described, 31
   electrical connections for
      single phase, 50, *51*
      three-phase, 50–52
   generator field, 34–36
   grounding, 52–53
   operating principles, 32–33
   protection
      described, 57
      differential, 61, 65–66
      ground fault, 67–68
      large generators, 76
      loss of excitation, 68
      medium size generators, 74–76
      negative sequence, 68–69
      overcurrent, 58–59, *60*
      overload, 58
      overtemperature, 59, 61
      reverse power, 69–70
      small generators, 72–74
      very small generators, 70, *73*
   testing
      bearing insulation, 104–7
      core loss, 96–101
      described, 79
      high-potential, 89–93
      insulation, 80–89
      insulation power factor, 93–96
      surge comparison, 102–4
   See also DC generators
AC motors
   classification of
      application, 126, 128
      electrical type, 128–29, *130–32*
      environmental protection/cooling and, 129, 131–32
      insulation, 133–34
      size, 126
      speed variability, 132–33
   control centers, 164–69
   four-digit frame number, 126, *127*
   induction, 117–20
      characteristics of, *130–31*
      electrical tests on, 205–14
      speed ratings for, *136–37*
   insulation, 133–34
   maintenance

magnetic center, 203–5
physical inspection, 200–202
rotor air gap, 203
nameplates
information on, 140–43
temperature, 139
ratings, 134–40
rotating magnetic field, 112–14
selecting correct, 151–54
single-phase, 120–21
speed, 121–22
control, 169, 171–76
starter tests, 206
starters, 164–69
troubleshooting, 169, *170*
starting circuits
across-the-line, 158
part winding, 161–63
resistance, 159–60
wye-delta, 163–64
stator electrical connections
single-phase, 148–49
split-phase, 149
three-phase, 144–47
synchronous, 114–16
characteristics of, *131*
electrical tests on, 214–16
troubleshooting, 216–24
voltage, autotransformer reduced, 160–61
see also DC motors
Across-the-line starting circuits, 158
Adjustable
speed motor, 133
varying-speed motor, 133
Air-cooled generators, 46
Air/water-cooled generators, 46
Alternating current, 11, 13
illustrated, *14*
insulation testing and, 82
testing, AC generator, 91–92
Alternating voltage, described, 10
Amortisseur windings, 115
Ampere, Andre M., 24

Amperes, described, 11
Ampere's law, 3, 24
Anchorage
AC motors, inspecting, 200
DC machinery, inspecting, 286
Apparent power, 18
Armature
AC generators, 34–36
DC,
generators, 237–38
motors, 264–65
Atoms, 9–10
Automatic gas pressure transformers, 340
Autotransformer reduced voltage, 160–61
Autotransformers, 330

B

Batteries, small AC generators and, 38–39
Bearing, insulation, 104–7
Breakdown torque, 135
Brush holders,
DC
generators, 239–40
motors, 266–67
Brushes
DC
generators, 239–40
machinery, 291–93
motors, 266–67
Bushing insulation resistance, 418
Bushings
transformer, 342, 345
inspecting, 373–74

C

Capacitive charting current, insulation testing, 81–82
Capacitor
run motor, 150
start motors, 150
Capacitors, troubleshooting, 224

Chattock potentiometer, core loss testing and, 99–101
Circuit breakers, molded-case, 189
Class
    A motor insulation, 133
    B motor insulation, 133
    F motor insulation, 134
    H motor insulation, 134
Clutch, eddy current, 171
Coils, 23–24
    reversed, troubleshooting, 222
    transformer, 342
Commutator
    generators, 230–31, 238
    machinery, 293–300
    motors, 265
Compound-connected
    generators, 250–52
    motors, 280
Condensation, transformers and, 371–72
Connections
    compound
        generators, 250–52
        motors, 280
    inspecting
        AC motors, 200
        DC machinery, 286, 288
Conservator/expansion tank transformer, 338–39
Constant speed motor, AC, 132
Control centers, motor, 164–69
Controls, large AC generators, 48–50
Cooling
    fans, transformers and, 349
    systems, large generators, 46, 48
Core
    excitation, current testing, 459–64
    form magnetic transformers, 320, *321*
    loss
        described, 96
        testing, 99–101
    transformer, 342
Coulomb, Charles A., 7
Coulomb's law, 7
Current
    described, 11
    flow, illustrated, *14*
    running, checking, 210
    testing, core excitation, 459–64
    transformers, 319
Cycle, electrical, 13

D

Damper windings, 115
DC generators
    compound-connected, 230–31, 238
    construction
        armature, 237–38
        brushes/brush holders, 239–40
        commutator, 238
        field windings, 234–37
        frame, 234
    operating theory of, 230–34
    permanent magnet, 244–47
    ratings, 240–44
    self-excited, 248–52
    separately-excited, 244–47
    see also DC generators; DC machinery
DC machinery
    tests, electrical, 300–301
    troubleshooting, 302–11
    visual inspection of, 286–300
DC motors
    characteristics, 280–81, *282*
    compound-connected, 280
    construction
        armature, 264–65
        brushes/brush holders, 266–67
        commutator, 265
        field windings, 261–63

frame, 260–61
operating theory, 256–60
ratings, 267–76
rotation direction, 283
self-excited, 276–78
series-connected, 278
shunt-connected, 278, 280
speed control, 281
see also AC motors
Dew point, defined, 372
Diamagnetic, described, 4
Dielectric
  absorption
    ratio, 418
    tests, 88–89
  current, insulation testing, 82
Differential
  protection, AC generators, 61, 65–66
  relay
    checks, 206
    protection, 191–93
Direct current, 11, 13
  illustrated, *14*
  insulation testing, 80–82
  source, troubleshooting, 218
  testing, AC generator, 92–93
Direct voltage, described, 10
Distribution transformers, 317
Drip-proof machines, described, 129
Drives, variable frequency, 176
Dry cell, cross section of, *12*
Dry-type transformers, 336

E

Eddy current
  clutch, 171
  loss
    described, 96
    testing, 99–101
Electric current, magnetism and, 19–23
Electrical tests
  AC motors, synchronous, 214–16

  circuit, troubleshooting, 217–18
  DC machinery, 300–301
  induction motors, 205–14
Electricity, 9–19
Electromagnetic induction, 25–27
Electromagnetism, 19–24
Electron theory, 9–10
Electronic protection, 191
Electrons, 9
Encapsulated windings machines, 132
Excitation systems
  AC generators
    large, 43–46
    loss of protection by, 68
    small, 37–41
Exciters
  rotating, 43
  static, 44–46
Explosion-proof machines, 131

F

Fans, transformers and, 349
Faraday, Michael, 25
Faraday's law, 25–26
Fault gas detector, 353
Ferromagnetic, described, 4
Field
  application relays, testing, 215
  windings
    DC motors, 261–63
    generator, 234–37
Flux density, described, 8
Frame
  DC generators, 234
  DC motors, 260–61
Free breathing transformers, 337
Frequencies, AC motors, 134
Frequency, described, 11, 13
Full-load torque, 135
Fuses, 183, 185–86

G

Gas-in-oil analysis, transformers, 382–88

Gas-oil seal transformers, 339
Generator
   field, 34–36
   induction, 41
   ripple, described, 233
Generators *See* AC generators; DC generators
Ground fault protection
   AC generators, 67–68
      maximum ratings for, *184*
Grounded windings, troubleshooting, 219
Grounding
   AC generators, 52–53
   inspecting
      AC motors, 200
      DC machinery, 286
   transformers, 332
Growler, troubleshooting, 219
Guarded machines, described, 129

H

Heaters, motor space, 206
Hertz (Hz), electrical, 13
High-potential testing, 89–93
Horsepower, AC motors, 134–35, *136–38*
Humidity, insulation testing and, 82, 84
Hydrogen-cooled generators, 46, 48
Hysteresis loss
   described, 96
   testing, 99–101
Hz (Hertz), electrical, 13

I

Impedance, described, 13, 15
Induction
   disk overcurrent relays, 190–91
   generators, 41
   motors, 117–20
      polyphase, characteristics of, *130–31*
      speed, 121–22
      speed ratings for, *136–37*

Inductors, 23–24
Induction motors, speed ratings for, *136–37*
Instrument transformers, 317, 319
Insulation
   AC motors and, 133–34
   bearing, 104–7
   minimum resistance, 415
   resistance, 87–88
      checking, 210
   surge protectors resistance of, 205–6
   testing, 205
      alternating current, 82
      capacitive charging current, 81–82
      dielectric current, 82
      direct current, 80–82
      leakage current, 82
      physical construction, 80
      power factor, 93–96
      resistance, 84, 86–89, 411–12
      temperature/humidity, 82, 84
   transformers, 336–40
Interpoles, 236–37

L

Leakage current, insulation testing, 82
Left-hand rule, 21, *22*
Lenz, H.F. Emil, 26
Lenz's law, 3, 26–27
Lightning arresters, 374–75
Liquid level indicators
   transformers, 349
      visual inspection of, 368
Liquid-filled transformers, 337–40
Load current, transformers, visual inspection of, 368
Load voltage, transformers, visual inspection of, 368
Locked-rotor
   code letters for, 138, *139*
   currents, *132*
   torque, 135

Lodestones, 4
Lubrication
  inspecting
    AC motors, 202
    DC machinery, 288

M

Magnetic
  center, AC motors inspection of, 203–5
  fields, 7
    rotating, 112–14
  flux, 8
  force, described, 6
  overload relay, 188
  poles, 4–5
    force between, 7
Magnetism
  described, 4–6
  electric current and, 19–23
Magnets, described, 4
Maintenance, visual inspection, 200–202
Megohmmeter, 84
Melting alloy relays, 186–88
Molded-case circuit breakers, 189
Motor
  insulation, 133–34
  space heaters, 206
Motors
  See AC motors; DC motors
Mounting
  inspecting
    AC motors, 200
    DC machinery, 286
Multi-speed motor, 133

N

Negative sequence protection, AC generators, 68–69
Neutral point
  generators and, 232–33
  motors, 258
Neutrons, 9

Nucleus, 9

O

Oersted, Hans Christian, 19
Ohm's law, 15
Open rotor bars, troubleshooting, 222–24
Open tank transformers, 337
Open windings
  described, 129
  troubleshooting, 220
Overcurrent
  described, 180
  protection, 180–83
    AC generator, 58–59, *60*
    devices for, 183, 185–91
Overload
  AC generator protection, 58
  described, 181
  pickup settings, *181*
  units required, *182*
Overtemperature
  AC generator protection, 59, 61
  stator winding, 193–95

P

Paramagnetic, described, 4
Part winding starters, 161–63
Permanent magnets
  excitation of, 39–40
  generators, 39–40, 244–47
Pickup tests, 210
Polarity
  transformers, 323
    single-phase, 324
Polarization index, 415–16
Polyphase induction motors
  characteristics of, *130–31*
  speed ratings for, *137–38*
Potentiometer, core loss testing and, 99–101
Power
  apparent, 18
  DC motors, 258–60

described, 16–19
factor
  defined, 18
  insulation testing and, 93–96
  relays, testing, 215–16
  sampling, transformers, 380, 382
  tests, winding, 448–59
reactive, 17–18
real, 16–17
three-phase systems and, 19
transformers, 317
Pressure
  devices
    relief, 348
    sudden, 347–48
  vacuum indicators
    transformers, 353
      visual inspection of, 368
Primary resistance starter, 159
Protection
  differential relay, 191–93
  electronic, 191
  overcurrent, 180–83
    devices for, 183, 185–91
Protective devices, 206–7, 210
  DC machinery, 301
Protons, 9
Pull-up torque, 135

R

Ratings
  AC motors, 134–40
  DC
    generators, 240–44
    motors, 267–76
Reactive power, 17–18
Real power, defined, 16–17
Relays
  differential, 191–93
  field application, testing, 215
  induction disk overcurrent, 190–91
  magnetic overload, 188

melting alloy, 186–88
power factor, testing, 215–16
Relief devices, pressure, transformers and, 348
Residual magnetism, described, 6
Resistance
  busing insulation, 418
  described, 13, 15
  insulation
    checking, 210
    testing, 84, 86–89
  starters, 159–60
  temperature detectors, 206
  testing, windings, 418–26
Retentivity, defined, 6
Reverse power, AC generator protection by, 69–70
Reversed coils, troubleshooting, 222
Rotating
  exciters, 43
  magnetic fields, 112–14
Rotation test, 211
Rotor bars, open, troubleshooting, 222–24
Rotors
  inspecting, air gap, 203, 289
  large generators and, 42–43
  locked
    code letters for, 138, *139*
    currents, *132*
    torque, 135
  rounded, 43
  squirrel cage, 117–18
  wire wound, 118–20
Running current, checking, 210

S

Salient-pole rotor, 42–43
Sealed
  tank transformers, 337–38
  windings machines, 132
Secondary resistance
  speed control, 171
  starter, 160

Self-excitation systems, 40–41
  DC generators, 248–52
  DC motors, 276–78
Separately-excited DC generators, 244–47
Series-connected
  generators, 248
  motors, 278
Shaded pole motor, 150–51
Shell form magnetic transformer, *321*, 322
Short circuit
  protection, 181, 183
    maximum ratings for, *184*
Shorted windings, troubleshooting, 221
Shunt-connected
  generators, 248–49
  motors, 278, 280
Single-phase systems, 15–16
  AC motors, 120–21, 148–49
  electrical connections for, 50, *51*
  transformers, 323–24
    polarity of, 324
Speed
  AC motors, 134–35, *136–37*
    control, 169–76
  DC motors, 258–60
    control, 281
  secondary resistance, control, 171
Splash-proof machines, described, 129
Split-phase motor, 149
Squirrel cage rotors, 117–18
Starter tests, AC motors, 206
Starters
  comparison of five methods of, *165*
  motor, 164–69
  part winding, 161–63
  resistance, 159–60
  standard sizes of, *168*
  troubleshooting, 169, *170*
  wye-delta, 163–64

Starting
  circuits, across-the-line, 158
  windings, 115
Static exciters, 44–46
Stator
  electrical connections
    single-phase, 149
    three-phase, 144–47
  winding, overtemperature, 193–95
Step-voltage test, 89, 418
Sudden pressure devices, transformers and, 347–48
Surge
  arrestor, connections, *201*, *287*
  comparison testing
    methods/results of, 102–4
    principles of, 102
  protectors, insulation resistance of, 205–6
Synchronous motors
  AC, 114–16
    characteristics of, *131*
  electrical tests on, 214–16
  speed, 121

T

Tank, transformer, 341–42
Tap changers
  inspecting, 374
  transformer, 345
Temperature
  correction factors, *85*
  detectors, resistance, 206
  indicators, 351, *352*
  insulation testing and, 82, 84
Terminal markings, transformers, 322, 326
Test
  circuits, troubleshooting, 217–24
  equipment, 206–10
Three-phase systems, 15–16
  electrical connections for, 50–52
  power in, 19

Index 481

transformers, 325–29
Time ratings, AC motors, 138
Timing, checking, 210
Torque
  motors
    AC, 135
    DC, 258–60
Totally-enclosed
  fan-cooled machines, described, 131
  machines, described, 129
  nonventilated machines, described, 129–30
Transformers
  AC testing of, 426–59
  characteristics/ratings of, 354–58
  components/indicators of, 340, 342–53
  construction, electrical, 322–32
  cooling classifications of, 340 *341*
  core excitation current testing, 459–64
  DC testing of, 410–26
    safety precautions, 410–18
    winding resistance, 418–26
  diagnostic inspections of, 371–75
  gas-in-oil analysis, 382–88
  grounding, 332
  insulation of, 336–40
  liquid, 388–89
  liquid sampling
    precautions, 378–79
    procedures, 380
    purpose of, 378
    sampling jar preparation, 379–80
  liquid testing
    acid neutralization number test, 396–98
    color testing, 394–95
    dielectric breakdown voltage test, 394
    interfacial tension test, 398–401
    moisture content test, 402–10
    visual examination of, 395
  magnetic structures of, 320–22
  maintenance for, 375–76
  nameplates, 358–64
  power factor sampling, 380, 382
  principles of, 316
  testing, 376–77
  turns ratio, 426–48
  types of, 316–19
  visual inspection of, 368–77
  winding power factors test, 448–59
Troubleshooting
  AC motors, 216–24
  DC machinery, 302–11
  starters, 169, *170*
Turbo-rotors, 43

V

Variable frequency drives, 176
Varying-speed motor, 132
Ventilation
  inspecting
    AC motors, 202
    DC machinery, 289
Vibration analysis, DC machinery, 301
Voltage
  autotransformer reduced, 160–61
  cell, cross section of, *12*
  described, 10
  droop, DC generators and, 233
  rating, AC motors, 134
  regulation, DC generators and, 233
  test, step, 89
  transformers, 317, 319

W

Windings
  field
    DC motors, 261–63

generator, 234–37
grounded, troubleshooting, 219
open, troubleshooting, 220
power factor tests, 448–59

resistance testing, 418–26
shorted, troubleshooting, 221
Wire wound rotors, 118–20
Wye-delta starters, 163–64